Wave Optics

Jan Petykiewicz

Institute of Physics,
Warsaw University of Technology, Poland

Wave Optics

KLUWER ACADEMIC PUBLISHERS
DORDRECHT / BOSTON / LONDON

PWN – POLISH SCIENTIFIC PUBLISHERS
WARSZAWA

Library of Congress Cataloging-in-Publication Data

Petykiewicz, Jan.
Wave optics / by Jan Petykiewicz.
p. cm.
Includes bibliographical references.
ISBN 978-90-481-4065-7
1. Light, Wave theory of. I. Title.
QC403.P48 1990
535'.13 – dc20 90-30853

Published by Polish Scientific Publishers, PWN Ltd.
Miodowa 10, 00-251 Warszawa, Poland
in co-edition with Kluwer Academic Publishers,
P.O. Box 17, 3300 AA Dordrecht, The Netherlands

Distributors for the U.S.A. and Canada:
Kluwer Academic Publishers,
101 Philip Drive, Norwell, MA 02061, U.S.A.

Distributors for Albania, Bulgaria, Chinese People's Republic, Cuba, Czecho-Slovakia, Hungary, Korean People's Republic, Mongolia, Poland, Romania, the former U.S.S.R., Vietnam, and Yugoslavia:
Polish Scientific Publishers PWN Ltd., Miodowa 10, 00-251 Warszawa, Poland

Distributors for all remaining countries:
Kluwer Academic Publishers Group,
P.O. Box 322, 3300 AH Dordrecht, The Netherlands

Revised translation from the Polish original *Optyka Falowa*
published in 1986 by Państwowe Wydawnictwo Naukowe, Warszawa.

Translated by B. Bijak (Secs. 4.1–4.5 by M. Sadowski)

Contents

Introductory Remarks

The following work owes its inspiration to a course of lectures I held at the Department of Technical Physics and Applied Mathematics of the Warsaw University of Technology. It is devoted to the issues of wave optics. The material contained in it is based on the electromagnetic theory of light and includes those sections of optics which the classical laws of electrodynamics are sufficient to explain. It also includes a more or less successful compilation of theoretical descriptions of various optical phenomena scattered throughout various textbooks and monographic works, some of them already classics, most of them in foreign languages. Each chapter is devoted to a different section of optical phenomena.

The work also encompasses such areas of optics as elements of the theory of coherence and the diffraction of light, both in the scalar and vector approximations; elements of the theoretical description of light waves in anisotropic media including stimulated anisotropy; the effect of spatial and temporal dispersion on the propagation of light waves; a description of the effect of scattering in some selected cases, and those nonlinear phenomena in optics which it is possible to explain within a classical formulation. Such questions as the physical foundations of holography and the transition from wave optics to geometrical optics, with a substantiation of the basic laws of the latter, are also touched upon.

Among the important features of this work are: the first attempt to gather in textbook form the results of the work of one of the Polish leading physicists, Professor Adalbert Rubinowicz, in the field of diffraction; a discussion of the so-called Conjugate Phase

Optics, based on nonlinear optical effects; and a glance at electromagnetic shock waves.

The literature listed at the end of each chapter is meant to point out those works which were used preparing its contents, and by no means exhausts the subject. The reader will also find a fuller discussion of some topics there, merely mentioned in this work.

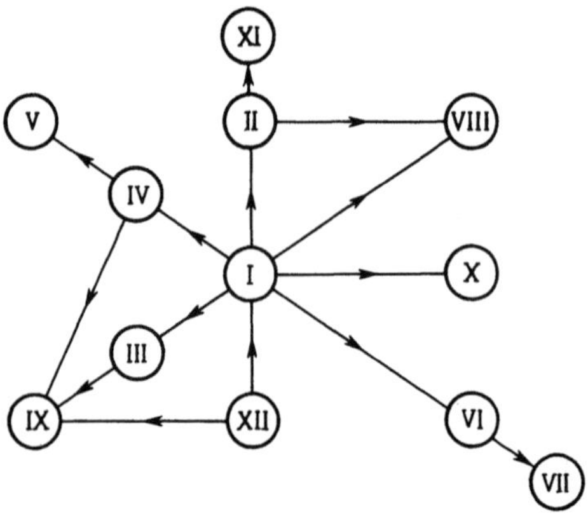

The selection and form of presentation of the problems are aimed at readers familiar with physics and mathematics at the level of basic courses taught at Universities of Technology. The central chapter of the book is Chapter One. The logical interdependences of the following chapters are shown by arrows in the figure above.

1. Fundamental Elements of the Classical Description of the Electromagnetic Field

1.1. Maxwell's Equations

If we are interested in the laws of propagation of electromagnetic waves with optical frequencies and not in their generation and registration, then in order to lay them down we can use the fundamental equations of the electromagnetic field, namely Maxwell's equations.

This system of differential equations connecting two pairs of vector fields, **E** and **B** with **H** and **D**, together with the constitutive relations, boundary conditions and initial conditions, plays in electrodynamics the same role as Newton's equation and the initial conditions in classical mechanics. Maxwell's equations, which are the quintessence of electrodynamics, allow us to determine in space that state of excitation which is produced in the presence of electric charges and their motion. This state of excitation is called an *electromagnetic field* and is described by the *electric field intensity vector* **E**, the *magnetic field intensity vector* **H** and by a second set of vectors, which take into account the presence of a material medium. These vectors are: the *current density vector* **j**, the *electric displacement vector* **D**, and the *magnetic induction vector* **B**.

The relations between time and spatial derivatives of the vectors mentioned above make up the content of Maxwell's equations. These equations are satisfied at any point around which the physical properties of the medium are continuous. Their explicit form is as follows

$$\operatorname{curl} \mathbf{H} - \dot{\mathbf{D}} = \mathbf{j}, \tag{1.1}$$

$$\operatorname{curl} \mathbf{E} + \dot{\mathbf{B}} = \mathbf{0}, \tag{1.2}$$

$$\operatorname{div} \mathbf{D} = \rho, \tag{1.3}$$

$$\operatorname{div} \mathbf{B} = 0. \tag{1.4}$$

The dot over a letter denotes time differentiation.

Equation (1.3) determines the electric charge density ρ. From equation (1.4) it follows that there is no free magnetic charge.

Equations (1.1)–(1.4) admit the possibility of occurrence of different types of electromagnetic fields. So, if

(a) $\mathbf{j} = \dot{\mathbf{D}} = \dot{\mathbf{B}} = \mathbf{0}$ – the electromagnetic field is static;

(b) $\mathbf{j} \neq \mathbf{0} = \dot{\mathbf{D}} = \dot{\mathbf{B}}$ – the electromagnetic field is stationary;

(c) in other cases the electromagnetic field is variable in time and space.

Further, particular attention is given to the areas where $\mathbf{j} = \mathbf{0}$, and $\rho = 0$, i.e., the areas free of sources.

In such areas electromagnetic fields propagating in space in a wave form can exist. The waves have the following properties:

(1) electromagnetic waves in free space are transverse waves ($\mathbf{E} \perp \mathbf{H}$ and both fields are perpendicular to the direction of propagation);

(2) the propagation velocity of electromagnetic waves depends on the properties of the medium.

Maxwell's equations must be supplemented by constitutive relations, which determine the relationship between vectors $\mathbf{D}$, $\mathbf{j}$ and $\mathbf{E}$, and between $\mathbf{B}$ and $\mathbf{H}$. Maxwell's theory does not give an insight into the microstructure of the medium. It formally describes the influence of the medium on the electromagnetic field by means of the phenomenological relations which are the content of the constitutive relations.

For different adjoining media it is necessary to formulate, in addition to Maxwell's equations and the constitutive relations, additional conditions imposed on the field vectors on the boundary.

1.2. Constitutive Relations for Monochromatic Fields

In order to obtain a univocal solution of Maxwell's equations, we should add to them relations describing the behaviour of matter under the influence of the field applied. These dependencies are called *constitutive relations*, and for a monochromatic field $(e^{i\omega t})$ they determine the functional relationship between vectors **D** and **E**, **B** and **H**, and **j** and **E**:

$$\begin{aligned} \mathbf{D} &= \boldsymbol{\varepsilon}(E_1, E_2, E_3), \\ \mathbf{B} &= \boldsymbol{\mu}(H_1, H_2, H_3), \\ \mathbf{j} &= \boldsymbol{\sigma}(E_1, E_2, E_3), \end{aligned} \tag{1.5}$$

where $\boldsymbol{\varepsilon}$, $\boldsymbol{\mu}$ and $\boldsymbol{\sigma}$ are vector functions, and E_l and H_l $(l = 1, 2, 3)$ are components of the appropriate vector fields.

The third equation is the generalization of Ohm's differential law. If the relations (1.5) are local in time and space, and the medium is at rest, then it is permissible to expand them in a Taylor series. As a result of this expansion we obtain for the individual components of the appropriate fields:

$$\begin{aligned} D_i &= \varepsilon_{ij}E_j + \frac{1}{2!}\varepsilon_{ijk}E_jE_k + \frac{1}{3!}\varepsilon_{ijkl}E_jE_kE_l + \dots, \\ B_i &= \mu_{ij}H_j + \frac{1}{2!}\mu_{ijk}\mathrm{H}_j\mathrm{H}_k + \dots, \\ j_i &= \sigma_{ij}E_j + \frac{1}{2!}\sigma_{ijk}\mathrm{E}_j\mathrm{E}_k + \dots . \end{aligned} \tag{1.6}$$

The coefficients ε_{ij}, μ_{ij}, σ_{ij}, ε_{ijk}, $\mu_{ijk}\dots$ are the tensors of the appropriate orders.

In notation (1.6) of the components of the appropriate vectors we have used the sum convention $(a_{ik}b_k \equiv \Sigma\, a_{ik}b_k)$.

Taking into account only first order terms of the series, we obtain the constitutive relations in the form

$$D_i = \varepsilon_{ij} E_j, \quad B_i = \mu_{ij} H_j, \quad j_i = \sigma_{ij} E_j \tag{1.7}$$

or

$$\begin{aligned} \mathbf{D} &= \varepsilon \cdot \mathbf{E}, \\ \mathbf{B} &= \mu \cdot \mathbf{H}, \\ \mathbf{j} &= \sigma \cdot \mathbf{E}. \end{aligned} \tag{1.8}$$

Here ε is the tensor of electric permittivity, μ – the tensor of magnetic permeability, and σ – the conductivity tensor. They are all second order tensors.

In optics, until quite recently, investigations were restricted to linear expansions, i.e., to what is called *linear optics.*

Considering the relations between vectors **D** and **E**, **B** and **H** in fields of high intensity, we must also take into account further terms in expansions (1.6). This is the situation which we face in the case of laser beams of high power. It is a new field of optics – *nonlinear optics.*

A medium is called *isotropic* if in linear approximation the tensors (1.8) have the form

$$\varepsilon = \mathfrak{J}\varepsilon, \quad \mu = \mathfrak{J}\mu, \quad \sigma = \mathfrak{J}\sigma, \tag{1.9}$$

where

$$\mathfrak{J} = \begin{pmatrix} 1 & 0 & 0 \\ 0 & 1 & 0 \\ 0 & 0 & 1 \end{pmatrix}.$$

Apart from the division of media into isotropic and anisotropic, it is possible to introduce the division into homogeneous and non-homogeneous media.

In homogeneous media the tensors ε, μ, and σ do not depend on the position vector, i.e., they are the same at every point of the medium. In non-homogeneous media ε, μ and σ may vary from point to point.

1.3. Dispersion of Electromagnetic Waves

If the frequency of a variable field is comparable with the frequency of free vibrations of ions or the frequency of free vibrations of electrons, which are connected with the electric and magnetic polarization of the medium, then the electric and magnetic permeability depends on frequency.

In the general case, the values of **D** and **B** at time t depend not only on the values of **E** and **H** at the same moment of time but also, because of the inertia of the medium (electric and magnetic polarization of the material lags behind the changes of electromagnetic field), on the situation in the past. The polarization of the medium stabilizes only after a certain period of time (which is determined by the relaxation processes taking place in the medium), and its temporary state depends not only on the value of the field at that moment but also on the whole history of the field. This leads to a phenomenon called *time dispersion.*

In the range of optical frequencies the inertia phenomenon concerns first of all the electric displacement vector **D** and the polarization vector **P**.

Because of the comparatively slow mechanisms of magnetic polarization, we neglect the time dispersion of the magnetic induction vector **B**, attributing to the media in question the magnetic properties of the vacuum μ_0.

In most cases of quick-variable fields and comparatively low intensities the relationship between **D** and **E** may be regarded as linear. The most general form of the linear relationship between $\mathbf{D}(t)$ and the values of the function $\mathbf{E}(t)$ at all previous moments of time can be written as

$$\mathbf{D}(t) = \varepsilon_0 \cdot \mathbf{E}(t) + \int_0^\infty f(\tau) \cdot \mathbf{E}(t - \tau) \mathrm{d}\tau, \tag{1.10}$$

where $f(\tau)$ is a function of time dependent on the properties of the medium and ε_0 is the permittivity of the vacuum.

For a monochromatic field the relationship between **D** and **E** takes the form

$$\mathbf{D} = \varepsilon(\omega)\mathbf{E},$$

where

$$\varepsilon(\omega) = \varepsilon_0 + \int_0^\infty f(\tau)\mathrm{e}^{\mathrm{i}\omega\tau}\mathrm{d}\tau. \tag{1.11}$$

Thus for periodic fields we can introduce the notion of permittivity denoting the coefficient of proportionality between **D** and **E**. However, this coefficient depends not only on the properties of the medium but also on the frequency of the field. The dependence of ε on the frequency is called *dispersion.* Usually, the function $\varepsilon(\omega)$ is a complex function. Denoting its real part by ε_r and imaginary part by ε_i and assuming $f(\tau)$ to be real, we may write:

$$\begin{aligned} \varepsilon_r &= \mathrm{Re}\,\varepsilon(\omega) = \varepsilon_0 + \int_0^\infty f(\tau)\cos(\omega\tau)\mathrm{d}\tau, \\ \varepsilon_i &= \mathrm{Im}\,\varepsilon(\omega) = \int_0^\infty f(\tau)\sin(\omega\tau)\mathrm{d}\tau. \end{aligned} \tag{1.12}$$

Thus $\varepsilon(\omega) = \varepsilon_r(\omega) + \mathrm{i}\varepsilon_i(\omega)$, and on the strength of (1.12) the components $\varepsilon_r(\omega)$ and $\varepsilon_i(\omega)$ have the properties:

$$\begin{aligned} \varepsilon_r(\omega) &= \varepsilon_r(-\omega) \\ \varepsilon_i(\omega) &= -\varepsilon_i(-\omega) \end{aligned} \tag{1.13}$$

Formulas (1.13) imply that ε_r is an even function and ε_i and odd function of frequency.

For small ω the function ε_r may be expended in a series with respect to ω. The expansion of ε_r has only the even powers, ε_i – the odd powers of ω. If $\omega \to 0$, then $\varepsilon(\omega) \to \varepsilon_0$. Thus in the expansion of ε_r the first term is ε_0 and the first term of $\varepsilon_i(\omega)$ is proportional to ω.

The complex character of permittivity is manifested with particular force in the areas of so-called *anomalous dispersion* (see Sec. 1.6), because the imaginary part of permittivity is closely connected with the phenomenon of resonant light absorption by the atoms of the medium.

In addition to time dispersion, also so-called *spatial dispersion* can occur in material media. This kind of dispersion is related to the fact that polarization at the point $\mathbf{r}$ depends not only on the value of the field at that point but also on the values of the field in its neighbourhood. The dimensions of this neighbourhood are connected with the time during which the thermodynamic processes taking place in the medium and leading to the stabilization of an equilibrium do not destroy the "memory" of the external effect (in gases it is the average length of the free path of the molecules).

In the most general case linear polarization $\mathbf{P}^{(L)}$ depends on the field $\mathbf{E}$ which induces it as follows:

$$P_j^{(L)}(\mathbf{r},t) = \int_{-\infty}^{t} \mathrm{d}t' \int_{\Omega} \mathrm{d}\mathbf{r}' \chi_{jl}(\mathbf{r},t;\mathbf{r}',t') \cdot E_l(\mathbf{r}',t'), \tag{1.14}$$

where $\chi_{jl}(\mathbf{r},t;\mathbf{r}',t)$ is a tensor characterizing the response of the medium at time t and at point $\mathbf{r}$ for the field effect at point $\mathbf{r}'$ and at time t'. It is determined for $t' \leq t$ (the response cannot occur earlier than at time t' which is the moment of switching the field on); $\mathrm{d}\mathbf{r}' = \mathrm{d}x'\mathrm{d}y'\mathrm{d}z'$.

Expanding the field $E_l(\mathbf{r}',t')$ in monochromatic plane waves (see Sec. 2.6), we obtain

$$E_l(\mathbf{r}',t) = \int_{-\infty}^{+\infty} \mathrm{d}\omega \int \mathrm{d}\mathbf{k} E_l(\omega,\mathbf{k})\, \mathrm{e}^{\mathrm{i}\mathbf{k}\cdot\mathbf{r}'-\mathrm{i}\omega t'}, \tag{1.15}$$

where $\mathrm{d}\mathbf{k} = \mathrm{d}k_1\,\mathrm{d}k_2\,\mathrm{d}k_3$. Inserting (1.15) in (1.14) and substituting for the integration variables

$$\tau = t - t', \qquad \boldsymbol{\rho} = \mathbf{r} - \mathbf{r}', \tag{1.16}$$

we obtain the relationship

$$P_j^{(L)}(\mathbf{r},t) = \int_{-\infty}^{+\infty} \mathrm{d}\omega \int \mathrm{d}\mathbf{k} \chi_{jl}(\mathbf{r},t;\omega,\mathbf{k}) E_l(\omega,\mathbf{k}) \mathrm{e}^{\mathrm{i}\mathbf{k}\cdot\mathbf{r}-\mathrm{i}\omega t}, \tag{1.17}$$

where

$$\chi_{jl}(\mathbf{r},t;\omega,\mathbf{k}) = \int \mathrm{d}\tau \int \mathrm{d}\boldsymbol{\rho}\, \chi_{jl}(\mathbf{r},t;t-\tau,\mathbf{r}-\boldsymbol{\rho}) \exp(-\mathrm{i}\mathbf{k}\cdot\boldsymbol{\rho} + \mathrm{i}\omega\tau). \tag{1.18}$$

For media which are homogeneous and stable in time the *electric susceptibility tensor* $\chi_{jl}(\mathbf{r}, t; \omega, \mathbf{k})$ is independent of $\mathbf{r}$ and t. In this case it is merely a function of variables ω and $\mathbf{k}$. This dependence of the electric susceptibility tensor upon ω and $\mathbf{k}$ is different in different spectrum areas.

Within the optical frequency range media with non-local response, i.e. with spatial dispersion, are at the same time optically active – they can change the plane of light polarization. In the latter case the relationship between the linear polarization of medium $P_j^{(L)}$ and the field E_l for a medium which is homogeneous and stable in time can be expanded as

$$P_j^{(L)} = \chi_{jl}(\omega) E_l(\mathbf{r}, t) + \alpha_{jlm}(\omega) \frac{\partial}{\partial x_m} E_l(\mathbf{r}, t) + \dots \tag{1.19}$$

The justification of this relationship is as follows. The change of the integration variables (1.16) in the integral (1.14) gives:

$$P_j^{(L)} = \int \mathrm{d}\tau \int \mathrm{d}\boldsymbol{\rho}\, \chi_{jl}(\mathbf{r}, t; \mathbf{r} - \boldsymbol{\rho}, t - \tau) E_l(\mathbf{r} - \boldsymbol{\rho}, t - \tau). \tag{1.20}$$

If we confine ourselves to the case of a monochromatic field, i.e. to

$$E_l(\mathbf{r} - \boldsymbol{\rho}, t - \tau) = E_l(\mathbf{r} - \boldsymbol{\rho}) \mathrm{e}^{\mathrm{i}\omega(t-\tau)},$$

and apply the expansion

$$E_l(\mathbf{r} - \boldsymbol{\rho}) = E_l(\mathbf{r}) - x_m \frac{\partial}{\partial x_m} E_l(\mathbf{r}),$$

where x_m is the appropriate Cartesian component of vector $\boldsymbol{\rho}$, (1.20) may be written in the form

$$P_j^{(L)} = \int \mathrm{d}\tau \int \mathrm{d}\boldsymbol{\rho}\, \chi_{jl}(\mathbf{r}, t; \mathbf{r} - \boldsymbol{\rho}, t - \tau) \mathrm{e}^{\mathrm{i}\omega(t-\tau)} E_l(\mathbf{r}) -$$
$$- \int \mathrm{d}\tau \int \mathrm{d}\boldsymbol{\rho}\, \chi_{jl}(\mathbf{r}, t; \mathbf{r} - \boldsymbol{\rho}, t - \tau) \mathrm{e}^{\mathrm{i}\omega(t-\tau)} x_m \frac{\partial}{\partial x_m} E_l(\mathbf{r}) + \dots \tag{1.21}$$

or

$$P_j^{(L)} = \chi_{jl}(\mathbf{r}, t; \omega) \cdot E_l(\mathbf{r}) \mathrm{e}^{\mathrm{i}\omega t} + \alpha_{jlm}(\mathbf{r}, t; \omega) \frac{\partial}{\partial x_m} E_l(\mathbf{r}) \mathrm{e}^{\mathrm{i}\omega t}. \tag{1.22}$$

For a medium which is homogeneous and stable in time this form reduces to the relationship (1.19).

In the case of a plane monochromatic wave field, i.e. if

$$E_l(\mathbf{r}, t) = E_{l0}\,\mathrm{e}^{\mathrm{i}(\omega t - \mathbf{k}\cdot\mathbf{r})},$$

the j-th component of the linear polarization vector can be written in the form

$$P_j^{(L)} = \chi_{jl}(\omega)\, E_l(\mathbf{r}, t) + \alpha_{jlm} k_m\, E_l(\mathbf{r}, t). \tag{1.23}$$

1.4. Dielectric Permittivity for Very High Frequencies

In the boundary case $\omega \to \infty$, $\varepsilon(\omega) \to \varepsilon_0$. This fact is obvious for simple physical reasons. When the changes of the field are quick enough, the polarization processes leading to the difference between **E** and **D** will have no time at all to act. This happens if the field frequency is much higher than the frequency of the electron free vibrations in the atoms of a given medium. If this condition is satisfied, the interaction between the electrons and between the electrons and the atomic nucleus can be disregarded and the electrons can be regarded as free.

The velocity $\mathbf{v}$ of the electrons in atoms is small in comparison with c. That is why v/ω, i.e. the distance covered by the electrons during the wave period, is small in comparison with the wavelength c/ω. In these conditions, in order to determine the velocity attained by the electrons in an electromagnetic wave field, we can assume that the field is homogeneous. Then the equation of motion for the electrons has the form

$$m\frac{\mathrm{d}\mathbf{v}}{\mathrm{d}t} = e\mathbf{E}_0\mathrm{e}^{-\mathrm{i}\omega t},$$

where e is the charge of the electron, and m – its mass. Hence, assuming a periodic character of field changes with frequency ω, we have

$$\dot{\mathbf{r}} = \mathbf{v} = \frac{\mathrm{i}e\mathbf{E}}{\omega m}$$

and the radius vector is

$$\mathbf{r} = -\frac{e\mathbf{E}}{m\omega^2}.$$

The polarization vector $\mathbf{P}$ is equal to the dipole moment of a volume unit of the material

$$\mathbf{P} = \sum e\mathbf{r} = -\frac{e^2}{m\omega^2} N\mathbf{E},$$

where N is the number of electrons in a volume unit of a given material. The vector $\mathbf{D}$ can be shown in the form

$$\mathbf{D} = \varepsilon \cdot \mathbf{E} = \varepsilon_0 \mathbf{E} + \mathbf{P}.$$

Finally, for high frequencies we obtain

$$\varepsilon(\omega) = \varepsilon_0 - \frac{Ne^2}{m\omega^2}, \tag{1.24}$$

which is true for far ultraviolet in the case of low-Z elements, and for X-rays in the case of high-Z elements.

1.5. Dielectric Permittivity for Optical Frequencies

A. Free Vibrations of Electrons

In a transparent, isotropic dielectric a field is described by vectors $\mathbf{E}$ and $\mathbf{H}$. Excluding magnetic material it can be assumed that $\mathbf{B} = \mu_0 \mathbf{H}$ and $\mathbf{D} = \varepsilon \mathbf{E} = \varepsilon_0 \mathbf{E} + \mathbf{P}$, where $\mathbf{P}$ is the polarization vector.

The field $\mathbf{E}$ displaces electrons from their equilibrium state by the radius vector $\mathbf{r}$. It is assumed that

$$\mathbf{P} = -Ne\mathbf{r},$$

where e is the electron charge, and N the number of dispersion electrons in a volume unit.

The influence of the field $\mathbf{E}$ on the electric charge $-e$ is shown in Fig. 1.1. The electric charge $-e$ has been pushed away from the ion

$+e$, with which it initially coincided, and its displacement caused by the field $\mathbf{E}$ is opposite to that of the positive ion. The charge $+e$ connected with the ion mass $M (M \gg m)$ can be regarded as stationary. Thus, an electric dipole corresponding to the field direction is formed.

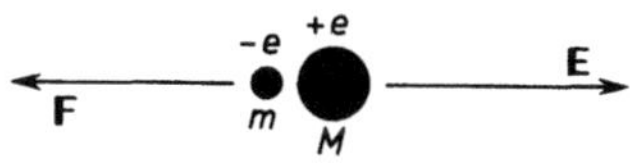

Figure 1.1

An electron is "quasi-elastically" connected with its equilibrium state. When subjected to the influence of the external field $\mathbf{E}$, it will tend to a new equilibrium state. This situation is expressed by the following differential equation:

$$m\ddot{\mathbf{r}} + f\mathbf{r} = -e\mathbf{E},$$

where $f\mathbf{r}$ is a "quasi-elastic" force.

The equation may be written in the form

$$\ddot{\mathbf{r}} + \omega_1^2 \mathbf{r} = -\frac{e}{m}\mathbf{E}, \tag{1.25}$$

where $\omega_1^2 = f/m$ is the square of the free vibration frequency of an electron. In the case of an ideal gas it is the frequency of the ultraviolet.

From (1.25) we obtain the equation for the polarization vector

$$\left(\frac{\partial}{\partial t^2} + \omega_1^2\right)\mathbf{P} = \frac{Ne^2}{m}\mathbf{E}. \tag{1.26}$$

Maxwell's equations

$$\operatorname{curl} \mathbf{E} = -\mu_0 \dot{\mathbf{H}}, \quad \operatorname{curl} \mathbf{H} = \varepsilon_0 \dot{\mathbf{E}} + \dot{\mathbf{P}},$$

yield

$$\varepsilon_0 \mu_0 \ddot{\mathbf{E}} + \mu_0 \ddot{\mathbf{P}} = -\operatorname{curl}(\operatorname{curl} \mathbf{E}) = \Delta \mathbf{E}. \tag{1.27}$$

Applying the operator $(\partial^2/\partial t^2 + \omega_1^2)$ to equation (1.27) and the operator $(\mu_0 \partial^2/\partial t^2)$ to equation (1.26), and adding the corresponding sides of the results, we obtain

$$\left(\frac{\partial^2}{\partial t^2} + \omega_1^2\right)\left(\frac{1}{c^2}\ddot{\mathbf{E}} - \Delta\mathbf{E}\right) + \frac{Ne^2}{m}\mu_0\ddot{\mathbf{E}} = 0. \qquad (1.28)$$

In the case of a plane monochromatic wave $E_y = E_0 e^{i(kx-\omega t)}$ propagating along the x-axis and polarized along the y-axis, (1.28) yields the relationship

$$(-\omega^2 + \omega_1^2)\left(-\frac{\omega^2}{c^2} + k^2\right) = \frac{\mu_0 Ne^2}{m}\omega^2.$$

Hence, we obtain

$$k^2 = \frac{\omega^2}{c^2}\left(1 + \frac{\mu_0 c^2 Ne^2/m}{\omega_1^2 - \omega^2}\right). \qquad (1.29)$$

The quantity $u = \omega/k$ is the *phase velocity* of a plane wave propagating in a dispersive medium. According to the definition, the refractive index of the medium with respect to the vacuum is

$$n = \frac{c}{u} = \frac{ck}{\omega}.$$

Combining this relationship with the dependence $\mu_0 c^2 = 1/\varepsilon_0$ we obtain

$$n^2 = 1 + \frac{Ne^2/m\varepsilon_0}{\omega_1^2 - \omega^2}. \qquad (1.30)$$

Thus, the refractive index depends on the frequency (dispersion).

For very high ω_1, by expanding (1.30) in a series with respect to ω and neglecting all terms except the first we obtain

$$n^2 = 1 + \frac{Ne^2}{m\varepsilon_0\omega_1^2}\left(1 + \frac{\omega^2}{\omega_1^2}\right). \qquad (1.31)$$

Substituting $\lambda = 2\pi c/\omega$, where λ is the wavelength in the vacuum, we obtain

$$n^2 = 1 + A\left(1 + \frac{B}{\lambda^2}\right), \qquad (1.32)$$

where A is the refractive index and B the dispersion coefficient.

B. Free Vibrations of Ions

The refractive index is influenced by the infrared free vibrations of the ions. The polarization **P** can be presented as the sum of two components: $\mathbf{P}_1$ (connected with the electrons) and $\mathbf{P}_2$ (connected with the ions)

$$\mathbf{P} = \mathbf{P}_1 + \mathbf{P}_2. \tag{1.33}$$

The differential equation of forced vibrations of electrons in a variable field **E** has the form

$$\left(\frac{\partial^2}{\partial t^2} + \omega_1^2\right)\mathbf{P}_1 = \frac{Ne^2}{m}\mathbf{E}. \tag{1.34}$$

The external field causes mutual displacement of the differently charged parts of an atom, and forced vibrations not only of the external electrons but also of the ions (the free vibration frequency of the ions is denoted by ω_2). For the description of ion vibrations we use the mutual displacement of the two parts of the atom and the reduced mass

$$\frac{1}{M^*} = \frac{1}{m} + \frac{1}{M},$$

where M is the mass of the ion.

Because of the electrical neutrality of the optical material, we assume that the ion number is N/p (the number of the electrons divided by the ionization multiplicity), and their charge is $Q = ep$.

For $\mathbf{P}_2$ we obtain the equation

$$\left(\frac{\partial}{\partial t^2} + \omega_2^2\right)\mathbf{P}_2 = \frac{Npe^2}{M^*}\mathbf{E}. \tag{1.35}$$

The relations (1.34), (1.35) and Maxwell's equations lead to the equation

$$\varepsilon_0\mu_0\ddot{\mathbf{E}} + \mu_0(\ddot{\mathbf{P}}_1 + \ddot{\mathbf{P}}_2) = \Delta\mathbf{E}. \tag{1.36}$$

Elimination of $\mathbf{P}_1$ and $\mathbf{P}_2$ from the equations (1.34), (1.35) and (1.36) gives us (for monochromatic field) sixth-order equations.

From them we obtain

$$k^2 - \frac{\omega^2}{c^2} = \mu_0 \omega^2 \left(\frac{Ne^2/m}{\omega_1^2 - \omega^2} + \frac{pNe^2/M^*}{\omega_2^2 - \omega^2} \right). \tag{1.37}$$

If we use the definition of the refractive index, the following equation is obtained:

$$n^2 = 1 + \frac{Ne^2/m\varepsilon_0}{\omega_1^2 - \omega^2} + \frac{pNe^2/M^*\varepsilon_0}{\omega_2^2 - \omega^2}. \tag{1.38}$$

In the presence of several free vibration frequencies of the ions and electrons, the number of terms on the right-hand side of (1.38) will be larger.

1.6. Anomalous Dispersion

The character of the dispersion is studied in the immediate neighbourhood of the free vibration frequency $\omega = \omega_1$ on the assumption that the last quantity lies in the area of visible frequencies. In order to avoid trouble with infinite amplitude, we act here as in all resonance cases. In the already considered equations for forced vibrations of electrons under the influence of variable electric field, we introduce a term $(g\omega_1 \dot{\mathbf{r}})$, characterizing the damping. Then, for the polarization vector the following equation is obtained:

$$\left(\frac{\partial^2}{\partial t^2} + g\omega_1 \frac{\partial}{\partial t} + \omega_1^2 \right) \mathbf{P} = \frac{Ne^2}{m} \mathbf{E}. \tag{1.39}$$

For the field $\mathbf{E}$ with the frequency ω equation (1.39) yields

$$\mathbf{P} = \frac{Ne^2}{m} \frac{\mathbf{E}}{\omega_1^2 - \omega^2 - ig\omega_1\omega}. \tag{1.40}$$

For the refractive index n we obtain the following expression:

$$n^2 = n_m^2 + \frac{Ne^2/m\varepsilon_0}{\omega_1^2 - \omega^2 - ig\omega_1\omega}, \tag{1.41}$$

where n_m describes the contributions of free vibrations with frequencies different from ω_1 within the visible part of the spectrum.

The refractive index is a complex quantity. It may be written in the form $n_0(1 + i\kappa)$ which enables us to separate the real part and the imaginary part:

$$n_0^2(1-\kappa^2) = n_m^2 + \frac{Ne^2}{m\varepsilon_0} \frac{\omega_1^2 - \omega^2}{(\omega_1^2 - \omega^2)^2 + g^2\omega_1^2\omega^2}, \tag{1.42}$$

$$2n_0^2\kappa = \frac{Ne^2}{m\varepsilon_0} \frac{g\omega_1\omega}{(\omega_1^2 - \omega^2)^2 + g^2\omega_1^2\omega^2}. \tag{1.43}$$

To simplify these expressions, we introduce the notation

$$\alpha^2 = \frac{Ne^2}{m\varepsilon_0\omega_1^2}, \qquad x = \frac{\omega^2 - \omega_1^2}{\omega_1^2},$$

$$y = \frac{n_0^2(1-\kappa^2)}{\alpha^2}, \qquad z = \frac{2n_0^2\kappa}{\alpha^2}.$$

Then formulas (1.42) and (1.43) take the form

$$y = \frac{n_m^2}{\alpha^2} - \frac{x}{x^2 + g^2(1+x)}, \tag{1.44}$$

$$z = \frac{g\sqrt{1+x}}{x^2 + g^2(1+x)}. \tag{1.45}$$

Calculating the extreme values of y:

$$\frac{\mathrm{d}y}{\mathrm{d}x} = \frac{1}{x^2 + g^2(1+x)}\left(1 - \frac{2x^2 + g^2 x}{x^2 + g^2(1+x)}\right) = 0,$$

we obtain

$$y = y_{\min} = \frac{n_m^2}{\alpha^2} - \frac{1}{2g + g^2} \qquad \text{for} \quad x = +g,$$

$$y = y_{\max} = \frac{n_m^2}{\alpha^2} + \frac{1}{2g - g^2} \qquad \text{for} \quad x = -g.$$

The diagram of the functional relationship between y (1.44) and z (1.45) and x is shown in Fig. 1.2. The curve y has its extreme values for $x = \pm g$, and intersects the line

$$y = \frac{n_n^2}{\alpha^2} \quad \text{for} \quad x = 0.$$

The curve z has the shape of a bell with the maximum for $x = 0$ (where $z = 1/g$) and the half width $= 2g$. The scale for each curve is different. The figure given above not only serves for a visual

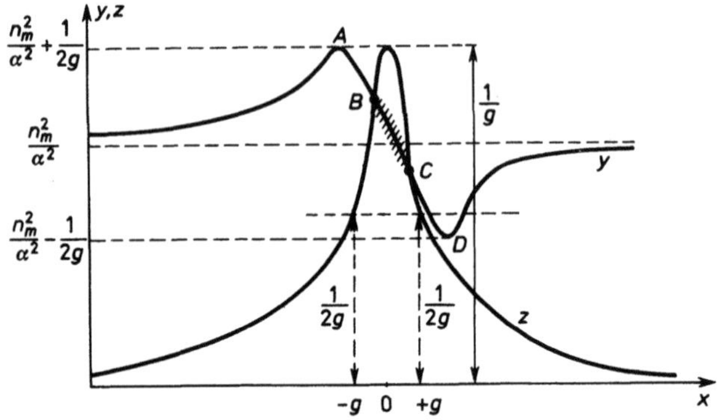

Figure 1.2

presentation of y and z but also shows the changes of n^2 and κ. In the neighbourhood of $x = 0$ these quantities differ slightly from their exact presentation given by formulas (1.44) and (1.45). The interesting part of the graph is the AD branch of the refractive index function. In the region up to the point A and away from the point D the curve ascends with the growth of ω. The AD branch of the refractive index function is called *anomalous dispersion*:

To shorter wavelength there corresponds weaker deflection than to longer wavelengths.

The interval BC of the descending part AD of the curve corresponds to strong absorption. That is why anomalous dispersion can only be observed in the intervals AB and CD.

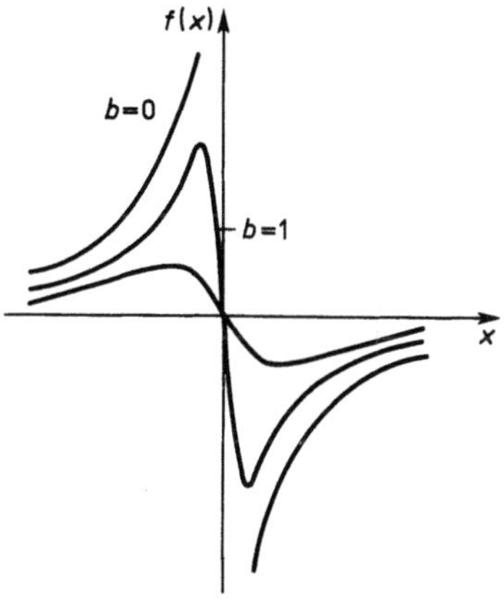

Figure 1.3

Mathematically, the character of the curve y is best given by the function

$$f(x) = -\frac{x}{x^2 + b^2}, \tag{1.46}$$

which for $b \to 0$ becomes a hyperbola $y = -\frac{1}{x}$. This corresponds to the dispersion curve for resonant frequencies in the absence of absorption (Fig. 1.3).

1.7. Boundary Conditions

Until now Maxwell's equations have been formulated for space areas within which the properties of the medium (characterized by ε and μ) are continuous. In optics we often come across a situation where those properties change rapidly in crossing one or more surfaces. Then, in cases where on the boundary surfaces in question the surface current j_{sur} and the surface charge ρ_{sur} are generated, vectors **E**, **H**, **B** and **D** may become discontinuous.

Let a surface S be the surface of discontinuity under consideration (Fig. 1.4). While crossing it the quantities ε and μ undergo

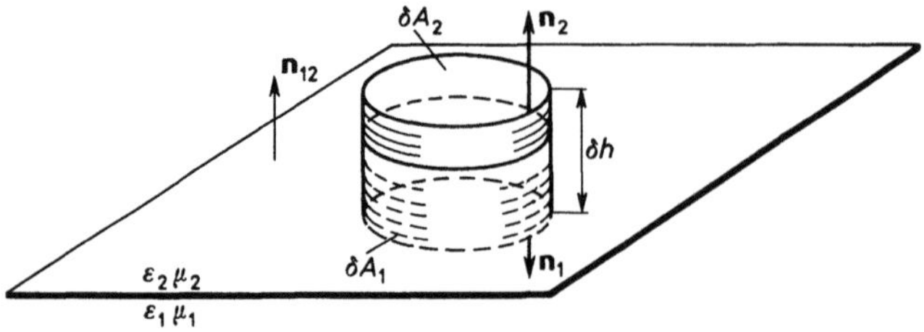

Figure 1.4

stepwise changes from the values ε_1, μ_1 on one side of the surface to the values ε_2, μ_2 on the other side of it. From the surface S we cut out a surface element dS and enclose it inside a closed surface composed of a cylindrical surface which is perpendicular to the surface S and has as its directrix the edge of the element dS, and two small areas δA_1 and δA_2, which lie on the different sides of the surface S and form "the ceiling" and "the bottom" of the cylinder constructed in this way. Integrating the equation $\operatorname{div} \mathbf{B} = 0$ over the volume of the cylinder and using Gauss' theorem, we obtain

$$0 = \int_V \operatorname{div} \mathbf{B} d\tau = \iint_S \mathbf{B} \cdot \mathbf{n} df. \tag{1.47}$$

The second interval is taken over the closed surface of the cylinder and $\mathbf{n}$ is a versor perpendicular to an element of the surface of integration. Since the choice of the cylindrical surface is fully arbitrary (the height of the cylinder may be very arbitrarily small), for small δA_1 and δA_2 (1.47) can be written in the form

$$(\mathbf{B}^{(1)} \cdot \mathbf{n}_1) \delta A_1 + (\mathbf{B}^{(2)} \cdot \mathbf{n}_2) \delta A_2 + \\ + \text{contribution from the side walls} = 0. \tag{1.48}$$

Neglecting the last term and assuming that $\delta A_1 = \delta A_2$, we obtain from (1.48) the equality

$$(\mathbf{B}^{(1)} \cdot \mathbf{n}_1 + \mathbf{B}^{(2)} \cdot \mathbf{n}_2) \delta A = 0. \tag{1.49}$$

If $\mathbf{n}_{12}$ is a versor directed from the first medium (ε_1, μ_1) to the second medium (ε_2, μ_2), then we find that $\mathbf{n}_{12} = -\mathbf{n}_1 = \mathbf{n}_2$. Hence the required condition for the behaviour of vector **B** in crossing the interface of two media has the form

$$\mathbf{n}_{12} \cdot (\mathbf{B}^{(1)} - \mathbf{B}^{(2)}) = 0. \tag{1.50}$$

It can be read as follows:

The normal component of vector **B** *is continuous in crossing the boundary surface between two media.*

The displacement vector **D** may be treated in a similar way. The integration of the equation $\operatorname{div} \mathbf{D} = \rho$ over the volume of the cylinder discussed above yields

$$\int_V \rho \, \mathrm{d}\tau = \int \operatorname{div} \mathbf{D} \, \mathrm{d}\tau = \iint \mathbf{D} \cdot \mathbf{n} \, \mathrm{d}f. \tag{1.51}$$

Passing with the height of the cylinder to zero and defining the surface density of the charge as

$$\lim_{h \to 0} \int \rho \, \mathrm{d}\tau = \iint \sigma_{\text{sur}} \mathrm{d}f, \tag{1.52}$$

we can write formula (1.51) in the form

$$\mathbf{D}^{(1)} \cdot \mathbf{n}_1 \delta A_1 + \mathbf{D}^{(2)} \cdot \mathbf{n}_2 \delta A_2 = \sigma_{\text{sur}} \mathrm{d}f. \tag{1.53}$$

Assuming that $\mathbf{n}_{12} = -\mathbf{n}_1 = \mathbf{n}_2$ and $\delta A_1 = \delta A_2 = \mathrm{d}f$ we obtain

$$\mathbf{D}^{(1)} \cdot \mathbf{n}_{12} - \mathbf{D}^{(2)} \cdot \mathbf{n}_{12} = \sigma_{\text{sur}}. \tag{1.54}$$

This means that:

In the presence of a surface charge at the boundary between two media the normal component of vector **D** *varies stepwise; the step in value is equal to the surface density of the charge.*

In order to examine the behaviour of the tangential components of **E**, let us consider a closed curve K (Fig. 1.5), which is immersed partially in the medium (ε_1, μ_1) and partially in the medium (ε_2, μ_2). Integrating the equation

$$\operatorname{curl} \mathbf{E} = -\frac{\partial \mathbf{B}}{\partial t},$$

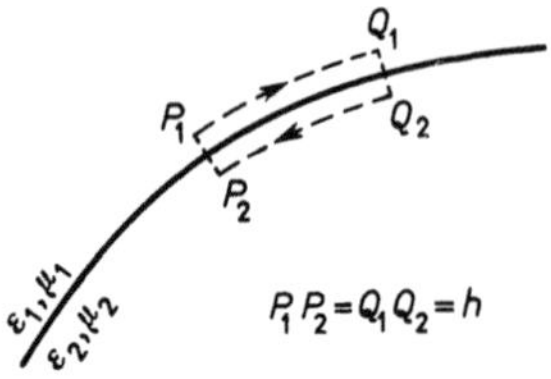

Figure 1.5

over the surface element enclosed within the curve, we obtain

$$\iint_S (\mathrm{curl}\,\mathbf{E})\cdot\mathbf{n}\mathrm{d}f = \oint_K \mathbf{E}\mathrm{d}\mathbf{l} = -\iint_S \dot{\mathbf{B}}\cdot\mathbf{n}\mathrm{d}f, \tag{1.55}$$

where $\mathbf{n}$ is a versor perpendicular to the surface element under consideration.

For small dimensions of the curve, in particular of its elements perpendicular to the boundary surface between two media, we obtain

$$(\mathbf{E}^{(1)}\cdot\mathbf{t}_1 + \mathbf{E}^{(2)}\cdot\mathbf{t}_2)\mathrm{d}l = -\dot{\mathbf{B}}\cdot\mathbf{n}h\mathrm{d}l. \tag{1.56}$$

Assuming the finiteness of $\dot{\mathbf{B}}$ and $h \to 0$, and taking into account the relations $-\mathbf{t}_1 = \mathbf{t}_2 = \mathbf{t}$, we obtain the following condition:

$$(\mathbf{E}^{(1)} - \mathbf{E}^{(2)})\cdot\mathbf{t} = 0. \tag{1.57}$$

The tangential component of the electric intensity vector $\mathbf{E}$ *is continuous in crossing the boundary surface between two media.*

The behaviour of the tangential components of the magnetic field at the discontinuity surface is found from the equation

$$\mathrm{curl}\,\mathbf{H} = \dot{\mathbf{D}} + \mathbf{j}.$$

As before, we integrate this equation over the surface enclosed within the curve K (Fig. 1.5) and apply Stokes' theorem. Hence

$$\oint_K \mathbf{H}\cdot\mathrm{d}\mathbf{l} = \iint_S (\dot{\mathbf{D}} + \mathbf{j})\cdot\mathbf{n}\mathrm{d}f. \tag{1.58}$$

For small dimensions of the curve, and for $h \to 0$, formula (1.58) takes the form

$$(\mathbf{H}^{(1)}\cdot\mathbf{t}_1 + \mathbf{H}^{(2)}\cdot\mathbf{t}_2)\mathrm{d}l = (\dot{\mathbf{D}}\cdot\mathbf{n}h\mathrm{d}l) + \mathbf{j}\cdot\mathbf{n}h\mathrm{d}l. \tag{1.59}$$

After substituting

$$-\mathbf{t}_1 = \mathbf{t}_2 = \mathbf{t}, \qquad \lim_{h\to 0}(\mathbf{j}\cdot\mathbf{n}h) = j_{\mathrm{sur}},$$

where j_{sur} stands for the density of the current flowing on the boundary surface, we obtain the equation

$$\mathbf{H}^{(1)}\cdot\mathbf{t} - \mathbf{H}^{(2)}\cdot\mathbf{t} = j_{\mathrm{sur}}. \tag{1.60}$$

This means that:

In the presence of a surface current j_{sur} at the interface of two media, the tangential component of vector **H** *changes stepwise by a value equal to the density of the surface current.*

1.8. The Law of Conservation of Energy for the Electromagnetic Field

From the point of view of electromagnetic theory, the light intensity may be interpreted as a flux of field energy.

We formulate the law of conservation of energy for the electromagnetic field. To this end, the notion of Poynting's vector is introduced. Poynting's vector is defined as follows

$$\mathbf{S} = \mathbf{E} \times \mathbf{H}. \tag{1.61}$$

In the system SI the unit of quantity (1.61) is $\mathrm{J/m^2 \cdot s}$ and so Poynting's vector represents the energy flow across a surface unit. Using (1.61) we obtain the following equality for $\operatorname{div}\mathbf{S}$:

$$\operatorname{div}\mathbf{S} = \mathbf{H}\cdot\operatorname{curl}\mathbf{E} - \mathbf{E}\cdot\operatorname{curl}\mathbf{H}. \tag{1.62}$$

After applying Maxwell's equations, equality (1.62) takes the following form:

$$\operatorname{div}\mathbf{S} = -\mathbf{H}\cdot\dot{\mathbf{B}} - \mathbf{E}\cdot\dot{\mathbf{D}} - \mathbf{E}\cdot\mathbf{j}. \tag{1.63}$$

Taking into consideration an area V (Fig. 1.6) confined by

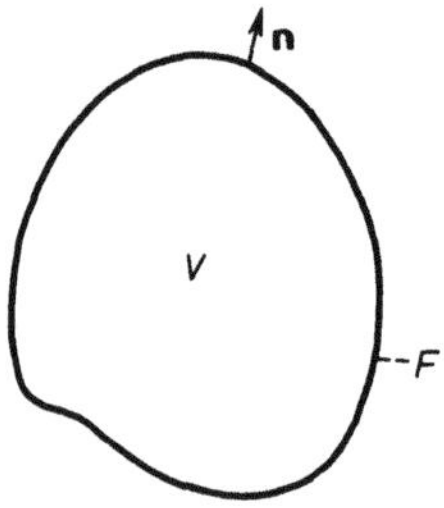

Figure 1.6

a closed surface F, and integrating formula (1.63) over that area, we obtain

$$\iint_F \mathbf{S}\cdot\mathbf{n}\mathrm{d}f = -\int_V (\mathbf{H}\cdot\dot{\mathbf{B}} + \mathbf{E}\cdot\dot{\mathbf{D}})\mathrm{d}\tau - \int_V \mathbf{j}\cdot\mathbf{E}\mathrm{d}\tau, \tag{1.64}$$

where $\iint_F \mathbf{S}\cdot\mathbf{n}\mathrm{d}f$ is the energy radiated from the area V and $\int_V \mathbf{j}\cdot\mathbf{E}\mathrm{d}\tau$ is Joule's heat emitted within the area V. (This interpretation may be given by using the so-called *differential form of Ohm's law*: $\mathbf{j} = \sigma\mathbf{E}$, where σ is the conductivity). The remaining term in the equality (1.64), after applying the relations $\mathbf{B} = \mu\mathbf{H}$ and $\mathbf{D} = \varepsilon\mathbf{E}$, is written in the form

$$-\int_V (\mathbf{H}\cdot\dot{\mathbf{B}} + \mathbf{E}\cdot\dot{\mathbf{D}})\mathrm{d}\tau = -\frac{\mathrm{d}}{\mathrm{d}t}\int_V \left(\tfrac{1}{2}\mathbf{H}\cdot\mathbf{B} + \tfrac{1}{2}\mathbf{E}\cdot\mathbf{D}\right)\mathrm{d}\tau. \tag{1.65}$$

This may be interpreted as the speed of the loss of energy of the field contained within the area V. The expressions $\frac{1}{2}(\mathbf{H}\cdot\mathbf{B})$ and $\frac{1}{2}(\mathbf{E}\cdot\mathbf{D})$ should then be regarded as the density of the energy of the magnetic field and the density of the energy of the electric field.

Thus, the *law of energy conservation* may be formulated in the form given below:

Loss of the field energy contained in the area V may be due to emission or to conversion into Joule's heat:

$$-\frac{\mathrm{d}}{\mathrm{d}t}\int_V\left(\tfrac{1}{2}\mathbf{H}\cdot\mathbf{B}+\tfrac{1}{2}\mathbf{E}\cdot\mathbf{D}\right)\mathrm{d}\tau=\iint_F(\mathbf{E}\times\mathbf{H})\cdot\mathbf{n}\,\mathrm{d}f+$$

$$+\int_V\mathbf{j}\cdot\mathbf{E}\,\mathrm{d}\tau. \tag{1.66}$$

Poynting's vector is closely related to the radiation intensity I. By that quantity we understand the energy carried in a time unit across a surface unit perpendicular to the direction of Poynting's vector.

The radiation intensity is equal to the value of Poynting's vector averaged in time:

$$I=\langle|\mathbf{E}\times\mathbf{H}|\rangle=\frac{1}{T}\left|\int_0^T(\mathbf{E}\times\mathbf{H})\mathrm{d}\tau\right|. \tag{1.67}$$

1.9. The Law of Conservation of Momentum for the Electromagnetic Field

An electromagnetic wave involves a momentum. In order to show this, we consider a physical system consisting of electrical charges endowed with mass (electrons) and of the electromagnetic field.

Every electron is acted upon by the Lorentz force directed from the electromagnetic field

$$\frac{\mathrm{d}}{\mathrm{d}t}\mathbf{p}_{\mathrm{el}}=e\mathbf{E}+e(\mathbf{v}\times\mathbf{B}). \tag{1.68}$$

The rate of momentum change for N electrons contained in the area V is

$$\frac{\mathrm{d}}{\mathrm{d}t}\sum_j^N\mathbf{p}_{\mathrm{el}}^j=\sum_j^N e\left[\mathbf{E}(\mathbf{r}_j)+\mathbf{v}_j\times\mathbf{B}(\mathbf{r}_j)\right]. \tag{1.69}$$

Now replacing the discrete distribution of charge and mass by a continuous distribution, i.e. writing

$$e=\rho_{\mathrm{el}}\mathrm{d}\tau,\qquad \mathbf{p}_{\mathrm{el}}=\mathbf{P}_{\mathrm{el}}\mathrm{d}\tau$$

(where ρ_{el} and $\mathbf{P}_{el}$ are, respectively, the charge density and the density of the electron momentum), we may replace sums by integrals over the area Ω of the occurrence of the electrons. Under these assumptions equality (1.69) can be rewritten in the form

$$\frac{\mathrm{d}}{\mathrm{d}t}\int\limits_{\Omega}\mathbf{P}_{el}\mathrm{d}\tau = \int\limits_{\Omega}(\rho_{el}\cdot\mathbf{E} + \rho_{el}\mathbf{v}\times\mathbf{B})\mathrm{d}\tau. \tag{1.70}$$

The use of Maxwell's equations permits a transformation of the above equality. Thus the j-th component of the above equality can be expressed in homogeneous media by the formula

$$\begin{aligned}\frac{\mathrm{d}}{\mathrm{d}t}&\int\limits_{\Omega}(\mathbf{p}_{el} + \mathbf{D}\times\mathbf{B})\cdot\mathbf{i}_j\mathrm{d}\tau\\ &= \int\limits_{\Omega}\operatorname{div}\left[E_j\mathbf{D} + H_j\mathbf{B} - \tfrac{1}{2}\mathbf{i}_j(\mathbf{E}\cdot\mathbf{D} + \mathbf{H}\cdot\mathbf{B})\right]\mathrm{d}\tau\\ &= \iint\limits_{S}\mathbf{n}\cdot\left[E_j\mathbf{D} + H_j\mathbf{B} - \tfrac{1}{2}\mathbf{i}_j(\mathbf{E}\cdot\mathbf{D} + \mathbf{H}\cdot\mathbf{B})\right]\mathrm{d}f,\end{aligned} \tag{1.71}$$

where $\mathbf{i}_j$ is the versor of the j-th axis of the Cartesian system, and S is the surface confining the area Ω occupied by the physical system under consideration. The quantity $(\mathbf{D}\times\mathbf{B})$ appearing in (1.71) may be interpreted as the density of the momentum of the electromagnetic field. The equation represents the law of conservation of momentum for the physical system in question.

The expression $\left[E_j\mathbf{D} + H_j\mathbf{B} - \tfrac{1}{2}\mathbf{i}_j(\mathbf{E}\cdot\mathbf{D} + \mathbf{H}\cdot\mathbf{B})\right]$, occuring under the surface integral, denotes a vector which is the j-th column of the tension tensor of electromagnetic field. The components of this tensor are

$$\pi_{lm} = \left[E_lD_m + H_lB_m - \tfrac{1}{2}\delta_{lm}(\mathbf{E}\cdot\mathbf{D} + \mathbf{H}\cdot\mathbf{B})\right], \tag{1.72}$$

where δ_{lm} is the Kronecker delta function.

The surface integral in the expression (1.71) represents the force causing the change of momentum of the system. It acts upon the surface of the area containing the physical system. In the absence of electrons, i.e. if $\mathbf{P}_{el} = \mathbf{0}$, the force is only connected with the change of momentum of the electromagnetic field.

Extending the area Ω over the whole space and assuming that the components of the tension tensor of the electromagnetic field decrease rapidly while moving away to the infinity, the surface integral becomes zero, and then

$$\frac{d}{dt}\int_{\Omega}(\mathbf{P}_{el} + \mathbf{D} \times \mathbf{B})d\tau = \mathbf{0},$$

which expresses the law of conservation of the total momentum of the physical system under consideration.

1.10. The Wave Equation

From Maxwell's equations for a homogeneous medium without dispersion, in the area where $\rho = 0$, we can obtain the wave equation.

Taking the curl of both sides of the equation $\operatorname{curl}\mathbf{E} = -\partial\mathbf{B}/\partial t$ and taking linearity and isotropy of a medium into account, i.e. assuming $\mathbf{B} = \mu\mathbf{H}$, $\mathbf{D} = \varepsilon\mathbf{E}$, we are given

$$\operatorname{curl}\operatorname{curl}\mathbf{E} = -\mu\frac{\partial}{\partial t}\operatorname{curl}\mathbf{H}. \tag{1.74}$$

Hence and considering the equation

$$\operatorname{curl}\mathbf{H} = \varepsilon\frac{\partial\mathbf{E}}{\partial t} + \mathbf{j}$$

we can eliminate the field $\mathbf{H}$ from (1.74); this gives

$$\operatorname{curl}\operatorname{curl}\mathbf{E} = -\mu\varepsilon\frac{\partial^2\mathbf{E}}{\partial t^2} - \mu\frac{\partial\mathbf{j}}{\partial t}. \tag{1.75}$$

Applying the identity $\operatorname{curl}\operatorname{curl}\mathbf{a} = \operatorname{grad}\operatorname{div}\mathbf{a} - \Delta\mathbf{a}$ and assuming the homogeneity of the medium and the absence of free charges: $\operatorname{div}\mathbf{D} = \varepsilon\operatorname{div}\mathbf{E} = 0$, we can write (1.75) in the form of the wave equation

$$\Delta\mathbf{E} - \varepsilon\mu\frac{\partial^2\mathbf{E}}{\partial t^2} = -\mu\frac{\partial\mathbf{j}}{\partial t}.$$

In particular, if $\partial\mathbf{j}/\partial t = \mathbf{0}$, then

$$\Delta\mathbf{E} - \frac{1}{v^2}\frac{\partial^2\mathbf{E}}{\partial t^2} = \mathbf{0}, \tag{1.76}$$

where $v = 1/\sqrt{\varepsilon\mu}$ is the velocity of propagation of the electromagnetic disturbance.

The wave equation is satisfied by any component of the electric vector also in the case where $\varepsilon = \varepsilon(\mathbf{r})$ provided that the spatial changes are slight within distances of the order of light wavelength (to be discussed below).

The physical sense of the wave equation becomes clear if we note that an arbitrary function

$$\psi = f\left(t - \frac{\mathbf{r}\cdot\mathbf{n}}{v}\right) \tag{1.77}$$

is a solution of this equation on condition that the function f has continuous second derivatives ($\mathbf{r}$ is the position vector of the observation point, $\mathbf{n}$ – the constant vector). In the solution of (1.77) v is not a function of frequency. The solution represents a plane wave propagating in space filled with a homogeneous medium $\varepsilon = \text{const}$. To prove this, it is enough to note that for a fixed value of argument

$$\eta = t - \frac{1}{v}\mathbf{n}\cdot\mathbf{r}$$

the function $f(\eta)$ is constant. For a fixed t we have $\eta = \text{const}$ on the plane $\mathbf{r}\cdot\mathbf{n} = \text{const}$. The vector $\mathbf{n}$ is perpendicular to this plane.

A particular case of the solution of the wave equation is the periodic function

$$g = A\cos\left(\omega t - \frac{\omega}{v}\mathbf{n}\cdot\mathbf{r}\right). \tag{1.78}$$

It may also be presented in the form

$$g = A\cos(\omega t - \mathbf{k}\cdot\mathbf{r}),$$

where **k** is defined as

$$\mathbf{k} = \frac{\omega}{v}\mathbf{n}$$

and is called the *wave vector*. Its length is

$$k = \frac{\omega}{v} = \omega\sqrt{\varepsilon\mu}. \tag{1.79}$$

In most media we have $v = v(\omega)$, and the waves propagate with different phase velocities for different ω. Hence the solution of equation (1.76) given in the form of plane wave (1.77) remains correct only in the vacuum.

For a non-homogeneous medium, i.e. where $\varepsilon = \varepsilon(\mathbf{r})$, equation (1.75) takes the form

$$\Delta\mathbf{E} - \nabla\left(\mathbf{E}\cdot\frac{\nabla\varepsilon}{\varepsilon}\right) - \mu\varepsilon\frac{\partial^2\mathbf{E}}{\partial t^2} = -\mu\frac{\partial\mathbf{j}}{\partial t}. \tag{1.80}$$

Equation (1.80) is useless in practice. The question arises in what non-homogeneous media the wave equation could still be applied? We find that it is possible in the case where the relative change of ε within distances of the order of the wavelength is considerably less than one. This condition is often fulfilled in most non-homogeneous media.

The justification for this theorem can be shown in the case of a plane monochromatic wave. We seek a condition under which the dominant terms in equation (1.80) are $\Delta\mathbf{E}$ and $\mu\varepsilon\partial^2\mathbf{E}/\partial t^2$. For a plane monochromatic wave these terms are

$$-\Delta \mathbf{E} = \left(\frac{2\pi}{\lambda}\right)^2 \mathbf{E},$$

$$-\mu\varepsilon \frac{\partial^2 \mathbf{E}}{\partial t^2} = \omega^2 \mu\varepsilon \mathbf{E} = \left(\frac{2\pi}{\lambda}\right)^2 \mathbf{E}, \tag{1.81}$$

i.e., they are equal.

For the estimation of term $\nabla(\mathbf{E}\cdot\nabla\varepsilon/\varepsilon)$ the operator ∇ is replaced by a directional derivative in space in a certain direction **s**. This permits us to write

$$\left|\nabla\left(\mathbf{E}\cdot\frac{\nabla\varepsilon}{\varepsilon}\right)\right| \approx \frac{\partial}{\partial s}\left(\mathbf{E}\cdot\frac{\nabla\varepsilon}{\varepsilon}\right) = \frac{2\pi}{\lambda}\mathbf{E}\cdot\frac{\nabla\varepsilon}{\varepsilon} + \mathbf{E}\frac{\partial}{\partial s}\left(\frac{\nabla\varepsilon}{\varepsilon}\right). \tag{1.82}$$

On the strength of (1.81) and (1.82) we obtain

$$\left|\frac{\nabla[\mathbf{E}\cdot(\nabla\varepsilon/\varepsilon)]}{\Delta\mathbf{E}}\right| \approx \frac{(2\pi/\lambda)\mathbf{E}\cdot(\nabla\varepsilon/\varepsilon)}{(2\pi/\lambda)^2 E} = \frac{\lambda}{2\pi}\frac{|\nabla\varepsilon|}{\varepsilon}. \tag{1.83}$$

In order to get (1.83) the term $\mathbf{E}\partial/\partial s(\nabla\varepsilon/\varepsilon)$ has been neglected. Approximating the gradient ε by the differential quotient

$$|\nabla\varepsilon| = \frac{\varepsilon(\mathbf{r}) - \varepsilon(\mathbf{r} + d\mathbf{r})}{|d\mathbf{r}|}$$

and assuming $|d\mathbf{r}| = \lambda$, we finally find the required estimation

$$\frac{|\nabla[\mathbf{E}\cdot(\nabla\varepsilon/\varepsilon)]|}{\Delta\mathbf{E}} \approx \frac{1}{2\pi}\frac{\varepsilon_1 - \varepsilon_2}{\varepsilon},$$

where $\varepsilon_1 = \varepsilon(\mathbf{r})$, $\varepsilon_2 = \varepsilon(\mathbf{r} + d\mathbf{r})$.

If the inequality

$$\frac{1}{2\pi}\frac{\varepsilon_1 - \varepsilon_2}{\varepsilon} \ll 1$$

is satisfied, then the omission of the second term in equation (1.80) is fully justified. This means that if the relative change of ε within a distance of the order of the wavelength is considerably less than one, then the use of the wave equation in such a medium is admissible.

The arguments mentioned above show that the wave equation may be applied also if ε is variable in space, but its changes must be insignificant within distances of the order of the wavelength. Unless the surface separates two media, this condition is almost always fulfilled. Thus the propagation of light in a non-homogeneous medium may be described by the wave equation.

In most practical cases, the difference between the wave equation and the equation (1.80) is very slight.

1.11. Kirchhoff's Integral Formula

Let us consider the scalar function $v(x, y, z, t)$ satisfying the wave equation

$$\Delta v - \frac{1}{c^2} \frac{\partial^2}{\partial t^2} v = -f, \tag{1.84}$$

where c is the velocity of propagation of the disturbance, and $f = f(x, y, z, t)$ is a given function describing the source of the disturbance field.

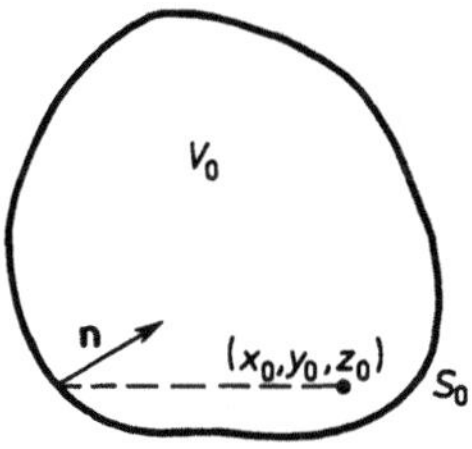

Figure 1.7

Let V_0 be a space area enclosed by a surface S_0 and containing a point (x_0, y_0, z_0) (Fig. 1.7). Then $v(x_0, y_0, z_0, t)$ is determined by the expression

$$v(x_0, y_0, z_0, t) = \frac{1}{4\pi} \int\limits_{V_0} \frac{1}{r} [f] \, \mathrm{d}v + \frac{1}{4\pi} \iint\limits_{S_0} \left\{ [v] \frac{\partial}{\partial n} \left(\frac{1}{r} \right) - \right.$$

$$\left. - \frac{1}{cr} \frac{\partial r}{\partial n} \left[\frac{\partial v}{\partial t} \right] - \frac{1}{r} \left[\frac{\partial v}{\partial n} \right] \right\} \mathrm{d}f, \tag{1.85}$$

where r is the distance between the chosen point (x_0, y_0, z_0) and a point (x, y, z) on the surface S_0; $\partial/\partial n = \mathbf{n} \cdot \nabla$ stands for differentiation with respect to the inward normal to the surface; the symbol $[\varphi]$ denotes a delay described in the following way $[\varphi] = \varphi(x, y, z, t - r/c)$.

If $f(x, y, z, t) = 0$ at all points inside V_0, then the first term in expression (1.85) equals zero.

The relation mentioned above is called *Kirchhoff's integral formula.* It enables us to express $v(x_0, y_0, z_0, t)$ by using the quantities v, $\partial v/\partial n$ and $\partial v/\partial t$ at all points of the surface S_0 at moments of time $\Delta t = r/c$ earlier than the time t under consideration. In order to prove Kirchhoff's integral formula, we use the following facts.

1. If a function $h(t)$ is continuous and

$$\int_a^b w(t) h(t) \mathrm{d}t = 0 \tag{1.86}$$

for any function $w(t)$, then $h(t) = 0$ within the interval (a, b).

2. If $w(t)$ is an arbitrary, twice differentiable function and (x_0, y_0, z_0) a fixed point situated outside the area Ω (Fig. 1.8), then the following function can be formed:

$$\psi(x, y, z, t) = \frac{w(t + r/c)}{r}, \tag{1.87}$$

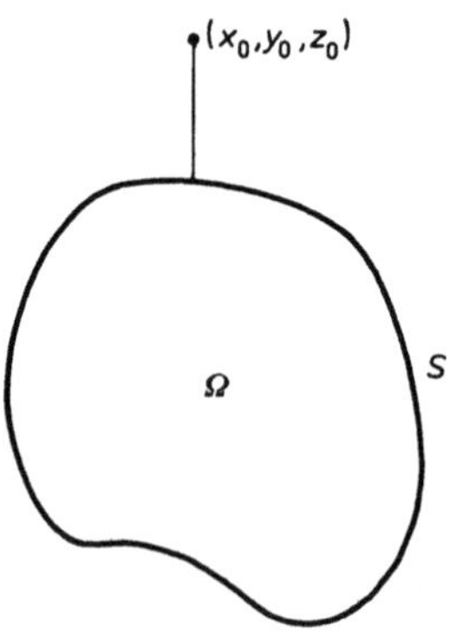

Figure 1.8

where $r = \sqrt{(x - x_0)^2 + (y - y_0)^2 + (z - z_0)^2}$. This function satisfies the homogeneous wave equation

$$\Delta\psi - \frac{1}{c^2}\frac{\partial^2 \psi}{\partial t^2} = 0 \quad \text{for} \quad r \neq 0.$$

3. If the functions u, v have continuous first and second derivatives within an area Ω enclosed by the surface S_0, then

$$\int\limits_{\Omega} (u\Delta v - v\Delta u)\,\mathrm{d}V = \iint\limits_{S_0} \left(v\frac{\partial u}{\partial n} - u\frac{\partial v}{\partial n}\right)\mathrm{d}f. \tag{1.88}$$

This is called *Green's theorem.*

Assuming that $u = \psi$ and demanding that v should satisfy equation (1.84), we obtain the following equality:

$$\psi\Delta v - v\Delta\psi = \psi\left(\frac{1}{c^2}\frac{\partial^2 v}{\partial t^2} - f\right) - v\frac{1}{c^2}\frac{\partial^2 \psi}{\partial t^2}. \tag{1.89}$$

After applying (1.89), Green's theorem may be written in the form

$$\frac{1}{c^2}\int\limits_{\Omega}\left(\frac{w}{r}\frac{\partial^2 v}{\partial t^2} - \frac{v}{r}\frac{\partial w^2}{\partial t^2}\right)\mathrm{d}\tau$$

$$= \iint\limits_{S_0}\left[-\frac{w}{r}\frac{\partial v}{\partial n} + v\frac{\partial}{\partial n}\left(\frac{w}{r}\right)\right]\mathrm{d}f + \int\limits_{\Omega}\frac{wf}{r}\,\mathrm{d}\tau, \tag{1.90}$$

where w and $\partial^2 w/\partial t^2$ are taken at the instant $t + r/c$. Denoting by r_m the maximum value of the distance between the fixed point (x_0, y_0, z_0) and the points of the area Ω, we impose the following condition on the function $w(t)$:

$$w(t) = 0 \quad \text{for} \quad |t| > \frac{2r_m}{c}. \tag{1.91}$$

For sufficiently large values of time $T = 3r_m/c$ at any point $(x, y, z) \in \Omega$, the equality

$$\int_{T}^{-T} \left(w \frac{\partial^2 v}{\partial t^2} - v \frac{\partial^2 w}{\partial t^2} \right) dt$$

$$= \left[w\left(t + \frac{r}{c}\right) \frac{\partial v}{\partial t} - v \frac{\partial w}{\partial t}\left(t + \frac{r}{c}\right) \right] \Bigg|_{-T}^{T} = 0 \qquad (1.92)$$

is satisfied, because $|-T + r/c| > 2r_m/c$.

Taking into account (1.92), we obtain from (1.90) the following equality:

$$\iint\limits_{S_0} \int_{-T}^{T} \left[-\frac{w}{r}\frac{\partial v}{\partial n} + v \frac{\partial}{\partial n}\left(\frac{w}{r}\right) \right] dt\, df + \int\limits_{\Omega} \int_{-T}^{T} \frac{w}{r} f\, dt\, d\tau = 0. \qquad (1.93)$$

The identity

$$\frac{\partial}{\partial n} \frac{w(t + r/c)}{r} = \left(\frac{\partial}{\partial n}\left(\frac{1}{r}\right) \right) w + \frac{1}{cr}\left(\frac{\partial r}{\partial n}\right) \frac{\partial w}{\partial t} \qquad (1.94)$$

and the condition (1.91) give

$$\int_{-T}^{T} v \frac{\partial w}{\partial t} dt = w\left(t + \frac{r}{c}\right) v \Bigg|_{-T}^{T} - \int_{-T}^{T} w \frac{\partial v}{\partial t} dt = - \int_{-T}^{T} w \frac{\partial v}{\partial t} dt \qquad (1.95)$$

If we take into account (1.94) and (1.95) in (1.93) and change the variables $(t' = t + r/c)$, the integration limits will be: $-T + r/c$, $T + r/c$. In view of the condition (1.91), the integration limits may be replaced by T and $-T$. Then we transpose the integration sequence, and remove $w(t')$, as independent of the spatial variables, from under the sign of integral taken over the spatial variables. This enables us to write (1.93) in the form

$$\int_{-T}^{T} w(t')\left\{\iint_S\left(-\frac{1}{r}\left[\frac{\partial v}{\partial n}\right]+[v]\frac{\partial}{\partial n}\left(\frac{1}{r}\right)-\frac{1}{cr}\frac{\partial r}{\partial n}\left[\frac{\partial v}{\partial t}\right]\right)\mathrm{d}f+\right.$$

$$\left.+\int_\Omega\frac{[f]}{r}\,\mathrm{d}\tau\right\}\mathrm{d}t'=0, \tag{1.96}$$

where the square bracket [ ] denotes that the appropriate expressions are taken for the time $t = t' - r/c$. Expression (1.96) is valid for any $|t'| < 2r_m/c$, i.e., for $|t| < r_m/c$. Since the function $w(t)$ is arbitrary, the expression in the curly brackets (see (1.86)) should equal to zero for any t and any point (x_0, y_0, z_0).

To finish the proof of Kirchhoff's integral formula (1.85), it is assumed that $S = S_0 + S_1$, as shown in Fig. 1.9. If S_1 is a spherical

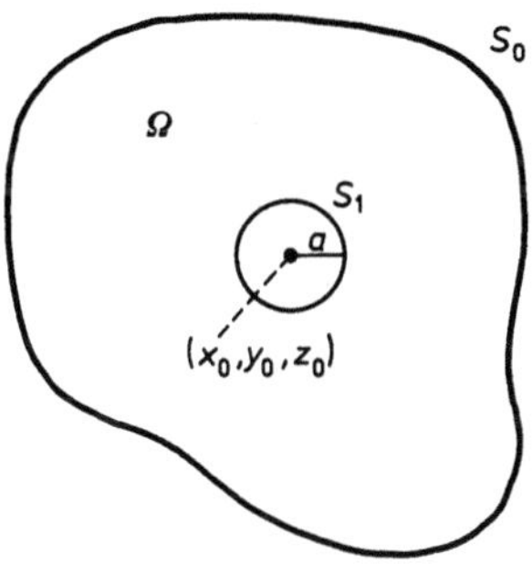

Figure 1.9

surface with the centre at the point (x_0, y_0, z_0), then at all points of the surface the following equalities are satisfied:

$$\frac{\partial r}{\partial n}=1,\qquad \frac{\partial(1/r)}{\partial n}=-\frac{1}{a},\qquad r=a.$$

Hence

$$\iint_{S_1}(\ldots)\mathrm{d}f=\iint_{S_1}\left(-\frac{1}{a}\left[\frac{\partial v}{\partial n}\right]-\frac{[v]}{a^2}-\frac{1}{ac}\left[\frac{\partial v}{\partial t}\right]\right)\mathrm{d}f\to -4\pi v,$$

since $\mathrm{d}f = a^2 \mathrm{d}\varphi$ (where $\mathrm{d}\varphi$ is a solid angle) and $v = v(x, y, z, t - a/c) \to v(x_0, y_0, z_0, t)$ for $a \to 0$. Thus the quantity in the parentheses under the integral taken over the surface S_1 tends to $4\pi v$, which proves the correctness of Kirchhoff's integral formula.

1.12. Radiation of Electromagnetic Waves

Let us assume that the whole space is filled with an isotropic medium described by the constitutive relations

$$\mathbf{D} = \varepsilon \mathbf{E}, \qquad \mathbf{B} = \mu \mathbf{H}.$$

It is also assumed that there are no free charges in the medium (i.e. $\rho = 0$) but only local currents of the form

$$\mathbf{j} = \mathbf{j}_0 \delta(\mathbf{r}) \mathrm{e}^{\mathrm{i}\omega t}, \tag{1.97}$$

where $\delta(\mathbf{r})$ is Dirac's delta, and the current $\mathbf{j}_0$ is directed along the z-axis of the chosen coordinate system (Fig. 1.10).

Under these assumptions Maxwell's equations lead to the following non-homogeneous wave equation for vector $\mathbf{H}$:

$$\Delta \mathbf{H} - \varepsilon \mu \frac{\partial^2 \mathbf{H}}{\partial t^2} = -\operatorname{curl} \mathbf{j}. \tag{1.98}$$

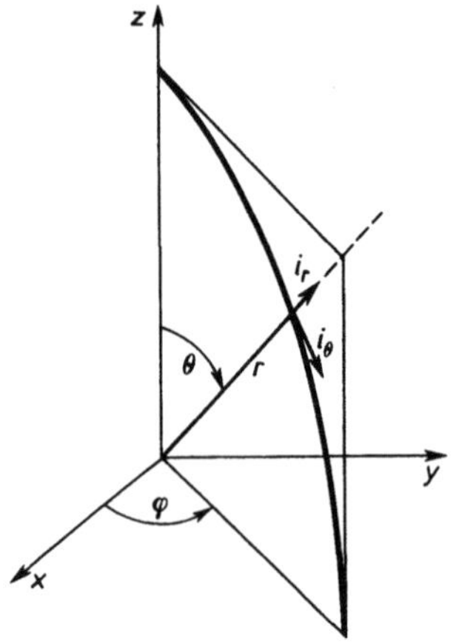

Figure 1.10

On the ground of Kirchhoff's integral (see formula (1.85)), the solution of equation (1.98) for unlimited space has the form

$$\mathbf{H} = \frac{1}{4\pi}\int \frac{[-\operatorname{curl}\mathbf{j}]}{r}\,\mathrm{d}\tau. \tag{1.99}$$

Hence, on taking into account the properties of the function $\delta(\mathbf{r})$, we get

$$\mathbf{H} = \frac{1}{4\pi}\frac{\mathbf{j}_0 \times \mathbf{r}_0}{r_0}\left[\mathrm{i}\frac{\omega}{vr_0} + \frac{1}{r_0^2}\right]\exp\left[\mathrm{i}\left(\omega t - \frac{\omega r_0}{v}\right)\right]. \tag{1.100}$$

Changing over to the spherical coordinates, using the versors $\mathbf{i}_r$, $\mathbf{i}_\varphi$, $\mathbf{i}_\theta$, and applying the relationship

$$\mathbf{i}_z \times \mathbf{i}_r = \mathbf{i}_\varphi \sin\theta,$$

we obtain for the spherical coordinates of vector $\mathbf{H}$ the expressions

$$\begin{aligned} H_\varphi &= j_0\left(\mathrm{i}\frac{\omega}{vr_0} + \frac{1}{r_0^2}\right)\exp\left[\mathrm{i}\left(\omega t - \frac{\omega r_0}{v}\right)\right]\frac{\sin\theta}{4\pi}, \\ H_r &= 0, \quad H_\theta = 0, \end{aligned} \tag{1.101}$$

and the expressions for vector $\mathbf{E}$ conjugated with it

$$\begin{aligned} E_r &= \left[\frac{j_0}{2\pi}\left(\frac{1}{v\varepsilon}\right)\left(\frac{1}{r_0^2} - \mathrm{i}\frac{v}{\omega r_0^3}\right)\cos\theta\right]\exp\left[\mathrm{i}\left(\omega t - \frac{\omega r_0}{v}\right)\right], \\ E_\varphi &= 0, \\ E_\theta &= \left[\frac{\omega j_0}{4\pi v\varepsilon}\left(\mathrm{i}\frac{1}{vr_0} - \frac{3}{\omega r_0^2} - \mathrm{i}\frac{3v}{\omega^2 r_0^3}\right)\sin\theta\right]\exp\left[\mathrm{i}\left(\omega t - \frac{\omega r_0}{v}\right)\right], \end{aligned} \tag{1.102}$$

respectively. It is worth noting that in the area where $r_0 \gg \lambda$ an essential contribution to the values of the field vectors is made by the terms proportional to r_0^{-1}.

Then the only non-zero components of the field vectors are

$$H_\varphi = \left(\frac{j_0\omega}{4\pi v r_0}\sin\theta\right)\exp\left[\mathrm{i}\left(\omega t - \frac{\omega r_0}{v}\right)\right],$$
$$E_\theta = \left(\frac{\omega j_0 \mu}{4\pi r_0}\sin\theta\right)\exp\left[\mathrm{i}\left(\omega t - \frac{\omega r_0}{v}\right)\right]. \tag{1.103}$$

The radiation field has in this case the character of a transverse wave, since vectors **E** and **H** are both perpendicular to the observation direction **r**, which is also the direction of wave propagation. From the practical point of view the area $r_0 \gg \lambda$ is of essential importance. It is called the *area of the wave field of radiation.*

1.13. Polarization of Plane Electromagnetic Waves

The electromagnetic field described by vectors **E** and **H**, whose directions may be given at any moment of time, is a polarized wave. If the direction of vector **E**, and correspondingly that of vector **H**, vary in space in a random way, the field is not polarized. The simplest example of a polarized wave is the field of a plane wave. The plane in which lies the direction of wave propagation and which is perpendicular to the vector **E** of the wave is called the *polarization plane.* If the location of the polarization plane with respect to a stationary coordinate system remains constant, then the plane wave has linear polarization. In this case vector **E** is always parallel to a given fixed axis. However, linear polarization is not the only possible kind that may occur in the case of a harmonic plane wave.

Let us consider two plane waves propagating along the z-axis, their electric vectors being given by

$$\mathbf{E}_1 = \mathrm{i}_x E_1 \cos(\omega t - kz),$$
$$\mathbf{E}_2 = \mathrm{i}_y E_2 \cos(\omega t - kz - \varphi),$$

where φ is the phase shift between them. The set of these waves may be looked upon as one plane wave with a more complex polarization structure. The electrical vector of the summary field $\mathbf{E} = \mathbf{E}_1 + \mathbf{E}_2$ has in the plane $z = 0$ the following components:

$$E_x = E_1 \cos \omega t, \qquad E_y = E_2 \cos(\omega t - \varphi).$$

It is necessary to study how this vector behaves in time. Since

$$\frac{E_x}{E_1} = \cos \omega t, \qquad \frac{E_y}{E_2} = \cos \omega t \cdot \cos \varphi + \sin \omega t \cdot \sin \varphi,$$

we obtain

$$\frac{E_x^2}{E_1^2} + \frac{E_y^2}{E_2^2} - 2\frac{E_x E_y}{E_1 E_2} \cos \varphi = \sin^2 \varphi. \tag{1.104}$$

Equation (1.104) is the equation of an ellipse whose axes form a certain angle with the axes x and y of the coordinate system. It is the geometric locus of points at which the end-point of vector $\mathbf{E}$ is found at different moments of time. Vector $\mathbf{E}$ rotates in the plane $z = \text{const}$. The time it takes to come full circle is $T = 2\pi/\omega$ and its size varies accordingly.

An electromagnetic wave of this type is a wave with *elliptic polarization*. The polarization plane of such a wave rotates continuously in space, the sense of rotation depending on the sign of angle φ.

We distinguish waves whose polarization planes rotate to the right and to the left. In the former case the sense of rotation of the polarization plane observed from the propagation direction is clockwise, in the latter it is counterclockwise.

In the case where $\varphi = 0$, equation (1.104) takes the form

$$\frac{E_x}{E_1} - \frac{E_y}{E_2} = 0. \tag{1.105}$$

It is the equation of the straight line. The wave has *linear polarization.*

For $\varphi = \pm\pi/2$ and $E_1 = E_2 = E$ equation (1.104) becomes the equation of the circle

$$E_x^2 + E_y^2 = E^2, \tag{1.106}$$

and the wave has *circular polarization.*

1.14. Poynting's Vector for Plane Electromagnetic Waves

According to formula (1.77), the field of a plane wave depends on the time and on the spatial coordinates through the quantity $\eta = (t - \mathbf{n}\cdot\mathbf{r}/v)$, i.e., each component of the field vectors **E** and **H** is a function of this quantity. Thus

$$\operatorname{curl}\mathbf{E} = \frac{1}{v}\frac{\partial}{\partial\eta}(\mathbf{E}\times\mathbf{n}), \qquad \frac{\partial\mathbf{H}}{\partial\eta} = \frac{\partial\mathbf{H}}{\partial t}. \tag{1.107}$$

Relations (1.107) imply

$$\mathbf{H} = \frac{1}{v\mu}(\mathbf{n}\times\mathbf{E}), \qquad \mathbf{E} = \frac{1}{v\varepsilon}(\mathbf{H}\times\mathbf{n}),$$

where **n** is the versor determining the propagation direction of the surface of the constant phase. It may be seen from (1.107) that the three vectors **E**, **H** and **n** are mutually perpendicular, which signifies the transverse character of the electromagnetic wave. Poynting's vector for this wave has the form

$$\mathbf{S} = \mathbf{n}v\mathbf{E}\cdot\mathbf{D} = \mathbf{n}v(\mathbf{H}\cdot\mathbf{B}) = \mathbf{n}v\left\{\tfrac{1}{2}\mathbf{E}\cdot\mathbf{D} + \tfrac{1}{2}\mathbf{H}\cdot\mathbf{B}\right\}, \tag{1.108}$$

i.e. it is expressed either by the electric vector or by the magnetic vector. This fact is of great importance for the scalar description of the phenomena of light.

Bibliography

1. Ingarden, R. S. and Jamiołkowski, A.: (1981), 'Classical Electrodynamics' (in Polish), PWN, Warszawa.
2. Kompaneets, A. S.: (1978), 'A Course of Theoretical Physics'. Mir Publishers, Moscow.
3. Landau, L. and Lifshits, E. M.: (1980), 'Field Theory' (in Polish), PWN, Warszawa.
4. Landau, L. and Lifshits, E. M.: (1960), 'Electrodynamics of Continuous Media', Pergamon Press, Oxford, London.
5. Sommerfeld, A.: (1949), 'Electrodynamik', Akad. Verlag, Leipzig.
6. Suffczyński, M.: (1980), 'Electrodynamics' (in Polish), PWN, Warszawa.

2. Electromagnetic Description of Light

2.1. Spectral Distribution of Light

One of the most important characteristics of light is its spectrum. Under this term we understand the complete set of monochromatic waves into which the electromagnetic radiation can be expanded.

Analysing the wave equation, we find that its solutions may include the functions

$$\mathbf{E}_\omega(\mathbf{r}, t) = \mathbf{E}_0(\mathbf{r}, \omega)\mathrm{e}^{\mathrm{i}\omega t}$$

with monochromatic time dependence. The factor $\mathbf{E}_0(\mathbf{r}, \omega)$ is the solution of the Helmholtz equation

$$\Delta\mathbf{E}_0(\mathbf{r}, \omega) + \frac{\omega^2}{v^2}\mathbf{E}_0(\mathbf{r}, \omega) = 0.$$

A monochromatic wave of this kind can be polarized linearly, elliptically or circularly.

In a medium of weak dispersion also a more general function

$$\mathbf{E}(\mathbf{r}, t) = \frac{1}{2\pi}\int_{-\infty}^{+\infty}\mathbf{E}_0(\mathbf{r}, \omega)\mathrm{e}^{\mathrm{i}\omega t}\mathrm{d}\omega \tag{2.1}$$

may be regarded as a solution of the wave equation.

Expression (2.1) is the Fourier integral, which presents an arbitrary time function as a superposition of harmonic waves from a continuous region of the spectrum.

The general theorem concerning Fourier transforms states that the functions in question should have finite discontinuities and

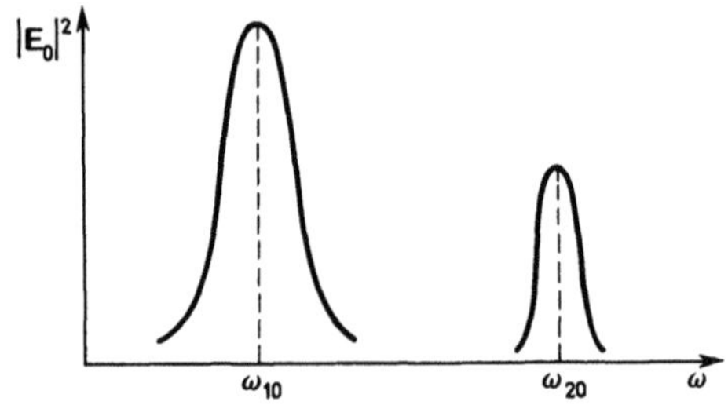

Figure 2.1

a convergent integral of their absolute values taken over the interval from $-\infty$ to $+\infty$.

Under these assumptions

$$\mathbf{E}_0(\mathbf{r}, \omega) = \int_{-\infty}^{+\infty} \mathbf{E}(\mathbf{r}, t) e^{-i\omega t} dt. \tag{2.2}$$

The reverse Fourier transform makes it possible to calculate the amplitude of a wave from the frequency interval $(\omega, \omega + d\omega)$; it is $\mathbf{E}_0(\mathbf{r}, \omega) d\omega$.

The theorem given above enables us in every case to obtain the spectrum distribution, i.e. the amplitude of the harmonic component as a function of frequency. If this amplitude has strongly marked maxima for frequencies ω_{n0} (Fig. 2.1), we can speek about quasi-monochromatic waves. Their number may be finite or infinite.

Every quasi-monochromatic wave is a superposition of monochromatic waves from the finite region

$$\omega_{n0} - \frac{\Delta\omega}{2} \leq \omega \leq \omega_{n0} + \frac{\Delta\omega}{2} \tag{2.3}$$

concentrated around a certain average frequency ω_{n0}; its width fulfils the condition

$$\frac{\Delta\omega}{\omega_{n0}} \ll 1. \tag{2.4}$$

2.2. The Wave Packet

A particular realization of the physical situation formulated above is the superposition of plane monochromatic waves propagating along a fixed direction (e.g. along the x-axis) and having the same direction of oscillation of vector **E** for all waves.

This superposition gives

$$\mathbf{E}(x, t) = \int_{-\infty}^{+\infty} \mathbf{A}(\omega) \mathrm{e}^{\mathrm{i}(\omega t - kx)} \mathrm{d}\omega, \tag{2.5}$$

where the amplitudes of the waves $\mathbf{A}(\omega)$ are defined as follows:

$$\mathbf{A}(\omega) = \begin{cases} \mathbf{A}_0 = \text{const} & \omega_0 - \dfrac{\Delta\omega}{2} \leq \omega \leq \omega_0 + \dfrac{\Delta\omega}{2}, \\ 0 & \text{for other } \omega. \end{cases}$$

The wave vector **k** is a function of frequency. Assuming that the width of permissible frequencies is small as compared with the average frequency $(\Delta\omega/\omega_0 \ll 1)$, we may expand $k = k(\omega)$ in a power series with respect to $(\omega - \omega_0)$ and consider the first two terms of the series

$$k(\omega) = k(\omega_0) + (\omega - \omega_0)\left(\frac{\mathrm{d}k}{\mathrm{d}\omega}\right)_{\omega_0}.$$

In view of the above formula, the resultant wave $\mathbf{E}(x, t)$ may be written in the form

$$\mathbf{E}(x, t) = \mathbf{A}_0 \exp[\mathrm{i}(\omega_0 t - k_0 x)] \times$$

$$\times \int_{\omega_0 - \Delta\omega/2}^{\omega_0 + \Delta\omega/2} \exp\left[+\mathrm{i}(\omega - \omega_0)\left(t - \left.\frac{\mathrm{d}k}{\mathrm{d}\omega}\right|_{\omega_0} x\right)\right] \mathrm{d}\omega. \tag{2.6}$$

Introduction of a new integration variable $\eta = \omega - \omega_0$ enables us to calculate the integral appearing above:

$$E(x,t) = A_0 \frac{2\sin\left[\dfrac{\Delta\omega}{2}\left(t - \left.\dfrac{\mathrm{d}k}{\mathrm{d}\omega}\right|_{\omega_0} x\right)\right]}{t - \left.\dfrac{\mathrm{d}k}{\mathrm{d}\omega}\right|_{\omega_0} x} \exp\left[\mathrm{i}\left(\omega_0 t - k_0 x\right)\right]. \tag{2.7}$$

It is a homogeneous plane wave of average frequency ω_0. Its amplitude varies in time and space as $(\sin\xi)/\xi$. It has an infinite number of maxima, one of them being the highest and appearing if $t - (\mathrm{d}k/\mathrm{d}\omega)_{\omega_0} x = 0$. This means that it occurs for $x = (\mathrm{d}k/\mathrm{d}\omega)_{\omega_0}^{-1} t$; it has no fixed location, but moves with the velocity

$$v_{\mathrm{gr}} = \frac{1}{(\mathrm{d}k/\mathrm{d}\omega)_{\omega_0}} \tag{2.8}$$

called the *group velocity.*

This type of disturbance limited in space is called a *wave packet.*

Velocity (2.8), describing the propagation of a wave packet in a medium without dispersion ($\mathrm{d}\varepsilon/\mathrm{d}\omega = 0$), is the same as the velocity of a monochromatic wave with carrier frequency ω_0 and is equal to $v_{\mathrm{gr}} = \omega_0/k_0 = v_0$. In a medium with dispersion ($\mathrm{d}\varepsilon/\mathrm{d}\omega \neq 0$), on applying the equality $k = 2\pi/\lambda$ (where λ is the wavelength), we obtain the following expression for the group velocity:

$$v_{\mathrm{gr}} = v_0 - \lambda_0 \frac{\mathrm{d}v}{\mathrm{d}\lambda}. \tag{2.9}$$

In the region of normal dispersion (where $\mathrm{d}v/\mathrm{d}\lambda > 0$) the group velocity is less than the velocity of a monochromatic wave with carrier frequency ω_0, i.e. less than the phase velocity $v_0 = \omega_0/k_0$.

2.3. Monochromatic Homogeneous and Non-homogeneous Plane Waves

The simplest solution of the wave equation has the form of monochromatic plane wave

$$\mathbf{E}(r,t) = \mathbf{A}\exp(\mathrm{i}\varphi_0)\exp\left[\mathrm{i}\left(k_1 x + k_2 y + k_3 z\right)\right]\exp(-\mathrm{i}\omega t). \tag{2.10}$$

The factor proceeding $\exp[-\mathrm{i}\omega t]$ expresses the *complex amplitude of vibrations*. Its absolute value is the field amplitude, and its argument is the phase of the vibrations. The complex amplitude gives full information about the plane wave which is completely determined by its frequency ω (the length of the wave), the propagation direction $\mathbf{k}$, the amplitude $\mathbf{A}$ and the initial phase of vibrations φ_0.

The components of the wave vector $\mathbf{k}$, determining the propagation direction of the wave, can be expressed by the angles θ and φ defining the direction in space

$$\begin{aligned} k_1 &= k \sin\theta \cos\varphi, \\ k_2 &= k \sin\theta \sin\varphi, \\ k_3 &= k \cos\theta. \end{aligned} \tag{2.11}$$

From the relations given above it follows that for a given value of the wave vector only two of its components are independent, the third being given by the relation

$$\begin{aligned} k_3 &= \pm\sqrt{k^2 - k_1^2 - k_2^2} \\ &= \pm\sqrt{k^2 - k^2 \sin^2\theta} = \pm k \cos\theta, \end{aligned} \tag{2.12}$$

which does not specify its sign.

Thus, the complex amplitude of a plane monochromatic wave can be written in the form

$$\mathbf{A}\exp(\mathrm{i}\varphi_0)\exp\left(\pm\mathrm{i}z\sqrt{k^2 - k_1^2 - k_2^2}\right)\exp[\mathrm{i}(k_1 x + k_2 y)]. \tag{2.13}$$

In the above expression we have in an explicit form the following wave parameters: the amplitude $\mathbf{A}$, the wavelength $\lambda = 2\pi/k$, the phase φ_0, and the propagation direction (k_1 and k_2). Although formally expression (2.13) will satisfy the wave equation for any k_1 and k_2 varying from $+\infty$ to $-\infty$, a propagating plane wave arises only if the condition

$$k_1^2 + k_2^2 \leq k^2 \quad \text{(homogeneous plane wave)}$$

is fulfilled.

If this condition is not satisfied, we have a *non-homogeneous wave*, which decays exponentially along one of the axes of space variables. A wave of this kind is also a solution of the wave equation.

2.4. Plane Wave Distribution of a Monochromatic Wave

The wave equation is a linear equation, and so the superposition principle applies to it.

Since a field in the form of a plane, monochromatic, homogeneous or non-homogeneous wave is a solution of the wave equation, the sum of such waves with arbitrary amplitudes, phases and propagation directions is also a solution of that equation. Allowing continuous changes of the parameters mentioned above, the solution may be written as the integral of the function (2.13) taken over the independent parameters of a plane wave. The integral takes the form

$$\mathbf{E}(x, y, z, t) = \exp(-\mathrm{i}\omega t)\frac{1}{4\pi^2}\iint\limits_{-\infty}^{+\infty}\mathbf{g}(k_1, k_2)\exp\left(\pm\mathrm{i}z\sqrt{k^2-k_1^2-k_2^2}\right)\times \exp(\mathrm{i}k_1x+\mathrm{i}k_2y)\mathrm{d}k_1\,\mathrm{d}k_2. \tag{2.14}$$

Here, $\mathbf{g}(k_1, k_2)$ is a complex function describing the amplitude and the phase of a particular plane wave propagating in the direction determined by variables k_1 and k_2. Naturally, formula (2.14) embraces also so-called plane non-homogeneous waves.

Formula (2.14) is a generalization of the notion of complex amplitude for the case of non-plane monochromatic waves and is known as the *Rayleigh formula*. It is the general solution of the wave equation. Having (2.14) at our disposal, we can give the precise solution of the wave equation satisfying given boundary conditions on the plane $z = 0$. Expression (2.14) represents an analytic function in the whole space, and such functions have univocal analytic continuations inside a compact area.

An example of the application of formula (2.14) is given below.

We seek a solution of the wave function for $z \geq 0$ which would take on the plane $z = 0$ the values of a given wave field. Let us take a closed surface consisting of the plane $z = 0$ and a hemisphere which has an infinitely large radius for $z > 0$ and is supported on the plane $z = 0$. We look for a solution of the wave equation which would satisfy the following conditions:

(a) for $z = 0$ it is equal to the given field,

(b) for a hemisphere with $R \to \infty$ it is equal to zero (Sommerfeld's condition).

On the plane $z = 0$ the given field has a complex amplitude $\mathbf{E}(x, y, 0)$. It is required that the general solution should coincide with the given field for $z = 0$. Then

$$\mathbf{E}(x,y,0) = \frac{1}{4\pi^2} \iint\limits_{-\infty}^{+\infty} \mathbf{g}(k_1,k_2) \exp[\mathrm{i}(k_1 x + k_2 y)] \, \mathrm{d}k_1 \mathrm{d}k_2. \quad (2.15)$$

Hence, taking the reverse Fourier transform, we can determine $\mathbf{g}(k_1, k_2)$. We get

$$\mathbf{g}(k_1, k_2) = \iint\limits_{-\infty}^{+\infty} \mathbf{E}(x,y,0) \exp[-\mathrm{i}(k_1 x + k_2 y)] \, \mathrm{d}x \mathrm{d}y. \quad (2.16)$$

Defined in this way, function $\mathbf{g}(k_1, k_2)$ makes it possible to find a general solution. Since the sign of the k_3 component can be chosen from the two possible signs, the solution is ambiguous. This ambiguity is eliminated by using the boundary condition on the surface of an infinite sphere, on which the wave field should vanish. Which of the signs should be chosen may be established from the behaviour of the non-homogeneous waves when z is increased (these waves occur if $k_1^2 + k_2^2 > k^2$). If we demand that for $z > 0$ and an increasing z non-homogeneous waves should vanish, we should choose the sign "+": for $z < 0$ the sign "−" should be chosen. Thus we obtain for $z > 0$

$$\mathbf{E}(x, y, z) = \frac{1}{4\pi^2} \iint\limits_{-\infty}^{+\infty} \mathbf{g}(k_1, k_2) \exp\left(\mathrm{i}z\sqrt{k^2 - k_1^2 - k_2^2}\right) \times$$
$$\times \exp[\mathrm{i}(k_1 x + k_2 y)] \, \mathrm{d}k_1 \mathrm{d}k_2 \quad (2.17)$$

and for $z<0$

$$\mathbf{E}(x, y, z)=\frac{1}{4\pi^{2}} \iint_{-\infty}^{+\infty} \mathbf{g}(k_{1}, k_{2}) \exp\left(-\mathrm{i} z \sqrt{k^{2}-k_{1}^{2}-k_{2}^{2}}\right) \times$$
$$\times \exp\left[\mathrm{i}\left(k_{1} x+k_{2} y\right)\right] \mathrm{d} k_{1} \mathrm{d} k_{2}, \tag{2.18}$$

where $\mathbf{g}(k_1, k_2)$ is defined by formula (2.16). The function $\mathbf{g}(k_1, k_2)$ determining the distribution of amplitudes and phases of plane waves according to the direction of propagation is called the *spatial spectrum* or the *angular spectrum of the field.* The term angular spectrum reflects the relationship between k_1, k_2 and the angles determining the direction of propagation of plane waves.

2.5. The Dual Physical Sense of Spatial Frequencies

Formulas (2.15), (2.16) are Fourier integrals for two pairs of variables: k_1, k_2 and x, y. The coordinates x, y are expressed in a length unit, variables k_1, k_2 in a unit being the inverse of a length unit and are called *spatial frequencies.*

The justification for the term is as the following. For a function of time $f(t)$, the Fourier transforms take the form

$$\begin{aligned} f(t) &= \frac{1}{2\pi} \int_{-\infty}^{+\infty} C_{f}(\omega) \exp[\mathrm{i} \omega t] \, \mathrm{d} \omega, \\ C_{f}(\omega) &= \int_{-\infty}^{+\infty} f(t) \exp[-\mathrm{i} \omega t] \, \mathrm{d} t. \end{aligned} \tag{2.19}$$

Formulas (2.15) and (2.16) differ from those given above only in number of variables and symbols used to denote the quantities appearing in them. Thus they are also Fourier transforms but twice repeated. The relations between $\mathbf{E}(x, y, 0)$ and $\mathbf{g}(k_1, k_2)$ differ from the relations between $f(t)$ and $C_f(\omega)$ only in the unit and in the notation used.

Therefore, it can be stated that the properties of the variables k_1 and k_2 are identical with the well-known properties of frequency ω. In this sense k_1 and k_2 are spatial frequencies. The function $\mathbf{g}(k_1, k_2)$, analogous to the function $C_f(\omega)$, can be called the *spatial spectrum of the function* $\mathbf{E}(x, y, 0)$. It has all the properties of its time analog – the function $C_f(\omega)$. The corresponding mathematical relations are true for $\mathbf{g}(k_1, k_2)$, as well.

In respect of their physical meaning, k_1 and k_2 show duality. In relations (2.15) and (2.16) these quantities appear as spatial frequencies, in relations (2.11) and (2.12) they determine propagation directions of plane waves. Thus k_1 and k_2 have a double physical sense: (a) on the one hand, they are spatial frequencies, (b) on the other hand, they determine the propagation directions of the plane waves into which we can resolve the wave field (angular spectrum).

The variables k_1 and k_2 can be interpreted in two ways: as spatial frequencies or as angular variables. However, the specific assignment of an angle to a spatial frequency depends on the length of the radiation wave.

In relations (2.15) and (2.16), defining k_1 and k_2 as spatial frequencies, the wavelength does not appear. Provided the geometry of the source distribution does not change, the values of the spatial frequencies remain unchanged independently of the frequency of the radiating sources and the velocity of wave propagation in space.

The angular spectrum of plane waves corresponding to a given spatial spectrum depends on the length of the radiation wave. If the wavelength changes, the angular spectrum expressed by the angle values undergoes deformation.

2.6. Complete Field Representation

As shown in Sec. 2.1., a field with any time dependence can be presented as a superposition of monochromatic waves

$$\mathbf{E}(x, y, z, t) = \int_{-\infty}^{+\infty} \mathbf{E}(x, y, z, \omega) \exp[\mathrm{i}\omega t] \,\mathrm{d}\omega, \tag{2.20}$$

where $\mathbf{E}(x, y, z, \omega)$ is the amplitude of the different components of monochromatic waves. This amplitude, according to formula (2.14), can be presented as a superposition of plane waves, i.e., in the form

$$\mathbf{E}(x,y,z,\omega) = \frac{1}{4\pi^2}\iint_{-\infty}^{+\infty} \mathbf{g}(k_1,k_2)\exp\left(\pm \mathrm{i}z\sqrt{k^2 - k_1^2 - k_2^2}\right) \times$$

$$\times \exp(\mathrm{i}k_1 x + \mathrm{i}k_2 y)\mathrm{d}k_1\,\mathrm{d}k_2 \tag{2.21}$$

Substituting into formula (2.20) instead of $\mathbf{E}(x, y, z, \omega)$ the form given by formula (2.21), we obtain as a result the distribution of a wave field with any time dependence into monochromatic plane waves:

$$\mathbf{E}(x, y, z, t)$$

$$= \frac{1}{4\pi^2}\iiint_{-\infty}^{+\infty} \mathbf{g}(k_1,k_2)\exp[\mathrm{i}\omega t]\exp\left[\pm \mathrm{i}z\sqrt{k^2 - (k_1^2 + k_2^2)}\right] \times$$

$$\times \exp(\mathrm{i}k_1 x + \mathrm{i}k_2 y)\mathrm{d}\omega\,\mathrm{d}k_1\,\mathrm{d}k_2. \tag{2.22}$$

2.7. Monochromatic Wave at the Interface of Two Dielectrics

The interface between two dielectrics, provided we consider only a small fragment of it (of the order of a few hundred wavelengths), can be regarded as being flat. This interface is taken as the plane $y = 0$ in a right-handed coordinate system.

The following questions should be answered:

1. How does the electromagnetic wave behave at the interface of two dielectrics?

2. What are the relationships between the amplitudes and the phases of an incident wave, a reflected wave and a refracted wave?

For monochromatic plane waves these relationships are called *Fresnel's formulas*. They are obtained from the boundary conditions for the electromagnetic field on the interface surface S of two isotropic dielectrics:

$$\begin{aligned} \mathbf{n} \times \mathbf{E}_1 &= \mathbf{n} \times \mathbf{E}_2, \\ \mathbf{n} \times \mathbf{H}_1 &= \mathbf{n} \times \mathbf{H}_2, \\ \mathbf{D}_1 \cdot \mathbf{n} &= \mathbf{D}_2 \cdot \mathbf{n}, \\ \mathbf{B}_1 \cdot \mathbf{n} &= \mathbf{B}_2 \cdot \mathbf{n}, \end{aligned} \tag{2.23}$$

where the indices 1 and 2 determine, respectively, the field vectors in two different media and $\mathbf{n}$ is a unit vector normal to the surface.

Let us assume that a plane monochromatic wave

$$\mathbf{E} = \mathbf{E}_0 \exp[\mathrm{i}(\mathbf{k} \cdot \mathbf{r} - \omega t)],$$

where $\mathbf{E}_0 = \text{const}$ is the amplitude and $\mathbf{k} \cdot \mathbf{r} - \omega t$ the phase, falls from the half-space $y > 0$ onto the interface surface. In an isotropic medium ($\mathbf{D} = \varepsilon \mathbf{E}$) the Maxwell equation $\operatorname{div} \mathbf{D} = 0$ yields the equality $\mathbf{k} \cdot \mathbf{E} = 0$. This confirms perpendicularity of vector $\mathbf{E}$ with respect to the wave vector lying in the incident plane. Any monochromatic plane wave can be presented in the form of two monochromatic waves of the same frequency. Vectors $\mathbf{E}$ of these waves oscillate along two perpendicular directions. Therefore, in order to know the behaviour of a plane wave at the interface of two media, it is enough to consider separately the plane wave whose vector $\mathbf{E}_{\mathrm{I}}$ oscillates in the incident plane and the plane wave whose vector $\mathbf{E}_{\perp}$ oscillates perpendicularly to the incident plane.

It is assumed that the two media are unlimited on both sides of the interface. That is why we can neglect multiple reflections and assume that at the interface only three waves exist: an incident wave, a reflected wave and a refracted wave.

1. Let us consider the case of a wave with vector $\mathbf{E}_{\mathrm{I}}$ lying in the incident plane. Denote the electric vectors of the incident wave, reflected wave and refracted wave by $\mathbf{E}_{\mathrm{I}}$, $\mathbf{E}_{\mathrm{I}}^{r}$, $\mathbf{E}_{\mathrm{I}}^{t}$, respectively. Suppose that the propagation directions of the above waves form with the normal to the interface of the two dielectrics the angles α, β, γ, respectively (Fig. 2.2).

The equations for these waves can be written in the form

$$E_{\mathrm{I}} = E_{\mathrm{I}0} \exp[\mathrm{i}(k_1 x + k_2 y - \omega t)],$$

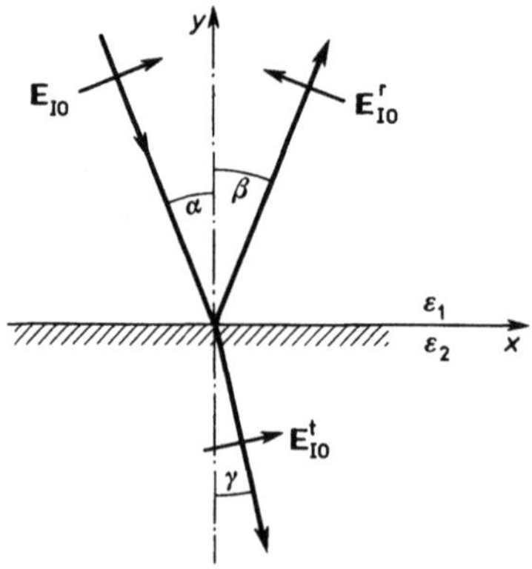

Figure 2.2

$$E_{\mathrm{I}}^{r} = E_{\mathrm{I}0}^{r} \exp\left[\mathrm{i}\left(k_1^* x + k_2^* y - \omega t\right)\right], \tag{2.24}$$

$$E_{\mathrm{I}}^{t} = E_{\mathrm{I}0}^{t} \exp\left[\mathrm{i}\left(k_1'' x + k_2'' y - \omega t\right)\right],$$

where

$$\mathbf{k} = \mathbf{i}k_1 + \mathbf{j}k_2, \quad \mathbf{k}^* = \mathbf{i}k_1^* + \mathbf{j}k_2^*, \quad \mathbf{k}'' = \mathbf{i}k_1'' + \mathbf{j}k_2'' \tag{2.25}$$

are the wave vectors for the incident wave, reflected wave and refracted wave, respectively (**i** and **j** are the versors of the axes of the chosen coordinate system), and $E_{\mathrm{I}0}$, $E_{\mathrm{I}0}^{r}$, $E_{\mathrm{I}0}^{t}$ the amplitudes of these waves. By (2.23), we can write for $y = 0$ the following condition:

$$\exp\left[\mathrm{i}\left(k_1 x - \omega t\right) E_{\mathrm{I}0} \cos\alpha - \exp\left[\mathrm{i}\left(k_1^* x - \omega t\right)\right] E_{\mathrm{I}0}^{r} \cos\beta = \exp\left[\mathrm{i}\left(k_1'' x - \omega t\right)\right] E_{\mathrm{I}0}^{t} \cos\gamma, \tag{2.26}$$

resulting from the continuity of the tangent components of an electric field.

On applying Maxwell's equation

$$\operatorname{curl}\mathbf{E} = -\mu \frac{\partial \mathbf{H}}{\partial t},$$

we can assign to the electric field (2.24) the appropriate magnetic fields (we assume $\mu_1 = \mu_2 = \mu$):

$$\mathbf{H}_\mathrm{I} = \frac{1}{\omega\mu}(\mathbf{k} \times \mathbf{E}_\mathrm{I}),$$

$$\mathbf{H}_\mathrm{I}^r = \frac{1}{\omega\mu}(\mathbf{k}^* \times \mathbf{E}_\mathrm{I}^r), \tag{2.27}$$

$$\mathbf{H}_\mathrm{I}^t = \frac{1}{\omega\mu}(\mathbf{k}'' \times \mathbf{E}_\mathrm{I}^t).$$

The continuity of the tangent component of vector **H** (for $y = 0$) yields the equation

$$\begin{aligned}\exp\left[\mathrm{i}(k_1 x - \omega t)\right]\left(-k_2 \mathbf{E}_{\mathrm{I}0}\cos\alpha + k_1 \mathbf{E}_{\mathrm{I}0}\sin\alpha\right) + \\ + \left(k_1^* E_\mathrm{I}^r \sin\beta + k_2^* E_{\mathrm{I}0}^r \cos\beta\right)\cdot\exp\left[\mathrm{i}(k_1^* x - \omega t)\right] \\ = \exp\left[\mathrm{i}(k_1'' x - \omega t)\right](k_1'' E_{\mathrm{I}0}^t \sin\gamma - k_2'' E_{\mathrm{I}0}^t \cos\gamma).\end{aligned} \tag{2.28}$$

Both equalities, (2.26) and (2.28), should be satisfied for any moment of time and for any value of x. This is possible only if

$$k_1 = k_1^* = k_1''. \tag{2.29}$$

Using the absolute values of the wave vectors, we get from (2.29) the following equality:

$$k\sin\alpha = k^*\sin\beta = k''\sin\gamma, \tag{2.30}$$

where

$$k = \omega\sqrt{\mu\varepsilon_1}, \quad k^* = \omega\sqrt{\mu\varepsilon_1}, \quad k'' = \omega\sqrt{\mu\varepsilon_2}.$$

From equality (2.29) follow the conditions

$$\sin\alpha = \sin\beta, \tag{2.31}$$

and thus the reflection law, i.e. $\alpha = \beta$, and

$$\frac{\sin\alpha}{\sin\gamma} = \sqrt{\frac{\varepsilon_2}{\varepsilon_1}}, \tag{2.32}$$

i.e. the *refraction law* (*Snell's formula*).

Considering (2.29) in (2.26) and in (2.28) leads to the equations connecting the amplitude of the incident wave with the amplitudes of the reflected wave and the refracted wave, i.e. to the equalities

$$E_{I0}\cos\alpha - E^r_{I0}\cos\alpha = E^t_{I0}\cos\gamma,$$

$$k(E_{I0} + E^r_{I0}) = E^t_{I0}k'',$$

which implies

$$\frac{E^r_{I0}}{E_{I0}} = \frac{\tan(\alpha-\gamma)}{\tan(\alpha+\gamma)} \tag{2.33a}$$

and

$$\frac{E^t_{I0}}{E_{I0}} = \frac{2\cos\alpha\sin\gamma}{\sin(\alpha+\gamma)\cos(\alpha-\gamma)}. \tag{2.33b}$$

2. The case of a wave with vector $\mathbf{E}_\perp$ perpendicular to the incident plane is computed similarly, and the result is

$$\frac{E^r_{\perp 0}}{E_{\perp 0}} = \frac{\sin(\gamma-\alpha)}{\sin(\gamma+\alpha)}, \tag{2.34a}$$

$$\frac{E^t_{\perp 0}}{E_{\perp 0}} = \frac{2\sin\gamma\cos\alpha}{\sin(\gamma+\alpha)}. \tag{2.34b}$$

The resulting formulas (2.33) and (2.34) are known as *Fresnel's formulas.* They permit computing the amplitude values of the reflected wave and the refracted wave for a given incident angle and a given amplitude of the incident wave.

From Fresnel's formulas it follows that if $\alpha + \gamma = \pi/2$, i.e. if the reflected ray and the refracted ray are mutually perpendicular, then for the wave with vector $\mathbf{E}_I$ lying in the incident plane there is no reflected wave. The incident angle at which the sum of the incident angle and the refracted angle equals $\pi/2$ is called the *Brewster angle.*

This angle satisfies the equation

$$\tan\alpha_B = \sqrt{\frac{\varepsilon_2}{\varepsilon_1}}. \tag{2.35}$$

For this incident angle only the wave of vector $\mathbf{E}_\perp$ (perpendicular to the incidence plane) is possible in the reflected wave.

2.8. Total Internal Reflection and the Goose–Hänchen Effect

Let us consider a plane monochromatic wave falling on the interface between two media, described by the material constants ε_1, μ_0 and ε_2, μ_0 respectively, from the side of the medium of higher optical density $\varepsilon_1 > \varepsilon_2$. Then, in accordance with the reflection law (2.32), for the value of the angle $\alpha \geq \arcsin\sqrt{\varepsilon_2/\varepsilon_1}$, the refraction angle γ assumes imaginary values (since $\sin\gamma > 1$). From Fresnel's formulas (2.33a) and (2.34a) for this value of the incident angle we obtain information that for any polarization state of the incident wave the modules of the amplitudes of the electric vector for the incident and the reflected waves are equal.

This phenomenon of light reflection from the medium of lower optical density is called *total internal reflection.*

The two components of vector $\mathbf{E}_0$ of the incident wave, namely the component perpendicular to the incidence plane $E_{\perp 0}$ and the component lying in the incidence plane $E_{\mathrm{I}0}$, undergo in total internal reflection different phase jumps. The components of the vector $\mathbf{E}^r$ of the reflected wave are bound with the corresponding components of the incident wave through the relations

$$E_{\mathrm{I}}^r = E_{\mathrm{I}0}\exp(i\theta_{\mathrm{I}}), \qquad E_\perp^r = E_{\perp 0}\exp(i\theta_\perp), \tag{2.36}$$

where

$$\begin{aligned} \tan\frac{\theta_{\mathrm{I}}}{2} &= \frac{\sqrt{\sin^2\alpha - \varepsilon_2/\varepsilon_1}}{(\varepsilon_2/\varepsilon_1)\cos\alpha}, \\ \tan\frac{\theta_\perp}{2} &= \frac{\sqrt{\sin^2\alpha - \varepsilon_2/\varepsilon_1}}{\cos\alpha}. \end{aligned} \tag{2.37}$$

These phase jumps, different for different components, cause a change of the polarization state of the reflected light in relation to the polarization state of the incident wave.

In spite of total reflection, the electromagnetic field in the medium of lower optical density, described by (ε_2, μ_0), is by no means equal to zero. In fact, if we consider in the refraction law that the angle γ has an imaginary value for $\alpha > \arcsin\sqrt{\varepsilon_2/\varepsilon_1}$, then the Cartesian coordinates k_1'' and k_2'' of the wave vector of the refracted wave (see (2.25)) are

$$k_1'' = \frac{\omega}{c}\sqrt{\varepsilon_1}\sin\alpha,$$

$$k_2'' = \frac{\omega}{c}\mathrm{i}\sqrt{\varepsilon_2}\sqrt{\frac{\varepsilon_1}{\varepsilon_2}\sin^2\alpha - 1}. \tag{2.38}$$

This results in concluding (see (2.24)) that the refracted wave in the medium of lower optical density has the phase factor

$$\exp\left[-\mathrm{i}x\left(\frac{\omega}{c}\sqrt{\varepsilon_1}\sin\alpha\right)\right]\exp\left[-y\left(\frac{\omega}{c}\sqrt{\varepsilon_1\sin^2\alpha - \varepsilon_2}\right)\right], \tag{2.39}$$

which means that it propagates along the boundary plane, and vanishes exponentially with the growth of the distance from that plane.

On the other hand, there is no energy flow through the interface of two media in the case of total internal reflection. The energy of the wave flows along the interface and does not enter the other medium. This can be proved by calculating the average value of the normal component of Poynting's vector.

If upon the interface of two media from the side of the denser medium falls a spatially confined light beam at an angle greater than the critical angle, then the so-called *Goose–Hänchen effect* occurs. It consists in the point of incidence A and the point of reflection B of the light beam being spatially shifted in relation to each other (Fig. 2.3).

Every light beam confined in space can be presented as a superposition of plane waves propagating in different directions (see Sec. 2.4). A light beam confined in space is a laser beam which can be treated as a Gauss beam.

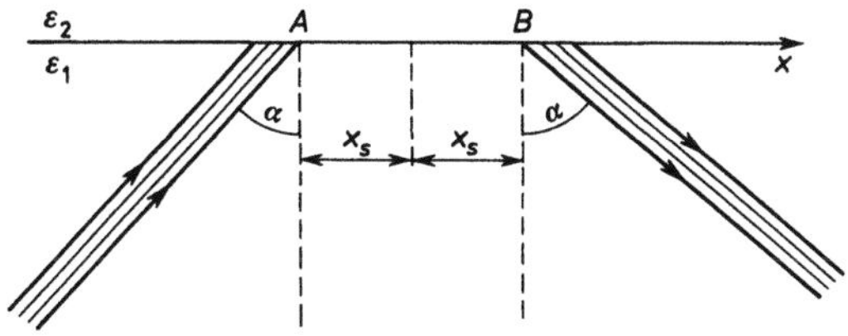

Figure 2.3

To explain the Goose–Hänchen effect we can take a simpler beam arising from a superposition of two plane waves falling on the interface of two media at slightly different angles.

Thus, if the x-th components of corresponding wave vectors equal respectively $k + \Delta k$ and $k - \Delta k$, then on the boundary $y = 0$ the conjugate resultant amplitude $A(x)$ of the incident wave may be written in the form

$$A(x) = \left(e^{i\Delta kx} + e^{-i\Delta kx}\right)e^{-ikx} = 2\cos(\Delta kx)e^{-ikx}. \tag{2.40}$$

The phase shift θ occurring in total internal reflection is a function of the incident angle α and the wave number k. For small variations of Δk we may write

$$\theta(k + \Delta k) = \theta(k) + \frac{d\theta}{dk}\Delta k. \tag{2.41}$$

On the boundary $y = 0$ the amplitude of the refracted wave packet $B(x)$ has the form

$$\begin{aligned} B(x) &= \left[e^{i(\Delta kx - \Delta\theta)} + e^{-i(\Delta kx - \Delta\theta)}\right]e^{-i(kx-\theta)} \\ &= 2\cos\left[\Delta k\left(x - \frac{d\theta}{dk}\right)\right]e^{-i(kx-\theta)}. \end{aligned} \tag{2.42}$$

Comparing the expression for the amplitude of the incident wave (2.40) with the expression for the amplitude of the reflected packet (2.42), we can notice that the quantity $d\theta/dk$, given in length units, represents the required spatial shift of the reflected beam in relation to the incident beam on the interface between the two media. The values of these shifts are different for a wave whose electric vector remains perpendicular to the incidence plane and for a wave whose electric vector lies in the incidence plane.

In the first case the phase jump in total reflection is

$$\theta_{\perp} = 2\arctan\left(\frac{\sqrt{\sin^2\alpha - \varepsilon_2/\varepsilon_1}}{-\cos\alpha}\right),$$

and hence the spatial shift is

$$2x_s^{\perp} = \frac{2}{\omega\sqrt{\varepsilon_1\mu}}\,\frac{\tan\alpha}{\sqrt{\sin^2\alpha - \varepsilon_2/\varepsilon_1}}. \tag{2.43}$$

In the second case the phase jump in total reflection is

$$\theta_{\mathrm{I}} = 2\arctan\frac{\sqrt{\sin^2\alpha - \varepsilon_2/\varepsilon_1}}{(\varepsilon_2/\varepsilon_1)\cos\alpha},$$

and the spatial shift

$$2x_s^{\mathrm{I}} = \frac{2x_s^{\perp}}{(\varepsilon_1/\varepsilon_2 + 1)\sin^2\alpha - 1}. \tag{2.44}$$

The occurrence of such spatial shifts of the point of incidence and the point of reflection seems to indicate that the process proceeds as if the light underwent total reflection in the medium of lower optical density at a depth not less than $y_s = x_s\cot\alpha$.

2.9. Electromagnetic Waves in a Conductor

First of all, it is worth noticing that the space charge density in a conductor decays exponentially with time. We can prove this by differentiating with respect to time the equation $\operatorname{div}\mathbf{D} = \rho$, and acting upon the equation $\operatorname{curl}\mathbf{H} = \sigma\mathbf{E} + \dot{\mathbf{D}}$ with the divergence operator. Considering the results of these operations and applying the constitutive relation $\mathbf{D} = \varepsilon\mathbf{E}$, we obtain the equation $\dot{\rho} = -\rho\sigma/\varepsilon$, whose solution is the exponential dependence of the space-charge density on time

$$\rho = \rho_0 e^{-t/\tau},$$

where $\tau = \varepsilon/\sigma$ (the ratio of electric permittivity to electric conductivity) is the relaxation time, which for waves from the optical range is so small in comparison with the period of light oscillations that the space-charge density in metals can be assumed to be zero.

As a consequence of this assumption, we can treat the electric field in a conductor as a field without sources ($\operatorname{div}\mathbf{E} = 0$). For monochromatic fields without sources, Maxwell's equations in a conductor can be replaced by the oscillation equation

$$\Delta\mathbf{E} + k^2\mathbf{E} = 0, \tag{2.45}$$

where

$$k^2 = \omega^2\mu\left(\varepsilon - \frac{\sigma}{\omega}\mathrm{i}\right) \tag{2.46}$$

is the square of the wave number.

Comparing this quantity with a similar expression for dielectric media (1.79), we can see that the role of electric permittivity is now taken over by the complex quantity

$$\hat{\varepsilon} = \varepsilon - \frac{\sigma}{\omega}\mathrm{i}. \tag{2.47}$$

Also, the so-called *complex refractive index* is introduced. It is defined as follows:

$$\hat{n} = n(1 + \mathrm{i}\kappa) \tag{2.48a}$$

and the following condition is imposed on it:

$$\hat{n}^2 = \mu\left(\varepsilon - \frac{\sigma}{\omega}\mathrm{i}\right)c^2. \tag{2.48b}$$

These two equations imply

$$\begin{aligned} n^2 &= \left(\tfrac{1}{2}\sqrt{\mu^2\varepsilon^2 + \frac{\mu^2\sigma^2}{\omega^2}} + \frac{\mu\varepsilon}{2}\right)c^2, \\ n^2\kappa^2 &= \left(\tfrac{1}{2}\sqrt{\mu^2\varepsilon^2 + \frac{\mu^2\sigma^2}{\omega^2}} - \frac{\mu\varepsilon}{2}\right)c^2. \end{aligned} \tag{2.49}$$

The simplest solution of equation (2.45) with a complex wave number $\hat{k} = \hat{n}\omega/c$, where $\hat{n}$ is the complex coefficient defined by (2.48a), is the following expression:

$$\mathbf{E} = \mathbf{E}_0 \exp\left[-\frac{n\kappa}{c}\omega(\mathbf{r}\cdot\mathbf{s})\right]\exp\left[\left(\frac{n}{c}(\mathbf{r}\cdot\mathbf{s}) - t\right)i\omega\right], \tag{2.50}$$

were $\mathbf{s}$ is a versor perpendicular to a constant phase plane.

As can be seen from the form of this solution, the electromagnetic wave in a conductor undergoes damping; its amplitude decays exponentially along the path of propagation.

From the explicit dependence of $(n\kappa)$ on σ (see (2.49)) it follows that the greater the conductivity the stronger the damping.

Since light intensity is proportional to the time-averaged square of vector $\mathbf{E}$ of a wave, it also decreases exponentially while it penetrates into the deep parts of a conductor according to the formula

$$I = I_0 \exp\left[-2\frac{n\kappa}{c}\omega(\mathbf{r}\cdot\mathbf{s})\right]. \tag{2.51}$$

The quantity $\chi = 2\frac{n\kappa}{c}\omega$ is called the *absorption coefficient.*

The law of the decay of light intensity in a conductor (2.51) is called *Bürger's law.*

Light intensity decreases to $1/e$ of its initial value after travelling the distance

$$d = \mathbf{s}\cdot\mathbf{r} = \frac{1}{\chi}.$$

This distance is different for different metals, and is strongly dependent on the wavelength of the incident light. It varies from 10^{-10} m for ultraviolet to 10^{-7} m for microwaves, and for $\sigma \to \infty$ it approaches zero.

2.10. Reflection and Refraction at a Conductor Surface

In conductors it is possible to use the complex refractive index (see (2.48)). The real part of the index describes the refraction of the electromagnetic wave, the imaginary part ($in\kappa$) is responsible for wave damping.

If in Snell's refraction law we formally replace the real refractive index by the complex refractive index, then the law will take the form

$$\frac{\sin\varphi_0}{\sin\varphi_t} = n(1 + i\kappa). \tag{2.52}$$

From this formula it can be seen that, for a real incident angle φ_0, the refraction angle φ_t for a wave in a conductor must be a complex quantity.

If we assume that the wave vector of a plane incident wave coincides with the direction of the versor $[\sin\varphi_0, \cos\varphi_0, 0]$, the spatial dependence of a wave phase in a conductor after refraction will be given by the factor $\hat{k}(\mathbf{s}^{(t)}\cdot\mathbf{r})$, where

$$s_x^{(t)} = \sin\varphi_t, \qquad s_y^{(t)} = \cos\varphi_t, \qquad s_z^{(t)} = 0.$$

It follows from the refraction law that

$$\begin{aligned} s_x^{(t)} &= \sin\varphi_t = \frac{1 - i\kappa}{n(1 + \kappa^2)} \sin\varphi_0, \\ s_y^{(t)} &= \sqrt{1 - \sin^2\varphi_t} = q e^{i\eta}, \end{aligned} \tag{2.53}$$

where

$$\begin{aligned} q^2 &= \frac{1}{n^2(1+\kappa^2)^2} \times \\ &\quad \times \sqrt{[n^2(1+\kappa^2)^2 + (\kappa^2 - 1)\sin^2\varphi_0]^2 + 4\kappa^2\sin^4\varphi_0}, \\ \tan 2\eta &= \frac{2\kappa\sin^2\varphi_0}{n^2(1 + \kappa^2)^2 + (\kappa^2 - 1)\sin^2\varphi_0}. \end{aligned} \tag{2.54}$$

Applying for $s_x^{(t)}$ and $s_y^{(t)}$ expressions (2.53), we can write the factor $\hat{k}(\mathbf{s}^{(t)}\cdot\mathbf{r})$ in the form

$$\hat{k}(\mathbf{s}^{(t)}\cdot\mathbf{r}) = \frac{\omega}{c}\left[x\sin\varphi_0 + ynq(\cos\eta - \kappa\sin\eta) + \right.$$
$$\left. + inqy(\kappa\cos\eta + \sin\eta)\right]. \tag{2.55}$$

From (2.55) it can be seen that under given conditions of wave incidence the planes of the constant amplitude are given by the equation $y = \text{const}$ and remain parallel to the interface of the conductor $y = 0$ (Fig. 2.4).

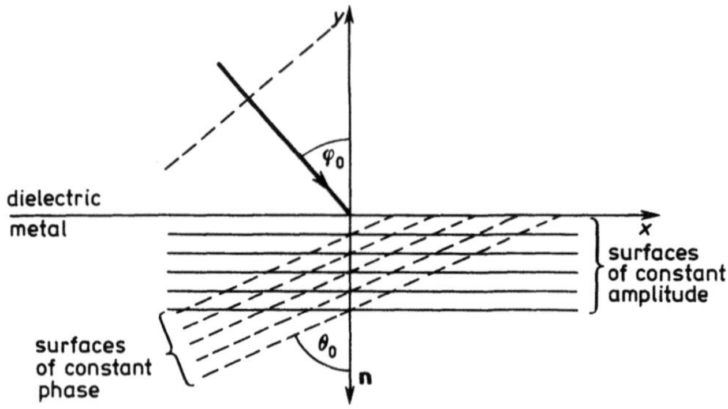

Figure 2.4

The surfaces of the constant phase are given by

$$x\sin\varphi_0 + ynq(\cos\eta - \kappa\sin\eta) = \text{const} \tag{2.56}$$

and form a real angle θ_0 with the normal to the interface surface of the medium (Fig. 2.4). This angle is found from the equation

$$\tan\theta_0 = \frac{\sin\varphi_0}{\eta q(\cos\eta - \kappa\sin\eta)}. \tag{2.57}$$

Taking this real angle into account, we may write the refraction law in the form

$$\frac{\sin\theta_0}{\sin\varphi_0} = \frac{1}{\sqrt{\sin^2\varphi_0 + n^2q^2(\cos\eta - \kappa\sin\eta)^2}}. \tag{2.58}$$

This form is analogous to Snell's law except that the refractive index is replaced by the quantity

$$n' = \sqrt{\sin^2\varphi_0 + n^2q^2(\cos\eta - \kappa\sin\eta)^2}\,, \tag{2.59}$$

which now depends not only on the medium properties but also on the incident angle.

Considering the rapid decay of the refracted wave in the conductor, of great significance in the analysis of the behaviour of light on the surface of the conductor is the reflected wave.

Using Fresnel's formulas (2.33) and (2.34), we can express the amplitudes of the reflected wave E^r by the incident angle φ_0 and the refractive index $\hat{n}$. The following relations are obtained:

$$\begin{aligned} \frac{E_{\mathrm{I}}^r}{E_{\mathrm{I}0}} &= \frac{\hat{n}^2\cos\varphi_0 - \sqrt{\hat{n}^2 - \sin^2\varphi_0}}{\hat{n}^2\cos\varphi_0 + \sqrt{\hat{n}^2 - \sin^2\varphi_0}}, \\ \frac{E_{\perp}^r}{E_{\perp 0}} &= \frac{\cos\varphi_0 - \sqrt{\hat{n}^2 - \sin^2\varphi_0}}{\cos\varphi_0 + \sqrt{\hat{n}^2 - \sin^2\varphi_0}}. \end{aligned} \tag{2.60}$$

Since the refractive index is now a complex quantity, the quantities $E_{\mathrm{I}}^r/E_{\mathrm{I}0}$ and $E_{\perp}^r/E_{\perp 0}$ are also complex. They can be written in the form

$$\frac{E_{\mathrm{I}}^r}{E_{\mathrm{I}0}} = \rho_{\mathrm{I}}\mathrm{e}^{\mathrm{i}\delta_{\mathrm{I}}}, \quad \frac{E_{\perp}^r}{E_{\perp 0}} = \rho_{\perp}\mathrm{e}^{\mathrm{i}\delta_{\perp}}. \tag{2.61}$$

In general, $\delta_{\mathrm{I}} \neq \delta_{\perp}$, and it means that there is a phase shift between the two orthogonal components of vector $\mathbf{E}^r$ of the reflectled wave, which leads to the elliptic wave polarization.

If a wave falling on the interface of a conductor is linearly polarized, then the azimuth angle β_0 which is formed by vector $\mathbf{E}_0$ and the incidence plane is found from the equality

$$\tan\beta_0 = \frac{E_{\perp 0}}{E_{\mathrm{I}0}}. \tag{2.62}$$

For the reflected wave the ratio of the appropriate components of vector $\mathbf{E}^r$ is then

$$\frac{E_{\perp}^{r}}{E_{\mathrm{I}}^{r}} = \frac{\rho_{\perp}}{\rho_{\mathrm{I}}} \mathrm{e}^{\mathrm{i}(\delta_{\perp} - \delta_{\mathrm{I}})} \tan \beta_0. \tag{2.63}$$

In the case of the perpendicular incidence of the wave on the interface of the conductor we obtain

$$\frac{E_{\perp}^{r}}{E_{\perp 0}} = \frac{1 - \hat{n}}{1 + \hat{n}} = \frac{1 - n - \mathrm{i}n\kappa}{1 + n + \mathrm{i}n\kappa}. \tag{2.64}$$

Since the *reflection coefficient* $\mathscr{R}$ is

$$\mathscr{R} = \left(\frac{E_{\perp}^{r}}{E_{\perp 0}}\right) \cdot \left(\frac{E_{\perp}^{r}}{E_{\perp 0}}\right)^{*}, \tag{2.65}$$

we can find the relationship between n, κ and the reflection coefficient $\mathscr{R}$. The explicit form of this relationship is as follows:

$$\mathscr{R} = \left|\frac{\hat{n} - 1}{\hat{n} + 1}\right|^2 = \frac{n^2(1 + \kappa^2) + 1 - 2n}{n^2(1 + \kappa^2) + 1 + 2n}. \tag{2.66}$$

Therefore, by examining the state of polarization of the reflected wave and the change of light intensity by reflection at a given incident angle, we can find the material constants for the reflecting conductor.

2.11. Pressure of Light

The pressure of light exerted on the surface of a black body is related to the transfer of the momentum of the electromagnetic field. In fact this pressure is due to a force arising during the momentum interaction between the field and the body in question. Let us assume that the direction of light propagation is perpendicular to a given surface ΔS of the body. Taking into account the transverse character of the electromagnetic wave and (1.71), we obtain for this interaction force the expression

$$\mathbf{F} = \int_{\Delta S} \mathbf{n} \cdot \tfrac{1}{2} \mathfrak{J} (\mathbf{H} \cdot \mathbf{B} + \mathbf{E} \cdot \mathbf{D}) \mathrm{d}f, \tag{2.67}$$

where $\mathfrak{J}$ is the unitary tensor, and $\mathbf{n}$ the normal versor.

If the surface of the body is a plane, and the field value at all its points can be regarded as constant, then

$$\mathbf{F} = \mathbf{n}\cdot\mathfrak{J}\tfrac{1}{2}(\mathbf{H}\cdot\mathbf{B} + \mathbf{E}\cdot\mathbf{D})\Delta S.$$

Hence, the value of the pressure exerted by light on the surface of a black body is

$$p = \frac{|\mathbf{F}|}{\Delta S} = \tfrac{1}{2}(\mathbf{H}\cdot\mathbf{B} + \mathbf{E}\cdot\mathbf{D}), \tag{2.68}$$

since $|\mathbf{n}\cdot\mathfrak{J}| = 1$.

This shows that the pressure of light exerted on the surface of a black body is equal to the energy density of the electromagnetic field.

For a totally reflecting surface the exchange of momentum between a field and a body is twice as great, and the pressure is

$$p = (\mathbf{H}\cdot\mathbf{B} + \mathbf{E}\cdot\mathbf{D}). \tag{2.69}$$

In partial reflection, characterized by the reflection coefficient $\mathscr{R} < 1$, the pressure will be given by the expression

$$p = (1 + \mathscr{R})\tfrac{1}{2}(\mathbf{H}\cdot\mathbf{B} + \mathbf{E}\cdot\mathbf{D}). \tag{2.70}$$

The pressure of light was experimentally detected in 1901 by P.N. Lebedev. He used an instrument whose essential part was a quartz thread hung vertically, to which two light plates a and b (Fig. 2.5) were attached. They were blackened on one side and silvered on the

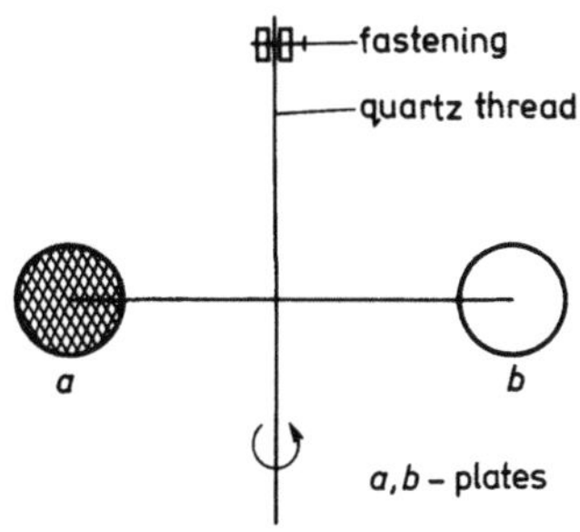

Figure 2.5

other. The instrument was placed in a vacuum. Owing to the different reflection conditions for each plate, the incident light caused the plates to rotate and the thread to twist.

To give an idea of the order of the pressure of light, it is worth to mention its value for light of the sun. It is about 4×10^{-3} N/m^2.

Bibliography

1. Born, M. and Wolf, E.: (1965), 'Principles of Optics', Pergamon Press, Oxford, London (third edition).
2. Goodman, J.: (1968), 'Introduction to Fourier Optics',McGraw-Hill, New York.
3. Kaliteevskii, N. I.: (1978), 'Wave Optics' (in Russian), Vysshaya Shkola, Moskva.
4. Korolev, F.: (1986), 'Theoretical Optics' (in Russian), Vysshaya Shkola, Moskva.
5. Lotsch, H. K. V.: (1968) *J. Opt. Soc. Am.*, **58**, 551.
6. Sommerfeld, A.: (1950), 'Optik, Vorlesungen über theoretische Physik', Bd IV, Wiesbaden.
7. Vainshtein, L. A.: (1957), 'Electromagnetic Waves' (in Russian), Sovet Radio, Moskva.

3. Coherence and Interference Phenomena

3.1. Introductory Remarks

In the geometrical optics approximation changes of intensity of a luminous flux can be described by changes of the cross-sectional area of a light beam (see Sec. 12.4). But in the case where two or more light fluxes are superposed, the intensity distribution cannot be described in such an easy way. If a light beam coming from a light source is first divided by a proper device into two beams which are next superposed, the light intensity in the area of their superposition changes from point to point from the maximum value, equal to the sum of the component beam intensities, to the minimum value, which can be zero. This phenomenon is called *interference.*

Light produced by real sources is never monochromatic, but its amplitude and phase undergo irregular fluctuations which cannot be detected by the human eye or by ordinary physical detectors. A light source consists of a great number of atoms. The waves emitted by such a source at a given moment of time are the sum of separate oscillations emitted by individual atoms. In fact, atoms radiate only during a limited period of time τ (in quantum mechanics it's called an *average lifetime*, in classical theory – the *decay time*). If we consider the radiation in the time interval considerably longer than τ, the oscillations observed at the initial moment die out whereas other atoms cause new oscillations not related to the initial oscillations. The oscillations produced by the sources are related to one another only if they are emitted within a short time interval (the time interval should be considerably shorter than τ). Interrelated oscillations are called *coherent oscillations*, the source which emits them is called a *coherent source.*

If two luminous fluxes are generated by the same source, the fluctuations existing in them are more or less correlated, and the correlation depends on the character of the source. If two fluxes come from two different sources, the fluctuations are completely incorrelated and the light beams are mutually incoherent. Their superposition causes no interference, and the resultant intensity is the sum of the individual beam intensities.

Below, we shall consider mainly the interference of fluxes generated by the same source. There exist two methods of obtaining several fluxes from one source:

1. A method caled *wave front partition*, which is exemplified by the Young experiment. The primary luminous flux is divided by the aperture configuration consisting of two slots in an opaque screen. The source emitting the primary flux has small angular dimensions.

2. The method called *amplitude division*. In this method also bulk sources can be used. The division of the primary beam is caused by reflection from partially reflecting surfaces. This method is applied in the Michelson interferometer.

3.2. Interference of Two Monochromatic Waves

Suppose that two monochromatic waves are mathematically described by expressions:

$$U_1(\mathbf{r}, t) = A_1(\mathbf{r})\exp[\mathrm{i}\omega t - \mathrm{i}g_1(\mathbf{r})],$$
$$U_2(\mathbf{r}, t) = A_2(\mathbf{r})\exp[\mathrm{i}\omega t - \mathrm{i}g_2(\mathbf{r})], \tag{3.1}$$

where $A_i(\mathbf{r})$ are the amplitudes, $g_i(\mathbf{r})$ — the phases of the component waves. The resultant fields is

$$U(\mathbf{r}, t) = A_1(\mathbf{r})\exp[\mathrm{i}\omega t - \mathrm{i}g_1(\mathbf{r})] + + A_2(\mathbf{r})\exp[\mathrm{i}\omega t - \mathrm{i}g_2(\mathbf{r})]. \tag{3.2}$$

Knowledge of the field at the point P determined by a position vector $\mathbf{r}$ enables us to find the wave intensity

$$I = UU^* = A_1^2 + A_2^2 + 2A_1A_2\cos(g_1 - g_2). \tag{3.3}$$

In equation (3.3) the symbol * denotes a complex quantity, conjugate to a given quantity.

Formula (3.3) shows that the resultant field intensity depends on the amplitudes of the component waves and on the difference of their phases. Introducing the symbols $I_1 = A_1^2$ and $I_2 = A_2^2$, formula (3.3) may be written as

$$I = I_1 + I_2 + 2\sqrt{I_1 I_2}\cos(g_1 - g_2). \tag{3.4}$$

As can be seen from this formula, at points P for which $\delta = g_1 - g_2$ takes the values: 0, 2π, 4π..., we obtain

$$I_{\max} = I_1 + I_2 + 2\sqrt{I_1 I_2}, \tag{3.5}$$

whereas at points, in which $\delta = g_1 - g_2$ is π, 3π..., we have the minimum intensity

$$I_{\min} = I_1 + I_2 - 2\sqrt{I_1 I_2}. \tag{3.6}$$

In the special case where $I_1 = I_2$, formula (3.4) reduces to the form

$$I = 4I_1 \cos^2\left(\tfrac{1}{2}\delta\right). \tag{3.7}$$

From (3.7) results that I is enclosed within the interval $0 \leq I \leq 4I_1$. The situation is presented in Fig. 3.1.

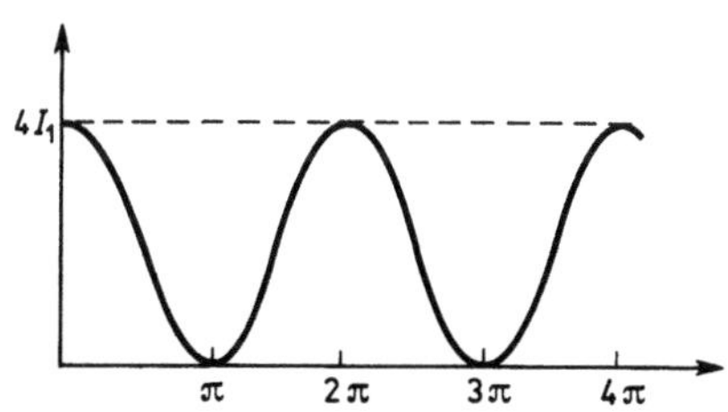

Figure 3.1

3.3. The Natural Width of a Spectral Line

A monochromatic wave is an ideal case. In reality, oscillations emitted by atoms are confined in time. Atoms radiate wave-trains whose duration and spectral composition (the frequency interval within which the spectral density is different from zero) are limited. The longer the wave-train, the narrower its frequency interval. The situation is presented in Fig. 3.2.

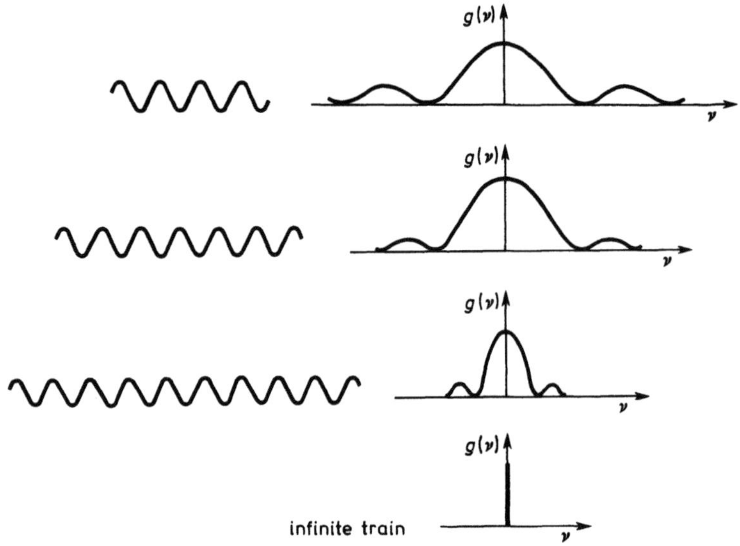

Figure 3.2

In fact, atoms radiate only within a limited time τ. The wave-train corresponding to this process presents quickly damping oscillations, and has the character shown in Fig. 3.3. Mathematical description of the oscillations under examination for $t > 0$ can be expressed by the formula

$$f(t) = A \exp\left(-\frac{t}{\tau}\right) \exp(i2\pi\nu_0 t)\sigma_0(t), \tag{3.8}$$

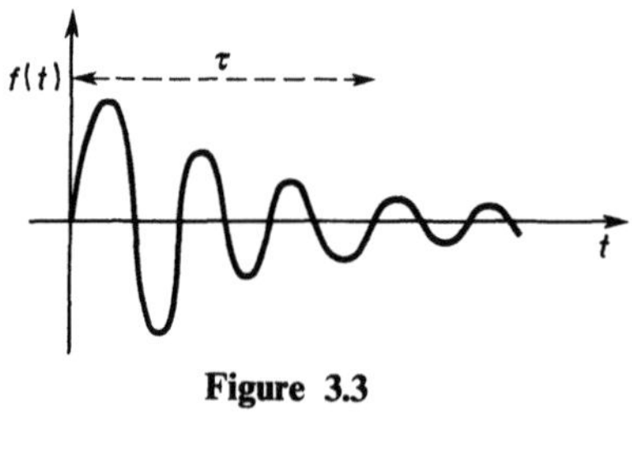

Figure 3.3

where

$$\sigma_0(t) = \begin{cases} 0 & \text{for} \quad t < 0 \\ 1 & \text{for} \quad t \geq 0, \end{cases}$$

and τ is an average lifetime. It characterizes the decay rate of the oscillations. If τ is long, the wave-train decays slowly and makes almost sinusoidal oscillations; the wave is almost monochromatic. Otherwise, if the oscillations decay quickly, then a wide frequency spectrum occurs; the wave is no longer monochromatic. As is shown below, in the case of decaying oscillations the frequency spectrum comprises a continuous interval of the width $\Delta\nu$. The relationship between $\Delta\nu$ and τ is

$$\Delta\nu \cdot \tau \geq \frac{1}{4\pi}. \tag{3.9}$$

The Fourier transformation of (3.8) has the form

$$g(\nu) = \int_{-\infty}^{+\infty} A \exp\left(\frac{t}{\tau}\right) \exp[-2\pi(\nu - \nu_0)it]\, \sigma_0(t) \mathrm{d}t \tag{3.10}$$

and is

$$g(\nu) = \frac{A}{1/\tau - \mathrm{i}2\pi(\nu - \nu_0)}. \tag{3.11}$$

In order to find the intensity distribution in the spectrum as a function of frequency, the square of module $g(\nu)$ is to be calculated. The quantity has the value

$$|g(\nu)|^2 = \frac{A^2}{4\pi^2(\nu_0 - \nu)^2 + 1/\tau^2}. \tag{3.12}$$

Introducing the pair of auxiliary symbols

$$x = \frac{\nu}{\nu_0}, \qquad \Gamma = \frac{1}{\nu_0 \tau}, \tag{3.13}$$

(3.12) can be written as

$$|g(x)| = \frac{A^2}{\nu_0^2} \frac{1}{4\pi^2(1-x)^2 + \Gamma^2}. \tag{3.14}$$

Formulas (3.12) and (3.14) express the intensity distribution dependence on the frequency, i.e. they give the spectrum picture of decaying oscillations. The graph of the function (3.14) is shown in Fig. 3.4.

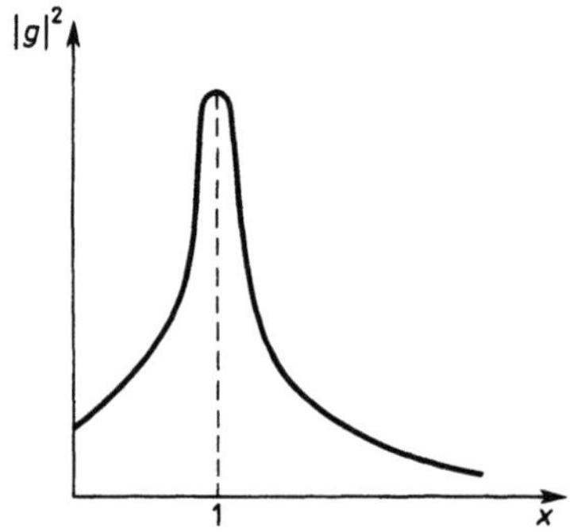

Figure 3.4

This function has a sharp maximum for $x = 1$, i.e for $\nu = \nu_0$, and for this frequency takes the value

$$|g(x)|^2_{\max} = \frac{A^2}{\nu_0^2 \Gamma^2} = A^2 \tau^2.$$

The values of x for which the function $|g(x)|^2$ is half its maximum value are found from the equation

$$\pm\Gamma = 2\pi(1-x).$$

Coming back to the original variables (3.13), the so-called *halfwidth of spectral line* $\Delta\nu_{1/2}$ can be found. It equals half of the distance between points at which $|g(x)| = \frac{1}{2} g_{\max}$. This quantity equals $\Delta\nu_{1/2} = \nu - \nu_0 = 1/2\pi\tau$, and hence

$$\Delta\nu_{1/2}\tau \geq \frac{1}{2\pi}.$$

The halfwidth of a spectral line is inversely proportional to the decay time, i.e., to the length of a wave-train. This result can also be achieved within the framework of quantum mechanics. If by τ we denote an average life time, the indeterminacy of energy ΔE is given by the Heisenberg relationship $(\tau\Delta E \geq \hbar)$, and hence

$$h\tau\Delta\nu \geq \hbar, \qquad \tau\Delta\nu \geq \frac{1}{2\pi}.$$

For example, for $\tau = 10^{-8}$ s and $\lambda = 0.5$ μm, we obtain the width of the line

$$\Delta\lambda = \frac{\lambda^2 \Delta\nu_{1/2}}{c} = \frac{\lambda^2}{2\pi c\tau} = 2.5 \times 10^{-5} \text{ nm}.$$

Even for the narrowest lines this value is smaller than the experimental value. In reality, spectral lines are wider because of the influence of other phenomena – the Doppler effect, pressure, etc.

3.4. Modification of a Spectral Line by the Doppler Effect

The Doppler effect occurs in the case of light emission by atoms of gas or vapour subjected to electric discharges. If atoms were at rest, they would emit waves of frequency determined by the relation

$$\nu_0 = \frac{E_m - E_0}{h}.$$

But atoms are in thermic motion. As a result of the Doppler effect, the frequency observed in the direction of x-axis is

$$\nu = \nu_0 + \frac{v_x}{c}\nu_0, \tag{3.15}$$

where v_x is the projection of thermic velocity of the atom on the x-axis. For the molecules of gas the velocity distribution, i.e. the probability of finding the velocity of a molecule within the range v_x, $v_x + dv_x$; v_y, $v_y + dv_y$; v_z, $v_z + dv_z$ is

$$f(v_x, v_y, v_z)\mathrm{d}v_x\mathrm{d}v_y\mathrm{d}v_z$$
$$= \left(\frac{m}{2kT\pi}\right)^{3/2} \exp\left[-\frac{m}{2kT}(v_x^2 + v_y^2 + v_z^2)\right] \mathrm{d}v_x\mathrm{d}v_y\mathrm{d}v_z. \tag{3.16}$$

The question may arise: what is the probability that a frequency observed in light emitted by atom during the transition from the level E_m to the level E_0 lies within the interval $(\nu, \nu + \mathrm{d}\nu)$?

From (3.15) we obtain

$$v_x = \frac{\nu - \nu_0}{\nu_0} c, \qquad \mathrm{d}v_x = \frac{c}{\nu_0}\mathrm{d}\nu. \tag{3.17}$$

Formula (3.16) allows us to state that this probability is proportional to

$$g(v_x)\mathrm{d}v_x = \left(-\frac{m}{2\pi kT}\right)^{3/2} \times$$
$$\times \iint\limits_{-\infty}^{+\infty} \exp\left[-\frac{m}{2kT}(v_y^2 + v_z^2)\right] \mathrm{d}v_y\mathrm{d}v_z \exp\left(-\frac{m}{2kT}v_x^2\right) \mathrm{d}v_x, \tag{3.18}$$

Substituting in (3.18) the variables (3.17) and calculating the integral, we obtain

$$\frac{c}{\nu_0} g\left(\frac{\nu - \nu_0}{\nu_0} c\right) \mathrm{d}\nu$$
$$= \frac{c}{\nu_0}\sqrt{\frac{m}{2kT\pi}} \exp\left[-\frac{m}{2kT}(\nu - \nu_0)^2 \frac{c^2}{\nu_0^2}\right] \mathrm{d}\nu. \tag{3.19}$$

The curve connected with the function

$$g_1(\nu - \nu_0) = \sqrt{\frac{m}{2kT\pi}} \exp\left[-\frac{mc^2}{2kT\nu_0^2}(\nu - \nu_0)^2\right] \tag{3.20}$$

is a Gauss curve. The maximum of the curve occurs for $\nu = \nu_0$, and its halfwidth, i.e. the width corresponding to half of the height, is (Fig. 3.5)

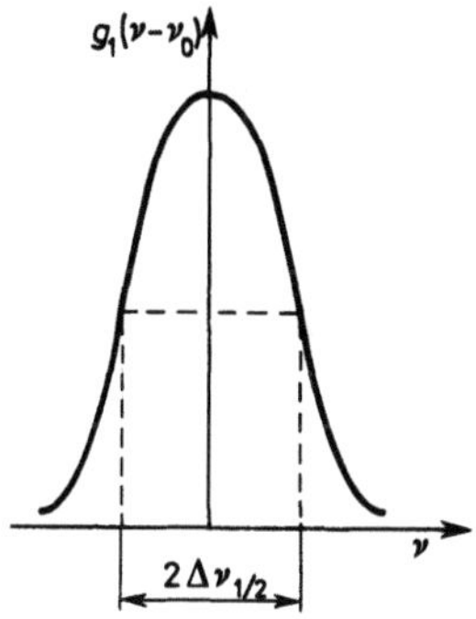

Figure 3.5

$$\Delta\nu_{1/2} = \nu_0 \sqrt{\frac{2kT}{mc^2}\ln 2}. \tag{3.21}$$

As can be seen from the above, light emitted by gaseous media is not a monochromatic wave, but always, even in a fixed transition between levels, it is quasi-monochromatic light, i.e., the light whose wavelength lies within the interval $-\frac{1}{2}\Delta\lambda + \lambda_0 \leq \lambda \leq \lambda_0 + \frac{1}{2}\Delta\lambda$, where the width of the interval $\Delta\lambda$ satisfies the condition

$$\frac{\Delta\lambda}{\lambda_0} \ll 1.$$

In the case in question this condition takes the form

$$\left|\frac{\Delta\lambda}{\lambda_0}\right| = 2\sqrt{\frac{2kT}{mc^2}\ln 2} \ll 1,$$

which takes place within the temperature range of our interest.

Continuing the consideration and applying formula (3.20), we can find the mathematical relation describing the propagation of waves emitted by atoms being in the state of thermic motion. Making use of the Fourier analysis for (3.20), we obtain

$$\begin{aligned} F(t) &= \int_{-\infty}^{+\infty} g_1(\nu - \nu_0)\exp(-2\pi i\nu t)\,d\nu \\ &= \exp\left(-\frac{2kT\nu_0^2\pi^2}{mc^2}t^2\right)\exp(-2\pi i\nu_0 t). \end{aligned} \tag{3.22}$$

Thus $F(t)$ is a modulated periodic wave with the amplitude as a Gaussian function of time. For this amplitude we can introduce the notion of halfwidth:

$$\Delta t_{1/2} = \sqrt{\frac{mc^2}{2kT\nu_0^2\pi^2}}\sqrt{\ln 2}.$$

Hence, on regarding (3.21), we have

$$\Delta t_{1/2}\Delta\nu_{1/2} = \frac{\ln 2}{\pi} \approx \frac{0.69}{\pi} \geq \frac{1}{2\pi}.$$

In the Fourier analysis it can be proved in general that the shorter in time a signal, the wider is its frequency spectrum. The time duration of the signal and the width of its spectrum are bound through the relation

$$\Delta t\,\Delta\nu \geq \frac{1}{4\pi}. \tag{3.23}$$

The examples given above confirm in individual cases this general relation.

3.5. Coherence Time, Coherence Length and their Influence on the Interference Phenomenon

Let us consider the Michelson interferometer illuminated by a point source of light. Suppose that the source S emits non-monochromatic light; it is presented in Fig. 3.6 as a wave-train between points O and 1. This limited wave-train can be described by a function $f(t)$ of the form

$$f(t) = \begin{cases} A_0 e^{i2\pi\nu_0 t} & \text{for } |t| \leq \dfrac{\Delta t}{2}, \\ 0 & \text{for } |t| > \dfrac{\Delta t}{2}. \end{cases} \tag{3.24}$$

The spectral function for this wave-train limited in time is

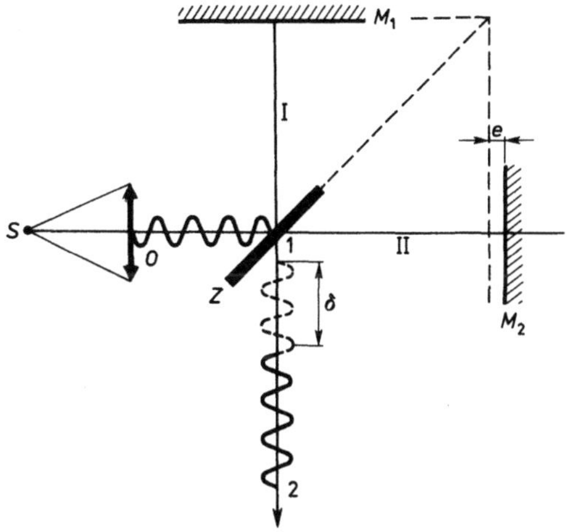

Figure 3.6

$$g(\nu) = A_0 \frac{\sin[\pi(\nu - \nu_0)\Delta t]}{\pi(\nu - \nu_0)\Delta t} \Delta t. \tag{3.25}$$

The region where $g(\nu)$ is a significant function is confined to the interval

$$-\pi \leq \pi(\nu - \nu_0)\Delta t \leq \pi, \tag{3.26}$$

i.e., to

$$|(\nu - \nu_0)\Delta t| \leq 1,$$

where Δt is the time duration of the wave-train.

The primary wave-train $O1$ is divided in two at the partially reflecting surface Z (the amplitude division takes place). One of them covers path I, the other path II. If the mirror M_2 is placed in the position shown in Fig. 3.6, then the wave-train going along path II covers a distance a little longer than that one going along path I. The two wave-trains meet at the output of the interferometer. The wave-train going along path II has a delay in comparison with that going along path I, and is shifted by $\delta = 2e$. If δ is smaller than the length of the wave-train, then the wave-trains practically cover and they can interfere. The interferential fringes are sharp, which is indicative of time coherence.

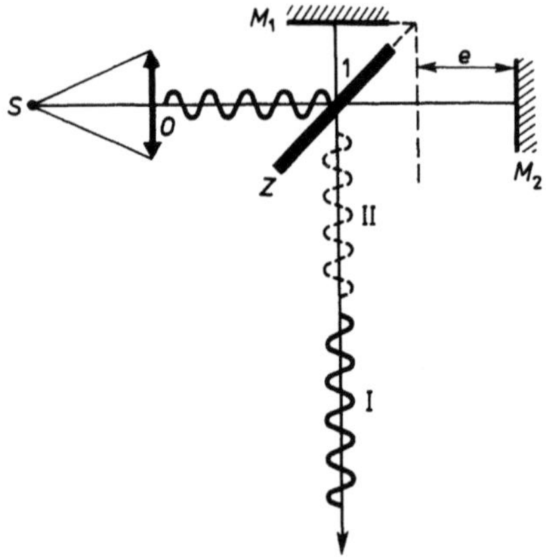

Figure 3.7

If $\delta = 2e$ is larger than the length of the wave-trains, then the two wave-trains I and II obtained from the same input wave-train cannot overlap at the output of the interferometer, and they cannot interfere (Fig. 3.7). However, also in this case it is possible to observe the interference phenomenon. It could be the result of overlapping of two wave-trains produced from different input wave-trains. Two wave-trains A and B can be emitted by the source S in different moments of time, and the wave-train B lags behind in comparison with A. They are shifted in space by δ_1 (Fig. 3.8).

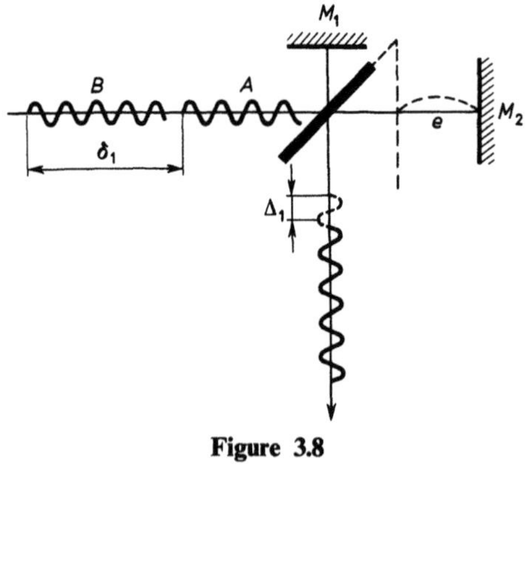

Figure 3.8

If the difference in path lengths caused by the interferometer is $\delta \geq c\Delta t$, the two wave-trains obtained from the wave-train A and travelling along the different paths I and II will not overlap. The same concerns the wave-trains generated by the wave-train B. However, it may happen that the wave-packet produced from the wave-train B and travelling along path I (of smaller length) may overlap the wave packet produced from the wave-train A and travelling along path II. The path difference created by the interferometer $\Delta = 2e$ may compensate the initial delay of the wave-train B in relation to A, which equals δ_1. Since the emission of wave-trains given by the source is a random phenomenon, $\delta_1, \delta_2, \delta_3 \ldots$ vary in time in a completely arbitrary way. The same concerns the differences Δ_1, Δ_2, $\Delta_3 \ldots$, which also are incidental.

During observation, a great number of overlapping momentary interference pattern follows in quick succession. These images cannot be distinguished – there is no time coherence.

If Δt is the time duration of a wave-train (coherence time), c – the velocity of light, then $\Delta l = c\Delta t$ is the coherence length.

From the above it follows:

(a) in order to observe interference pattern by means of a common light source, it is required that the path difference caused by the interferometer is less than the coherence length of wave-trains;

(b) the less is the path difference compared with the coherence length, the sharper is the interference pattern.

The problem presented above can be treated formally (mathematically). Let $F(t)$ be a light disturbance in the instant t connected with a single wave-train. It is assumed that F equals zero for $|t| \geq t_0$, and that it can be expressed by the Fourier integral

$$F(t) = \int_{-\infty}^{+\infty} f(\nu) e^{+2\pi\nu i t} d\nu, \tag{3.27}$$

where

$$f(\nu) = \int_{-\infty}^{+\infty} F(t) e^{-2\pi\nu i t} dt. \tag{3.28}$$

If N such packets go through the point during the observation time, then the total light disturbance is

$$V(t) = \sum_{n=1}^{N} F(t - t_n), \tag{3.29}$$

where t_n stands for the time of arriving of a successive packet. Light intensity averaged over the time interval $2T$ required for the observation is

$$I = \frac{1}{2T} \int_{-T}^{T} |V(t)|^2 \mathrm{d}t \sim \frac{1}{2T} \int_{-\infty}^{+\infty} |V(t)|^2 \mathrm{d}t. \tag{3.30}$$

T is long in comparison with half duration time t_0 for each wave-train.

The total disturbance $V(t)$ can be represented as the Fourier integral

$$V(t) = \int_{-\infty}^{+\infty} v(\nu) \mathrm{e}^{+2\pi i \nu t} \mathrm{d}\nu, \tag{3.31}$$

where

$$v(\nu) = f(\nu) \sum_{n=1}^{N} \mathrm{e}^{-2\pi i \nu t_n}. \tag{3.32}$$

On the ground of Parseval's theorem we can write

$$\int_{-\infty}^{+\infty} |V(t)|^2 \mathrm{d}t = \int_{-\infty}^{+\infty} |v(\nu)|^2 \mathrm{d}\nu$$

$$= \int_{-\infty}^{+\infty} |f(\nu)|^2 \sum_{n=1}^{N} \sum_{m=1}^{N} \exp[2\pi i(t_n - t_m)\nu] \mathrm{d}\nu. \tag{3.33}$$

Noting that

$$\sum_{n=1}^{N} \sum_{m=1}^{N} \exp[2\pi i \nu(t_n - t_m)] = N + \sum_{n \neq m} \exp[2\pi i \nu(t_n - t_m)]$$

$$= N + 2 \sum_{n<m} \cos 2\pi\nu(t_n - t_m), \tag{3.34}$$

and owing to the randomness of t_n, we can state that the double sum above equals N. This remark enables us to find the average intensity

$$I = \frac{N}{2T} \int_{-\infty}^{+\infty} |f(\nu)| \mathrm{d}\nu, \tag{3.35}$$

which is the integral of the intensities of the components of monochromatic waves.

By way of illustration let us consider a disturbance of the form

$$F(t) = \begin{cases} f_0 \mathrm{e}^{-2\pi i \nu_0 t} & \text{for} \quad |t| \leq \dfrac{\Delta t}{2}, \\ 0 & \text{for} \quad |t| > \dfrac{\Delta t}{2}, \end{cases}$$

where f_0 is a constant. Then

$$f(\nu) = f_0 \int_{-\Delta t/2}^{\Delta t/2} \mathrm{e}^{2\pi i(\nu - \nu_0)t} \mathrm{d}t = f_0 \Delta t \frac{\sin[\pi(\nu - \nu_0)\Delta t]}{\pi(\nu - \nu_0)\Delta t}.$$

$f(\nu)$ describes the intensity distribution of Fourier components. The region of frequency for which it has significant values is

$$\nu_0 - \frac{\Delta\nu}{2} \leq \nu \leq \nu_0 + \frac{\Delta\nu}{2},$$

where $\Delta\nu/2$ is the distance between ν_0 and zeros of the function $f(\nu)$ closest to ν_0. This equality is

$$\Delta\nu = \frac{1}{\Delta t}.$$

Thus the effective frequency region of the Fourier spectrum is of the order of the inverse of the time duration for a single wave-train (Fig. 3.9). The example given above, in which all wave-train are identical and have this simple form, is far from reality.

According to atomic theory, the energy emission performed by atoms produces attenuated wave-trains. The atoms are in continu-

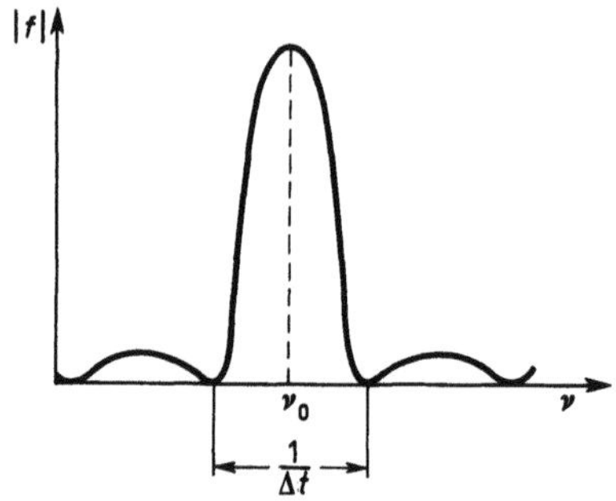

Figure 3.9

ous thermic motion (the Doppler effect). The radiating atoms interact, and for real light the terms "time duration of a wave-train" or "frequency region of a Fourier spectrum" have no such a simple meaning. However, for any disturbance $V(t)$ and its transform $v(\nu)$, it is possible to define the quantities Δt and $\Delta\nu$, which can be regarded as an average time duration and the effective frequency interval of the wave-trains $V(t)$. It can be shown that these averaged quantities satisfy the relation

$$\Delta t\,\Delta\nu \geq \frac{1}{4\pi}.$$

Time Δt is the coherence time, $\Delta l = c\Delta t$ is the coherence length, and

$$\Delta t \propto \frac{1}{\Delta\nu} = \frac{\lambda_0^2}{c\Delta\lambda_0},$$

where λ_0 is a mean wavelength.

3.6. Spatial Coherence

Until now, a point source of waves was taken into consideration while discussing time coherence. In nature such sources do not exist.

Let us consider once again the Michelson interferometer illuminated by a perfectly monochromatic point source S. Mirror M_2 is

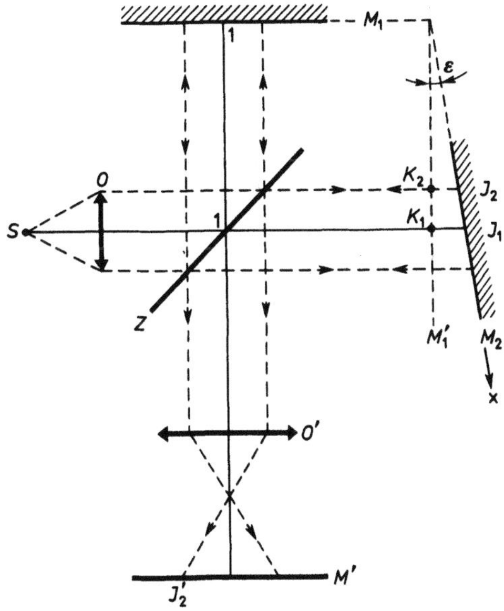

Figure 3.10

set up in such a way that it forms with plane M'_1 (M'_1 is symmetrical to M_1) a very small angle ε.

Mirrors M_1 and M_2 are perpendicular to the plane of the figure (the situation is shown in Fig. 3.10). Ray SO propagating along the optical axis of objective O is reflected from mirror M_2 at point J_1 and crosses plane M'_1 at point K_1. Since the angle ε is small, we can assume that the ray reflected at point J_1 practically covers the incident ray. The path difference between the ray reflected at point J_1 and the ray reflected from mirror M_1 is $\delta_1 = 2K_1J_1$. The path difference between the ray reflected at point J_2 and the ray reflected from mirror M_1 is $\delta_2 = 2K_2J_2 = 2\varepsilon x = 2e$.

Using lens O', we obtain image M' of mirror M_1 or M_2. Both mirrors are supposed to have the same reflecting power. The intensity I at point J'_2 corresponding to point J_2 is

$$I \propto 2\cos^2 \frac{2\pi x \varepsilon}{\lambda}. \tag{3.36}$$

It is caused by the fact that the resultant oscillations at the points of plane M' consist of two waves: one of them is the wave coming from mirror M_1 and given by the formula

$$U_1 = Ae^{i(\omega t - ks)},$$

whereas the other is coming from mirror M_2 and given by

$$U_2 = Ae^{i(\omega t - ks - 2ke)}.$$

The superposition of these two waves results in the wave field

$$U = U_1 + U_2,$$

whose intensity is

$$I \propto |U|^2 = 4A^2 \cos^2 \frac{2e\pi}{\lambda} = 4A^2 \cos^2 \frac{2x\varepsilon\pi}{\lambda}.$$

Along M', in a direction lying in the plane of the figure, the intensity varies likewise $\cos^2 x$ (Fig. 3.11). We obtain a pattern of interferential fringes; for $e = \varepsilon x$, the distance between the fringes is $\Delta x = \lambda/2\varepsilon$. The phenomenon proceeds so as if the rays were reflected from the planes M'_1 and M_2, i.e. from the aerial wedge with the vertical angle ε.

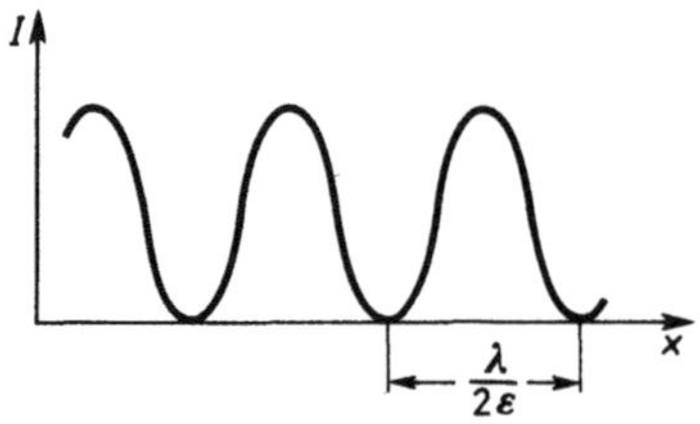

Figure 3.11

What will happen with the interferential fringes if the point source is located not at S but at S' (Fig. 3.12)? A ray going from S' to J_2 falls upon the plane M'_1 under the angle i. The path difference is now (Fig. 3.13) $\delta' = 2e/\cos i - 2e\tan i \sin i = 2e\cos i$. It has decreased, and the expression for the intensity in the plane of the image M' takes the form

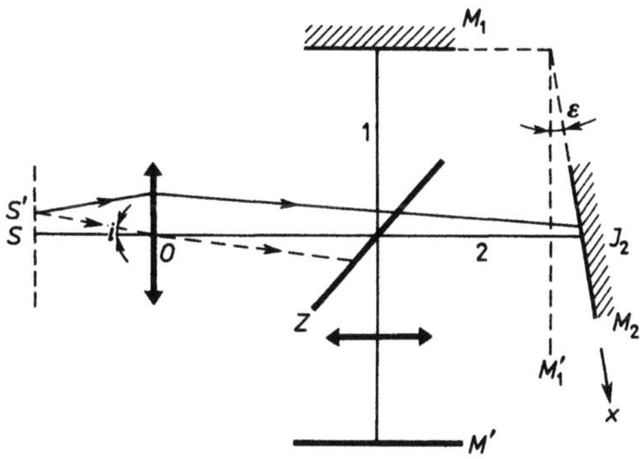

Figure 3.12

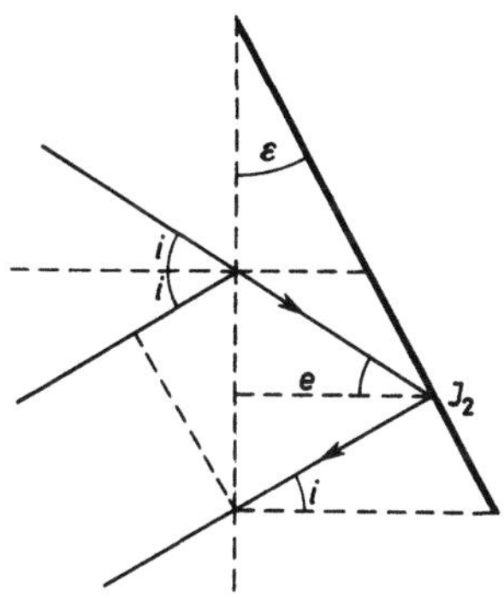

Figure 3.13

$$I \propto \cos^2\left(\frac{\pi\delta'}{\lambda}\right) = \cos^2\left(\frac{2\pi e \cos i}{\lambda}\right). \tag{3.37}$$

Thus the pattern of interferential fringes produced by the source S' is analogical to that given by the source S, but with larger distances between the fringes (Fig. 3.14). If the sources S and S' act simultaneously and are mutually independent, the resultant intensity given by the sources in the plane M' will be the sum of the component intensities produced by each source individually.

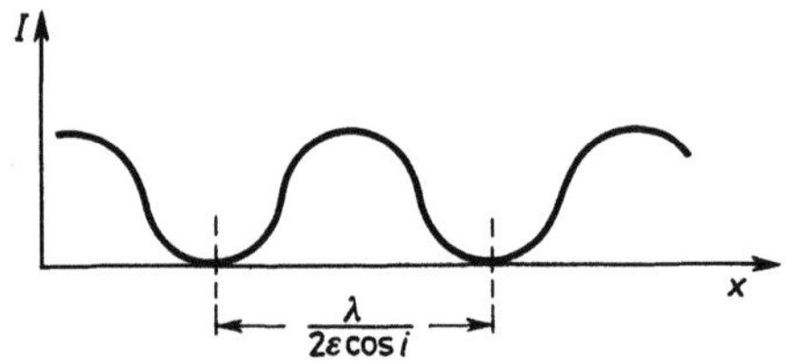

Figure 3.14

Owing to the difference in distances between the interferential fringes given by the sources S and S', the resultant pattern of the fringes will be less sharp. The bulk source can be regarded as a set of point sources lying close to one another. If source dimensions are small, then $\cos i \approx 1$. In this case the pattern of the interferential fringes obtained for S and that for S' are identical – the source is spatially coherent. Increasing the source diameter, from a certain moment we can no longer say that the condition $\cos i = 1$ holds, and that all points of the source give approximately the same pattern of the interferential fringes. The fringes become blurred, less sharp. In this case the source is said to have partial spatial coherence. With further increase of source dimensions, the image becomes more blurred, the fringes vanish and the image M' becomes homogeneous. The source is spatially incoherent.

Spatial coherence of a source depends on its angular dimensions:

– the source is coherent if it is of small dimensions and gives sharp interference pattern,

– increasing the source dimensions the fringes become less and less sharp, and the source becomes partially coherent,

– for large dimensions of the source the fringes vanish, and the source is incoherent.

Young's experiment with the interference of light produced by diffraction of two holes is a suitable starting-point for the presentations of partial spatial coherence of light introduced below.

If two holes are illuminated by a light source S, then as a result of diffraction of the incident wave, there arise at the holes T_1 and T_2 two secondary waves (Fig. 3.15). The superposition of these two waves gives in the observation plane Q oscillations, whose am-

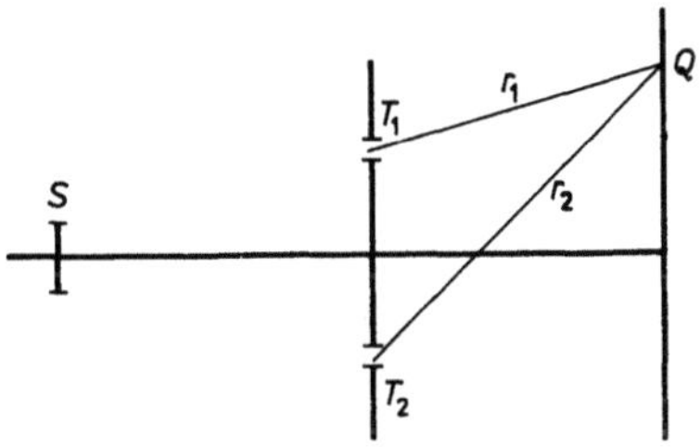

Figure 3.15

plitude will undergo a change from point to point. There will arise an interference pattern, in which, according to the relative intensities of the secondary waves, the total suppression of radiation may occur at certain of its points. The formation of sharp interference pattern is closely connected with the existence of the phase relations. In such cases the two secondary beams are said to be coherent. However, the interference pattern with the areas of the total suppression of radiation not always corresponds with an experimental situation. This is the case for ordinary thermal sources if only the angle under which the source is visible from a point of deflecting diaphragm is small enough. Moving a source of this kind with finite dimensions towards the diaphragm, the interference pattern usually blurs stepwise. In other words, the maxima and minima of intensity become less sharp.

To give a linear characteristic of sharpness of the interference pattern, we introduce the term of *visibility of the fringes* defind by the formula

$$\vartheta = \frac{I_{\max} - I_{\min}}{I_{\max} + I_{\min}}. \tag{3.38}$$

Using this term we can assert that as the source S approaches the diaphragm, the visibility ϑ approaches zero. In this case we obtain the image analogous to that when the source and the deflecting holes are replaced by two independent sources of thermal radiation positioned at the place of the deflecting holes. In other words, two points on the surface of the thermal source, at a sufficiently large distance between them, give the same interference pattern as that produced by two independent thermal sources. It is quite understandable, since also in a source atoms lying far enough from each other radiate independently, contributing to the total wave field; these contributions remain in different incidental phase ralations between themselves. If the source is placed near the diaphragm, or if the source and the holes are replaced by two independent thermal sources, the interference pattern vanishes as a result of complete lack of any phase relations between the radiations of the sources. During the time indispensable for the observation, a great number of optical oscillations take place, the momentary interference patterns undergo quick changes, giving as a result a blurred pattern. In this case the two wave-trains are incoherent.

Between these two extreme cases, i.e. the case with sharp interference patterns and the one with the lack of them, there exists an intermediate one, which is obtained with the use of *partially coherent light*. This situation is realized if the source S takes the intermediate position between the one close to the diaphragm and the other at a large distance from it. In this case the contrast of the fringes will depend on the distance between the source and the diaphragm, and it will oscillate between zero and a certain value, which decreases with decreasing distance. Approaching the source we can observe the disappearance of the fringes; with further approaching the source fringes may appear again – but they are of weaker contrast – and again disappear, and so on, till the final disappearance.

Although for a precise and exact description of the process it is necessary to know the behaviour of each radiating atom at every moment of time, simple problems can be examined with the limited knowledge about atoms. For example, a mean intensity at point T_1 may depend only on mean intensities of the individually radiating atoms, which can be assumed to be identical. In order to find the mean values of the intensities, we should apply statistical approach which makes use of the statistical description of optical fields.

3.7. Mean Values and Correlation Functions for Random Processes

As has already been mentioned in Sec. 3.5, each real source of light radiation consists of a large number of atoms, which in the case of thermal sources emit electromagnetic wave-trains of different time duration and in different time intervals. Therefore the total wave field produced by a source is a superposition of these at random appearing wave-trains. Thus the components of the vectors of the resultant electromagnetic field are random quantities dependent on time and spatial variables. The state of the field at a given point of space cannot be described by a precise mathematical relation. It is an undetermined quantity, which cannot be predicted for a moment of time on the ground of the previous state. Such behaviour of a physical quantity is a random process, and for its description the averaged statistical characteristics are required.

The result of one observation gives only one of numerous

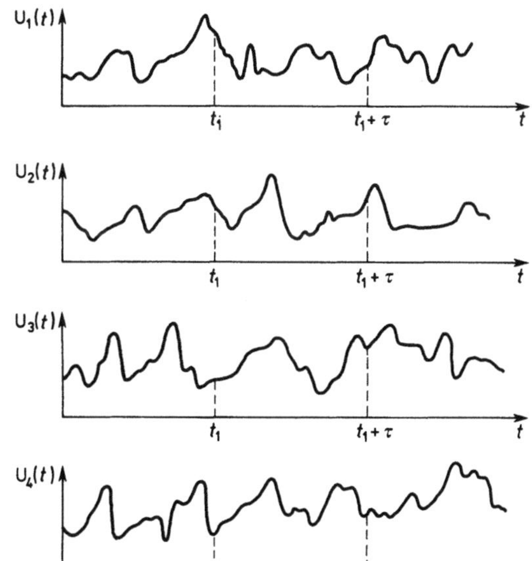

Figure 3.16

possible results, which can be written as the function $U = U(\mathbf{r}, t)$. ($U(\mathbf{r}, t)$ is generally a complex function and represents one of the components of the field vectors). It is called the *realizaton of a random process.* The set of all realizations which could be obtained by examining a given random phenomenon is called a *random process* or a *stochastic process* (Fig. 3.16).

From the point of view of stochastic processes, a physical phenomenon at a fixed point of space can be described in a given instant of time by averaging over a set of realizations representing a given random process. Thus the average value of a random process in the moment of time t_1 may be found by summing up the momentary values of each realization of the set in the moment of time t_1. Dividing the sum by the number of realizations, we get

$$U_u(t_1) = \lim_{N \to \infty} \frac{1}{N} \sum_{k=1}^{N} U_k(t_1). \tag{3.39}$$

On the other hand, the correlation between the values of a random process in two different moments of time is defined by averaging over the set of realizations the product of momentary values of a process in the moments t_1 and $t_1 + \tau$, i.e.

$$R_u(t_1, t_1 + \tau) = \lim_{N \to \infty} \frac{1}{N} \sum_{k=1}^{N} U_k^*(t_1) U_k(t_1 + \tau). \tag{3.40}$$

For a given stochastic process the quantity R_u gives the general relation between a process value in a given moment and a value in another moment of time, and it is called *autocorrelation.* In the first place it is applied while examining the relations between the values of the process in different moments of time. It is a mighty instrument for revealing determinate processes, which can be suppressed by the background noise.

In a general case, $U_u(t_1)$ and $R_u(t_1, t_1 + \tau)$ defined above change with the change of time t_1. It is said then that the process $\{U(\mathbf{r}, t)\}$ at a given point of space is *non-stationary.* If these quantities remain constant with the change of time t_1, the process is said to be *stationary*.

For majority of stationary random processes it is possible to describe them by averaging over time the individual realizations of the process, i.e. by introducing the quantities

$$U_u(k) = \lim_{T\to\infty} \frac{1}{T} \int_0^T U_k(t)\,\mathrm{d}t, \tag{3.41}$$

$$R_u(\tau, k) = \lim_{T\to\infty} \frac{1}{T} \int_0^T U_k^*(t)\, U_k(t+\tau)\,\mathrm{d}t. \tag{3.42}$$

If a random process $\{U(\mathbf{r}, t)\}$ is stationary and the quantities defined by formulas (3.41) and (3.42) are the same for different realizations, a process of this kind is called an *ergodic process*. For such a process the mean value (3.41) and the autocorrelation function (3.42) are equal to the appropriate mean values taken over the elements of the set $U_k(t)$, i.e.

$$U_u(k) = U_u, \quad R_u(\tau, k) = R_u(\tau). \tag{3.43}$$

Talking of the stationarity of a random process represented by one realization only, it is understood that its characteristics calculated for short, chosen at random segments are almost identical.

Above we have discussed a random process describing the state of the field at a fixed point of space. We admit also the change of the point of observation. Random processes depend then not only on time but also on spatial variables. Let $\{U(\mathbf{r}, t)\}$ (where $\mathbf{r}$ is the position vector, t — time) be a multidimensional random process. $U_1(\mathbf{r}, t)$ and $U_2(\mathbf{r}', t')$ stand for some realizations of this random process at two points $\mathbf{r}$ and $\mathbf{r}'$ of a certain area S. Over all possible realizations of a random process we can perform two kinds of averaging: in time and in space. If a field is stationary, the results of averaging over time do not depend on the moments of time t and t', and are merely the functions of the difference $\tau = t' - t$. The result of this averaging remains a function of $\mathbf{r}$ and $\mathbf{r}'$, i.e. spatial variables.

The time-averaged characteristics of a stationary field (the mean value and the time cross-correlation) are defined by

$$U_u(\mathbf{r}) = \lim_{T\to\infty} \frac{1}{2T} \int_{-T}^{T} U(\mathbf{r}, t)\mathrm{d}t,$$
$$R_{12}(\mathbf{r}, \mathbf{r}', \tau) = \lim_{T\to\infty} \frac{1}{2T} \int_{-T}^{T} U_1(\mathbf{r}, t)U_2(\mathbf{r}', t+\tau)\mathrm{d}t. \tag{3.44}$$

The *mean spatial value* and the *spatial cross-correlation function* are defined as follows:

$$u_u(\mathbf{r}) = \lim_{S\to\infty} \frac{1}{S} \int_S U(\mathbf{r}, t)\mathrm{d}S(\mathbf{r}),$$
$$R_{12}(\mathbf{r}, \mathbf{r}', t, t') = \lim_{S\to\infty} \frac{1}{S} \int_S U_1(\mathbf{r}, t)U_2(\mathbf{r}', t)\mathrm{d}S(\mathbf{r}),$$

where $\mathbf{r}' = \mathbf{r} + \boldsymbol{\xi}$.

If after averaging over space, we obtain a quantity dependent on $\mathbf{r}$ and $\mathbf{r}'$ only through the vector $\boldsymbol{\xi} = \mathbf{r}' - \mathbf{r}$, then a field of this kind is homogeneous. If the results of averaging over space depend only on $|\boldsymbol{\xi}|$ and do not depend on the orientation of the vector $\boldsymbol{\xi}$, then the field is *isotropic*. For homogeneous processes the results of averaging may depend on the choice of time t and t'. The results of averaging over time and space are independent of the choice of moment of time and of a coordinate system if a process is homogeneous and stationary.

3.8. Correlation Functions for Light Beams

Let us consider a wave field generated by a polychromatic bulk source σ. For a time being we neglect polarization effects and light disturbance is described by the analytic function $U(\mathbf{r}, t)$. Time variations of this quantity cannot be observed, since a detector registers only mean values over a time interval T much longer than the period of vibrations. What can be observed is the mean light intensity

$$I(\mathbf{r}) = \lim_{T\to\infty} \frac{1}{T}\int_0^T U(\mathbf{r},t)U^*(\mathbf{r},t)\mathrm{d}t = \langle U(\mathbf{r},t)U^*(\mathbf{r},t)\rangle. \quad (3.46)$$

For two points P_1 and P_2 (with the position vectors $\mathbf{r}_1$ and $\mathbf{r}_2$), we can find experimentally not only $I(\mathbf{r}_1)$ and $I(\mathbf{r}_2)$ but also interference effects produced by the superposition of vibrations coming from these points.

We place in a wave field studied an opaque diaphragm A (Fig. 3.17) with two rather small holes P_1 and P_2, and we observe the intensity distribution on the screen B, which is situated in a certain

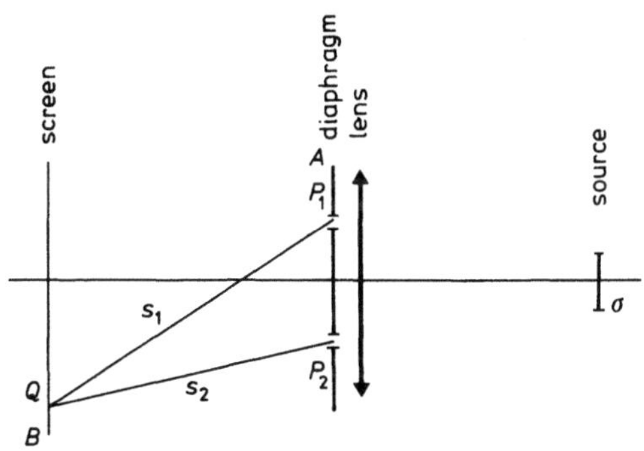

Figure 3.17

distance from A at the opposite, with relation to the diaphragm, side of the source. Let s_1 and s_2 denote the distances from point Q of the screen B to points P_1 and P_2, respectively. We can assume that these points are the centres of secondary disturbances, thus the disturbance at point Q can be written as

$$U(Q,t) = K_1 V(P_1, t - t_1) + K_2 V(P_2, t - t_2), \quad (3.47)$$

where $t_1 = s_1/c$ and $t_2 = s_2/c$ are times of light transition from P_1 to Q and from P_2 to Q, respectively. The coefficients K_i depend on the geometry of the system, are inversely proportional to s_i and imaginary.

The intensity at point Q is consequently

$$I(Q) = K_1K_1^*\langle V(P_1, t - t_1)V^*(P_1, t - t_1)\rangle + \\ + K_2K_2^*\langle V(P_2, t - t_2)V^*(P_2, t - t_2)\rangle + \\ + K_1K_2^*\langle V(P_1, t - t_1)V^*(P_2, t - t_2)\rangle + \\ + K_2K_1^*\langle V(P_2, t - t_2)V^*(P_1, t - t_1)\rangle. \qquad (3.48)$$

Under the assumption that the field is stationary, we can write the equality

$$\langle V(P_i, t - t_i)V^*(P_i, t - t_i)\rangle = \langle V(P_i, t)V^*(P_i, t)\rangle \\ = I(P_i) = I_i$$

(where $i = 1, 2$), which enables us to write expression (3.48) in the form

$$I(Q) = |K_1|^2 I_1 + |K_2|^2 I_2 + 2|K_1K_2|\Gamma^{(r)}(P_1, P_2, \tau), \qquad (3.49)$$

where $\Gamma^{(r)}(P_1, P_2, \tau)$ is the real part of the function

$$\Gamma(P_1, P_2, \tau) = \langle V(P_1, t + \tau)V^*(P_2, t)\rangle, \qquad (3.50)$$

where $\tau = (s_2 - s_1)/c$.

The quantity (3.50) is the time cross-correlation function, called in optics the *function of mutual coherence of light vibrations* at points P_1 and P_2. If the two points coincide, then

$$\Gamma(P_j, \tau) = \langle V(P_j, t + \tau)V^*(P_j, t)\rangle.$$

This is the autocorrelation function for vibrations of light at point P_j. If $\tau = 0$, these expressions represent intensities, i.e. $\Gamma(P_1, 0) = I_1$, and $\Gamma(P_2, 0) = I_2$.

The term $|K_i|^2 I_i$ in (3.49) represents intensity, which can be observed at point Q if there would exist only the hole P_j.

Introducing the notations

$$I^{(1)}(Q) = |K_1|^2 I_1 = |K_1|^2 \Gamma(P_1, 0),$$

$$I^{(2)}(Q) = |K_2|^2 I_2 = |K_2|^2 \Gamma(P_2, 0),$$

and

$$\gamma(P_1, P_2, \tau) = \frac{\Gamma(P_1, P_2, \tau)}{\sqrt{\Gamma(P_1, 0)}\sqrt{\Gamma(P_2, 0)}}, \tag{3.51}$$

equation (3.49) can be written in the form

$$I(Q) = I^{(1)}(Q) + I^{(2)}(Q) + $$

$$+ 2\sqrt{I^{(1)}(Q) I^{(2)}(Q)} \cdot \gamma^{(r)}(P_1, P_2, \tau), \tag{3.52}$$

where $\gamma^{(r)}(P_1, P_2, \tau)$ is the real part of (3.51). $\gamma(P_1, P_2, \tau)$ is called the *complex coherence degree* of vibrations of light.

Formula (3.52) represents the general interference law for stationary optical fields. It says that in order to find the intensity given by a superposition of two light beams, it is necessary to know the intensity of each beam and the real part of complex coherence degree. The quantities $|\gamma(P_1, P_2, \tau)|$ and $\arg \gamma(\mathrm{P}_1, P_2, \tau)$ can be found experimentally.

Starting from Schwartz' inequality, it can be shown that

$$|\Gamma(P_1, P_2, \tau)|^2 \leq \Gamma(P_1, 0) \Gamma(P_2, 0),$$

and hence

$$|\gamma(P_1, P_2, \tau)|^2 \leq 1.$$

Writing the complex coherence degree in the form

$$\gamma(P_1, P_2, \tau) = |\gamma(P_1, P_2, \tau)| \exp\{\mathrm{i}\alpha_{12}(\tau)\},$$

where $\alpha_{12} = \arg\gamma(P_1, P_2, \tau)$, we can go over from (3.52) to the formula

$$I(Q) = I^{(1)}(Q) + I^{(2)}(Q) + $$

$$+ 2\sqrt{I^{(1)}(Q) I^{(2)}(Q)} |\gamma(P_1, P_2, \tau)| \cos[\alpha_{12}(\tau)]. \tag{3.53}$$

If $|\gamma(P_1, P_2, \tau)| = 0$, the last term in (3.53) vanishes and we cannot observe any interference effects. In this case it is said that vibrations arriving at point Q from P_1 and P_2 are incoherent. For the second extreme case, where $|\gamma(P_1, P_2, \tau)| = 1$, we have maximum intensity oscillations; the contrast between maxima and minima is utmost.

Therefore the module of coherence degree is a measure of the contrast of the interference pattern. And so

$|\gamma(P_1, P_2, \tau)| = 1$ corresponds with total coherence,

$0 < |\gamma(P_1, P_2, \tau)| < 1$ corresponds with partial coherence,

$|\gamma(P_1, P_2, \tau)| = 0$ corresponds with the complete lack of coherence.

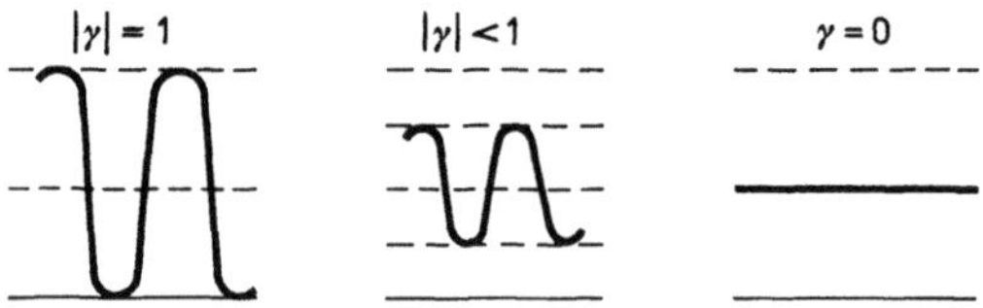

Figure 3.18

The amplitudes of the interference fringes for these three cases are shown in Fig. 3.18.

We can also determine the relationship between the visibility of the fringes and the coherence degree. From formula (3.53) we obtain the following extreme values of intensity:

$$I_{\max}(Q) = I_1(Q) + I_2(Q) + 2\sqrt{I_1(Q)I_2(Q)}\,|\gamma(P_1, P_2, \tau)|,$$
$$I_{\min}(Q) = I_1(Q) + I_2(Q) - 2\sqrt{I_1(Q)I_2(Q)}\,|\gamma(P_1, P_2, \tau)|. \tag{3.54}$$

Applying formula (3.38) and the above expressions for the extreme values, and assuming that $I_1(Q) = I_2(Q)$, we get

$$\vartheta(Q) = |\gamma(P_1, P_2, \tau)|. \tag{3.55}$$

This expression enables us to find experimentally the coherence degree of light; for this purpose the visibility of the fringes has to be determined.

The phase factor of the complex coherence degree

$$\alpha_{12}(\tau) = \arg\gamma(P_1, P_2, \tau)$$

determines the shift of the interference maxima in relation to the position of the maxima occurring if the diffraction system is illuminated by totally coherent light.

If a plane monochromatic wave falls perpendicularly on a flat diaphragm, the complex coherence degree of light at point Q is

$$\gamma(P_1, P_2, \tau) = e^{i\omega(s_1 - s_2)/c}.$$

Its real part

$$\gamma^{(r)} = \cos\left[\frac{\omega}{c}(s_1 - s_2)\right]$$

and the intensity distribution in the plane of the screen is

$$I(Q) = I_1(Q) + I_2(Q) + 2\sqrt{I_1(Q)I_2(Q)}\cos\left[\frac{\omega}{c}(s_1 - s_2)\right]. \tag{3.53a}$$

Choosing a reference system in this way that the coordinates of points P_1, P_2 and Q are respectively $(0, -\frac{1}{2}a, 0)$, $(0, \frac{1}{2}a, 0)$, $(d, \xi, 0)$, and considering a situation where

$$a,\ \xi \ll d,$$

we can make the approximations: $s_1 - s_2 \approx \xi a/d$.

The intensity maxima appear at points Q of the screen for which

$$\xi = \frac{d\lambda}{a} m \qquad (m = 0, \pm 1, \pm 2, \ldots).$$

For the same geometry of the diffraction system and for a quasi-monochromatic wave (see Sec. 2.2) falling on the diaphragm and described by the expression

$$V(\mathbf{r}, t) = V_0(\mathbf{r}, t)\exp\left[i\beta(\mathbf{r}, t) - i(kx + \omega t)\right],$$

where $V_0(\mathbf{r}, t)$ and $\beta(\mathbf{r}, t)$ are slow-changeable functions of its arguments, the complex coherence degree at the point of the screen $Q(d, \xi, 0)$ is

$$\gamma(P_1, P_2, \tau) = \frac{|\Gamma(P_1, P_2, \tau)|}{\sqrt{\Gamma(P_1, 0)\Gamma(P_2, 0)}}\exp\left[i\frac{2a\pi}{\lambda d}\xi + i\alpha(P_1, P_2, \tau)\right],$$

where

$$|\Gamma(P_1, P_2, \tau)| = |\langle V_0(P_1, t+\tau) V_0^*(P_2, t)\rangle|,$$

$$\alpha(P_1, P_2, \tau) = \arg\Gamma(P_1, P_2, \tau) - \frac{2a\pi}{\lambda d}\xi.$$

In this case the intensity distribution at the points of the screen is

$$I(Q) = I_1(Q) + I_2(Q) +$$
$$+ 2\sqrt{I_1(Q) I_2(Q)} |\gamma(P_1, P_2, \tau)| \cos\left[\frac{2a\pi}{\lambda d}\xi' + \alpha(P_1, P_2, \tau)\right]. \quad (3.53b)$$

The position of the intensity maxima can now be found from the condition

$$\frac{2a\pi}{\lambda d}\xi' + \alpha(P_1, P_2, \tau) = 2m\pi \qquad (m = 0, \pm 1, \pm 2, \ldots).$$

From this it follows that these maxima are shifted in relation to the positions of the maxima for a monochromatic wave by the value (Fig. 3.19)

$$\Delta\xi = \frac{\lambda d}{2a\pi}\alpha(P_1, P_2, \tau).$$

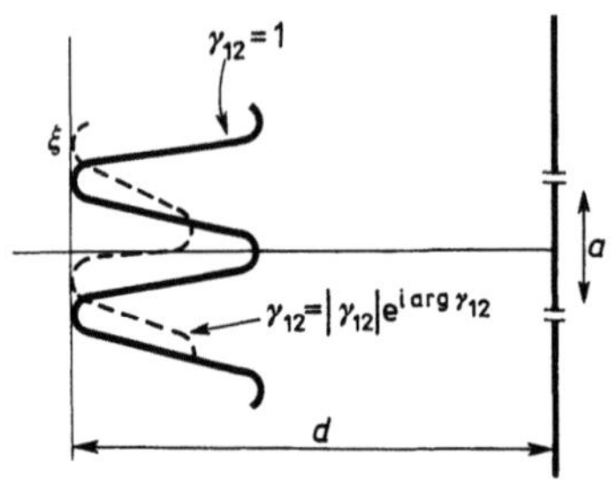

Figure 3.19

3.9. Correlation Functions for Fields Produced by Bulk Sources

Let a screen A be illuminated by a bulk quasi-monochromatic source σ (Fig. 3.20) lying in a plane parallel to the screen A. The

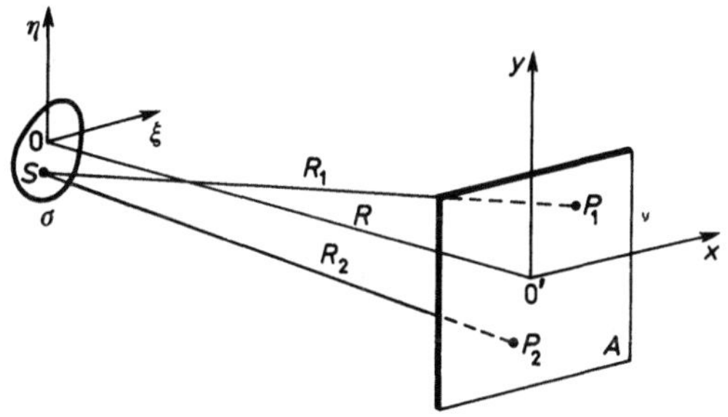

Figure 3.20

cross correlation function $\Gamma(P_1, P_2, 0)$ and the complex coherence degree $\gamma(P_1, P_2, 0)$ for the points P_1, P_2 lying in the screen A should be determined. The medium between the source and the screen is homogeneous. The dimensions of the source are small in comparison with the distance OO' between the source and the screen. Similarly, the angles betwen OO' and the lines connecting a point of the source σ with the points P_1, P_2 are small quantities.

In order to find the correlation function, the source is divided into elements $d\sigma_1, d\sigma_2, \ldots$ having the centres at points $s_1, s_2, \ldots$, and of small dimensions compared to the mean wavelength. If $V_m(P_1, t)$ and $V_m(P_2, t)$ are complex disturbances at the points P_1 and P_2 respectively, connected with the source element $d\sigma_m$, then the total disturbance at these points coming from all elements of the source is

$$V_1(t) = \sum_m V_m(P_1, t), \qquad V_2(t) = \sum_m V_m(P_2, t). \tag{3.56}$$

Hence the cross correlation function is

$$\Gamma_{12}(P_1, P_2, 0) = \langle V_1(t)\, V_2^*(t)\rangle$$

$$= \sum_m \langle V_m(P_1,t)\, V_m^*(P_2,t)\rangle + \sum_{m \neq n}\sum \langle V_m(P_1,t)\, V_n^*(P_2,t)\rangle. \tag{3.57}$$

We can assume that vibrations coming from different elements of the source are statistically independent, which gives (for $m \neq n$)

$$\langle V_m(P_1,t)\, V_n^*(P_2,t)\rangle = \langle V_m(P_1,t)\rangle \langle V_n^*(P_2,t)\rangle = 0.$$

If R_{m1} and R_{m2} are the distances from the source element $\mathrm{d}\sigma_m$ to P_1 and P_2 respectively, and if the source is quasi-monochromatic, then

$$V_{m1}(t) = A_m\left(t - \frac{R_{m1}}{c}\right)\frac{\exp[-2\pi\mathrm{i}\bar{\nu}(t - R_{m1}/c)]}{R_{m1}},$$

$$V_{m2}(t) = A_m\left(t - \frac{R_{m2}}{c}\right)\frac{\exp[-2\pi\mathrm{i}\bar{\nu}(t - R_{m2}/c)]}{R_{m2}},$$

where A_m is the complex wave amplitude from the m-th element of the source, c — the light velocity in vacuum. On the ground of the above we obtain

$$\langle V_m(P_1,t)V_m^*(P_2,t)\rangle$$
$$= \frac{\exp[2\pi\mathrm{i}\bar{\nu}(R_{m2}-R_{m1})/c]}{R_{m1}R_{m2}}\left\langle A_m(t)A_m^*\left(t - \frac{R_{m2}-R_{m1}}{c}\right)\right\rangle.$$

If $(R_{m2} - R_{m1})/c$ is small, it can be neglected in the argument of A_m^*. All this enables us to write the coherence function for disturbances at the points P_1 and P_2 in the form

$$\Gamma(P_1,P_2,0) = \sum_m \frac{\exp[2\pi\mathrm{i}\bar{\nu}(R_{m2}-R_{m1})/c]}{R_{m1}R_{m2}} I(S_m)\mathrm{d}\sigma_m, \tag{3.58}$$

where

$$\langle A_m(t)A_m^*(t)\rangle = I(S_m)\mathrm{d}\sigma_m \tag{3.59}$$

is the intensity of radiation coming from the source element $\mathrm{d}\sigma_m$.

Applying (3.59) and replacing the sum (3.58) by an appropriate integral, we get the formulas

$$\Gamma(P_1,P_2,0) = \int_\sigma I(S)\frac{\exp[\mathrm{i}\bar{k}(R_2-R_1)]}{R_1R_2}\,\mathrm{d}S, \tag{3.60}$$

$$\gamma(P_1,P_2,0) = \frac{1}{\sqrt{I(P_1)}\sqrt{I(P_2)}}\int_\sigma I(S)\frac{\exp[\mathrm{i}\bar{k}(R_2-R_1)]}{R_1R_2}\,\mathrm{d}S, \tag{3.61}$$

where

$$I(P_1) = \int_\sigma \frac{I(S)}{R_1^2}\,\mathrm{d}S, \qquad I(P_2) = \int_\sigma \frac{I(S)}{R_2^2}\,\mathrm{d}S$$

are the intensities at the points P_1 and P_2, respectively. It is the so-called *Van Cittert-Zernike theorem.* It says:

The complex coherence degree, which describes the correlation of vibrations at a fixed point P_2 and a variable point P_1 of a plane illuminated by a bulk quasi-monochromatic source, is equal to the complex amplitude of a certain diffraction pattern at an appropriate point P_1. The midpoint of this diffraction pattern lies at point P_2. This pattern is obtaind by replacing the source by a hole with the same dimensions and shape as the source, and by filling it with a spherical wave convergent to point P_2; the intensity distribution on the wave front in the hole is proportional to that in the source.

If by (ξ, η) denote the current coordinates of the source S, by (x_1, y_1) and (x_2, y_2) – the coordinates of the points P_1, P_2 in local reference systems (Fig. 3.20), then for parallel axes of both systems we can write

$$R_i^2 = (x_i - \xi)^2 + (y_i - \eta)^2 + R^2,$$

$$R_i \approx R + \frac{(x_i - \xi)^2 + (y_i - \eta)^2}{2R} \qquad (i = 1, 2),$$

which enables us to rewrite (3.61) in the form

$$\gamma(P_1, P_2, 0) = \frac{\int_\sigma I(\xi,\eta)\exp[\mathrm{i}k(p\xi + q\eta)]\,\mathrm{d}\xi\,\mathrm{d}\eta}{\int_\sigma I(\xi,\eta)\,\mathrm{d}\xi\,\mathrm{d}\eta}\,\mathrm{e}^{\mathrm{i}\psi}. \qquad (3.62)$$

In (3.62) the following notation has been introduced:

$$\psi = \frac{\bar{k}[(x_1^2 + y_1^2) - (x_2^2 + y_2^2)]}{2R},$$

$$p = \frac{x_1 - x_2}{R}, \quad q = \frac{y_1 - y_2}{R}.$$

From (3.62) it follows that if the linear dimensions of the source and the distance between P_1 and P_2 are small in comparison with the distance from these points to the source, the coherence degree $\gamma(P_1, P_2, 0)$ is equal to the value of the Fourier transform of a normalized function describing the intensity in the source.

For a homogeneous source in the form of circle of radius a and with the centre at point O, we obtain

$$\gamma(P_1, P_2, 0) = \left(\frac{2J_1(v)}{v}\right) e^{i\psi},$$

where $v = ka\sqrt{p^2 + q^2}$, J_1 – the first-order Bessel function. The expression $2J_1(v)/v$ decreases monotonously from unity for $v = 0$ to zero for $v = 3.83$, and then oscillates around zero value. As the distance between points P_1 and P_2 increases, we observe the decrease of the coherence degree. For $0.61 R\lambda/a \leq \sqrt{(x_1 - x_2)^2 + (y_1 - y_2)^2}$ we observe complete lack of coherence.

Assuming the Rayleigh criterion which admits 12% deviation of $|\gamma(P_1, P_2, 0)|$ from an ideal value equal unity, we obtain that the quasi-monochromatic source visible from the central point O' of the screen at the angle $\alpha = a/R$ illuminates coherently a field of diameter $0.16\bar{\lambda}/\alpha$. This result is useful for the estimation of the dimensions of a source required in experiments dealing with interference and diffraction.

3.10. Theoretical Description of Partial Polarization

In the sections above while considering a scalar disturbance $U(\mathbf{r}, t)$ at a point of space given by $\mathbf{r}$ and in the instant of time t, we have used the scalar light theory. But now, examining the phenomenon of polarization, we must take into account also the vectorial character of light.

The theoretical description of the state of polarization is connected with the coherence theory. We assume that a plane wave propagates in the positive direction of the z-axis. We make use of the complex description of the components of the vectors of the electromagnetic field. Thus, the projections of the vector **E** on the x- and y-axis at an arbitrary point of space **r** and in any time t are written in the form

$$E_x(\mathbf{r}, t) = E_x^{(r)}(\mathbf{r}, t) + \mathrm{i}E_x^{(i)}(\mathbf{r}, t),$$

$$E_y(\mathbf{r}, t) = E_y^{(r)}(\mathbf{r}, t) + \mathrm{i}E_y^{(i)}(\mathbf{r}, t).$$

Let us consider a non-polarized quasi-monochromatic field, i.e., a field in which every position of the electric vector in the xy-plane occurs with the same probability. In other words, the projection of the electric vector on the x-axis remains during the observation for the same period of time positive as negative. The same applies to the projection of the electric vector on the y-axis.

By this assumption we can expect that the averaging over a sufficiently long time will give

$$\langle E_x \rangle = 0, \quad \langle E_y \rangle = 0,$$

where the angular brackets denote time-averaging. The time-averaged light intensity is not equal zero

$$I_0 = \langle E_x E_x^* \rangle + \langle E_y E_y^* \rangle \neq 0.$$

The mean values of the intensities connected with each component of the vector **E** are equal

$$\langle E_x E_x^* \rangle = \langle E_y E_y^* \rangle. \tag{3.63}$$

Moreover, if there is no correlation between E_x and E_y, the following equalities are true:

$$\langle E_x E_y^* \rangle = \langle E_y E_x^* \rangle = 0. \tag{3.64}$$

It is convenient to identify the correlation functions (3.63) and (3.64) with the elements (2×2) of the matrix J:

$$J = \tfrac{1}{2} I_0 \begin{bmatrix} 1 & 0 \\ 0 & 1 \end{bmatrix}. \tag{3.65}$$

The matrix (3.65) is called the *coherence matrix* of a non-polarized field of radiation. In this case it is a unit matrix with the constant factor $\frac{1}{2}I_0$. We will consider now a quasi-monochromatic light wave with the mean frequency $\bar{\nu}$, propagating in the positive direction of the z-axis and having the components

$$\begin{aligned} E_x(t) &= a_1(t)\exp\left[\mathrm{i}\varphi_1(t) - 2\pi\mathrm{i}\bar{\nu}(t)\right], \\ E_y(t) &= a_2(t)\exp\left[\mathrm{i}\varphi_2(t) - 2\pi\mathrm{i}\bar{\nu}(t)\right]. \end{aligned} \tag{3.66}$$

For the quasi-monochromatic field (3.66) the coherence matrix

$$J = \begin{bmatrix} \langle E_x E_x^* \rangle & \langle E_x E_y^* \rangle \\ \langle E_y E_x^* \rangle & \langle E_y E_y^* \rangle \end{bmatrix} = \begin{bmatrix} J_{xx} & J_{xy} \\ J_{yx} & J_{yy} \end{bmatrix}, \tag{3.67}$$

after applying an explicit form of the field (3.66), takes the form

$$J = \begin{bmatrix} \langle a_1^2 \rangle & \langle a_1 a_2 \exp[\mathrm{i}(\varphi_1 - \varphi_2)] \rangle \\ \langle a_1 a_2 \exp[\mathrm{i}(\varphi_2 - \varphi_1)] \rangle & \langle a_2^2 \rangle \end{bmatrix}. \tag{3.68}$$

The diagonal elements of the matrix J are real and represent the intensities connected with the x-th and the y-th field component. Thus the trace of the coherence matrix gives the total intensity

$$I = \langle a_1^2 \rangle + \langle a_2^2 \rangle. \tag{3.69}$$

The non-diagonal elements, complex in general, are mutually conjugate, i.e. $J_{yx} = J_{xy}^*$. Using them, we can create the quantity

$$\mu_{xy} = \frac{J_{xy}}{\sqrt{J_{xx}}\sqrt{J_{yy}}}. \tag{3.70}$$

It plays the same role as the complex coherence degree introduced in Sec. 3.8. The introduced quantity has the property

$$|\mu_{xy}| \leq 1 \tag{3.71}$$

and is the measure of correlation between the x-th and the y-th component of the electric vector. The module $|\mu_{xy}|$ is the measure of coherence degree, and $(\arg \mu_{xy})$ – of phase difference.

Now we seek the intensity related to the oscillations of the vector **E** in the direction forming with the positive direction of the x-axis the angle θ. The projection of the vector $\mathbf{E}$ on this direction is expressed by the formula

$$E(t,\theta) = E_x \cos\theta + E_y \sin\theta. \tag{3.72}$$

The corresponding intensity is given by the expression

$$I(\theta) = \langle E(t,\theta)E^*(t,\theta)\rangle = J_{xx}\cos^2\theta + J_{yy}\sin^2\theta + \\ + (J_{xy} + J_{yx})\cos\theta\sin\theta. \tag{3.73}$$

Considering the complex character of the elements J_{xy} and J_{yx} and using (3.70), we can write the above formula in the form

$$I(\theta) = J_{xx}\cos^2\theta + J_{yy}\sin^2\theta + \\ + 2\sqrt{J_{xx}}\sqrt{J_{yy}}\,|\mu_{xy}|\cos(\arg\mu_{xy})\cos\theta\sin\theta. \tag{3.74}$$

This result allows us to discuss the dependence of the intensity upon the coherence degree of the field. First of all we can state that if

$$|\mu_{xy}| = 0$$

(which means the lack of any coherence between the components E_x and E_y), then (for $J_{xx} = J_{yy}$) we have $I(\theta) = \text{const}$. It means that the intensity connected with any component of the vector $\mathbf{E}$ perpendicular to the direction of propagation remains constant. Light of this kind is non-polarized.

The second extreme condition is the case where

$$|\mu_{xy}| = 1.$$

Assuming that a_1, a_2, φ_1, and φ_2 appearing in formula (3.66) do not depend on time, we can introduce a constant quantity $\delta = \varphi_1 - \varphi_2$, and then the coherence matrix and the complex coherence degree μ_{xy} are given by the formulas

$$J = \begin{bmatrix} a_1^2 & a_1 a_2 e^{i\delta} \\ a_1 a_2 e^{-i\delta} & a_2^2 \end{bmatrix}, \tag{3.75}$$

$$\mu_{xy} = e^{i\delta}. \tag{3.76}$$

In the particular case of linearly polarized light, as it was discussed in Sec. 1.13, $\delta = m\pi$ ($m = 0, \pm 1, \pm 2, \ldots$) and then

$$I(\theta) = J_{xx}\cos^2\theta + J_{yy}\sin^2\theta + 2(-1)^m \sqrt{J_{xx}}\sqrt{J_{yy}}\cos\theta\sin\theta. \tag{3.77}$$

From (3.77) it follows that

$$I(\theta) = \left(\sqrt{J_{xx}}\cos\theta \pm \sqrt{J_{yy}}\sin\theta\right)^2 = (a_1\cos\theta \pm a_2\sin\theta)^2, \tag{3.78}$$

where the sign "+" corresponds to even m, the sign "−" to odd m.

The coherence matrix takes here the form

$$J = \begin{bmatrix} a_1^2 & a_1a_2(-1)^m \\ a_1a_2(-1)^m & a_2^2 \end{bmatrix}, \tag{3.79}$$

and its determinant

$$|J| = 0.$$

In the case where $\delta = m\pi/2$ ($m = \pm 1, \pm 3, \pm 5, \ldots$) and $a_1 = a_2$, when according to the results from Sec. 1.13 the circular polarization occurs, the coherence matrix takes the form

$$J = a_1^2 \begin{bmatrix} 1 & \pm\mathrm{i} \\ \pm\mathrm{i} & 1 \end{bmatrix}, \tag{3.80}$$

and its determinant is zero, similarly as in the case of linear polarization of light. Therefore it can be stated that the zero determinant of the coherence matrix characterizes an electromagnetic wave with total polarization.

The third case is related with the condition

$$0 < |\mu_{xy}| < 1.$$

The coherence matrix has then the general form (3.67) and its determinant is

$$|J| = J_{xx}J_{yy} - J_{xy}J_{yx} \neq 0.$$

However, it is possible to write the coherence matrix as the sum of two matrices

$$J = J^{(1)} + J^{(2)}. \tag{3.81}$$

The first of them is of the form

$$J^{(1)} = \begin{bmatrix} A & 0 \\ 0 & A \end{bmatrix}, \tag{3.82}$$

and the second

$$J^{(2)} = \begin{bmatrix} B & D \\ D^* & C \end{bmatrix}. \tag{3.83}$$

The elements of these matrices are given by the expressions:

$$A = \tfrac{1}{2}(J_{xx} + J_{yy}) - \tfrac{1}{2}\sqrt{(J_{xx} + J_{yy})^2 - 4|J|},$$

$$B = \tfrac{1}{2}(J_{xx} - J_{yy}) + \tfrac{1}{2}\sqrt{(J_{xx} + J_{yy})^2 - 4|J|},$$

$$C = \tfrac{1}{2}(J_{yy} - J_{xx}) + \tfrac{1}{2}\sqrt{(J_{xx} + J_{yy})^2 - 4|J|},$$

$$D = J_{xy}, \qquad D^* = J_{yx}.$$

From formula (3.81) it follows that an electromagnetic wave can always be presented as a result of superposition of a polarized wave, described by the coherence matrix $J^{(2)}$, and a not-polarized one, described by the matrix $J^{(1)}$ (in this case $\mu^{(1)}(x, y) = 0$).

The ratio of the intensity of the polarized part to the total intensity of the wave is called the *polarization degree P*. It is expressed by the formula

$$P = \frac{I_{\text{pol}}}{I_{\text{total}}} = \frac{B + C}{J_{xx} + J_{yy}} = \sqrt{1 - \frac{4|J|}{(J_{xx} + J_{yy})^2}}. \tag{3.84}$$

From the above we can see that $0 \leq P \leq 1$, and for a polarized wave P equals one. If there is no correlation between the components E_x and E_y (i.e. if $J_{xy} = J_{yx} = 0$), the polarization degree is

$$P = \frac{J_{xx} - J_{yy}}{J_{xx} + J_{yy}} = 0. \tag{3.85}$$

The above considerations can be also used for quasi-monochromatic light.

Usually the polarization degree of light is defined by the formula

$$P = \frac{I(\theta)_{\max} - I(\theta)_{\min}}{I(\theta)_{\max} + I(\theta)_{\min}}, \tag{3.86}$$

where the extreme values of $I(\theta)$ are found either from formula (3.73) or experimentally with the use of the Nicol prism. It can be shown that the definitions (3.84) and (3.86) are fully consistent.

3.11. Interference Images in Plane-Parallel Glass Plates

Let us consider a plane-parallel glass plate (Fig. 3.21) with the refractive index n and thickness d, on which at the angle φ_1 falls

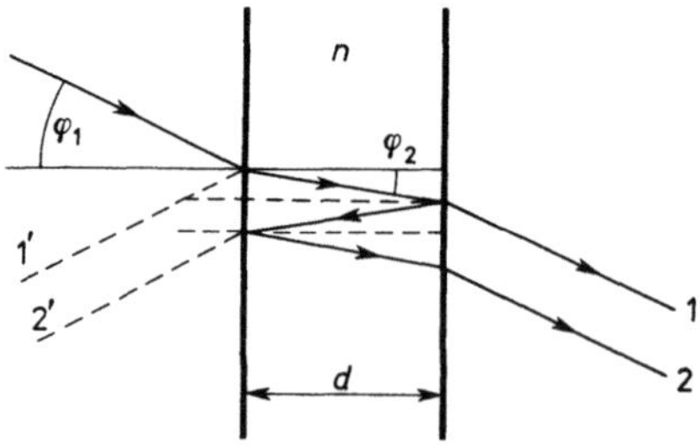

Figure 3.21

a plane monochromatic wave. On the first surface the wave splits into two: the reflected wave and the refracted one. The latter falls at the angle φ_2 on the second surface confining the plate and once again is divided into two plane waves – one of them leaving the plate, the second reflected. This process of dividing the wave remaining inside the plate is repeated many times.

At the interface of two dielectrics the wave amplitude undergoes changes. These changes are characterized by the reflection coefficient $\rho = |E^r/E_0|$ and transmittance $\tau = |E^t/E_0|$, which, after applying Fresnel's formula, are expressed by the angles φ_1, φ_2 and n. These coefficients will be different – for the first of the planes they are ρ, τ, for the second ρ', τ'.

Every time while crossing the interface between the two dielectrics the amplitude of the wave changes, decreasing τ- or τ'-fold; and reflecting from the interface it decreases ρ- or ρ'-fold. From Fresnel's equations it can be shown that for a given polarization we have

$$\rho = \rho'. \tag{3.87}$$

Thus the amplitudes of the waves 1, 2 passing through the plate are respectively

$$E_{00}\tau\tau', \quad E_{00}\tau\tau'\rho^2, \quad E_{00}\tau\tau'\rho^4, \dots \tag{3.88}$$

where E_{00} is the amplitude of the incident wave. The phase differences between "neighbouring" waves are

$$\delta = \frac{2\pi}{\lambda_0} 2nd\cos\varphi_2, \tag{3.89}$$

where λ_0 is the length of the wave in vacuum. This phase difference is taken into account by introducing the proper factor $e^{i\delta}$ for each of the transmitted waves. Therefore the expression for total amplitude of the transmitted wave can be written in the form

$$E_{20} = E_{00}\{\tau\tau' - \tau\tau'\rho^2\exp(i\delta) + \tau\tau'\rho^4\exp(2i\delta) + \dots$$
$$\dots + \tau\tau'\rho^{2(N-1)}\exp(i\delta[N-1])\}$$

or, after summation, in the form

$$E_{20} = E_{00}\frac{1 - \rho^{2N}\exp(i\delta N)}{1 - \rho^2\exp(i\delta)}\tau\tau'. \tag{3.90}$$

Considering that $\rho < 1$ and N is large enough, we can assume that

$$\lim_{N\to\infty} \rho^{2N} = 0.$$

This remark gives the needed formula for the amplitude of the wave passing through the plate:

$$E_{20} = E_{00}\frac{\tau\tau'}{1 - \rho^2\exp(i\delta)}, \tag{3.91}$$

and its intensity

$$I_{20} = I_{00} \frac{\tau^2 \tau'^2}{(1-\rho^2)^2 + 4\rho^2 \sin^2\left(\frac{1}{2}\delta\right)}. \tag{3.92}$$

Formula (3.92) shows that if the path difference between two succeeding partial waves is equal to an integer number of a wavelength, i.e., if the relation

$$2nd \cos\varphi_2 = m\lambda_0 \quad \text{or} \quad \tfrac{1}{2}\delta = m\pi \quad (m = 1, 2, 3, \dots)$$

is satisfied, then the intensity of the transmitted wave reaches its maximum values. For $\frac{1}{2}\delta = \left(l + \frac{1}{2}\right)\pi$ $(l = 1, 2, \dots)$ we have the reverse situation – minimum values occur.

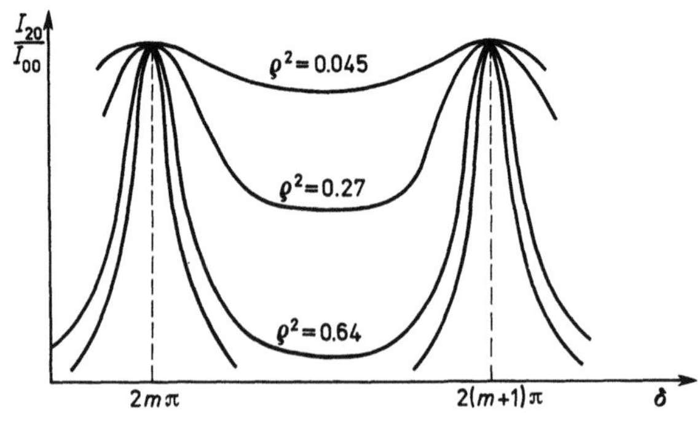

Figure 3.22

The diagram $I_{20} = I_{20}(\delta)$ (Fig. 3.22) shows the system of parallel maxima, which for sufficiently great values of $\rho\rho'$ are relatively sharp. They are separated by wide minima. Hence it can be seen that maxima become narrower with the increase of the reflecting power.

In a similar way we obtain the expression for the amplitude of the reflected wave

$$
\begin{aligned}
E_{10} &= E_{00}\{\rho + \tau\tau'\rho'\exp(\mathrm{i}\delta)[1 + \rho'^2\exp(\mathrm{i}\delta) + \ldots \\
&\quad \ldots + \rho'^{2(N-2)}\exp(\mathrm{i}\delta[N-2])\} \\
&= E_{00}\left\{\rho + \left[\frac{1-\rho'^{2(N-1)}\exp(\mathrm{i}\delta[N-1])}{1-\rho'^2\exp(\mathrm{i}\delta)}\right]\rho'\tau\tau'\exp(\mathrm{i}\delta)\right\}.
\end{aligned}
\tag{3.93}
$$

If the plate is sufficiently long, the number of partially reflected waves is large, and for $N \to \infty$ we obtain

$$
E_{10} = E_{00}\rho\frac{1-(\tau\tau' + \rho^2)\mathrm{e}^{\mathrm{i}\delta}}{1-\rho^2\mathrm{e}^{\mathrm{i}\delta}}. \tag{3.94}
$$

The intensity of the reflected wave is

$$
I_{10} = I_{00}\rho^2\frac{[1-(\tau\tau' + \rho^2)]^2 + 4(\tau\tau' + \rho^2)\sin^2\left(\frac{1}{2}\delta\right)}{(1-\rho^2)^2 + 4\rho^2\sin^2\left(\frac{1}{2}\delta\right)}. \tag{3.95}
$$

Using an explicit form for τ, τ' and ρ, expressed by the incident angle φ_1 and the refractive index, it turns out that

$$
\tau\tau' + \rho^2 = 1. \tag{3.96}
$$

This relation simplifies considerably the expression for the intensity of the reflected wave giving

$$
I_{10} = I_{00}\frac{4\rho^2\sin^2\left(\frac{1}{2}\delta\right)}{(1-\rho^2)^2 + 4\rho^2\sin^2\left(\frac{1}{2}\delta\right)}. \tag{3.97}
$$

As can be noticed from (3.97), the reflected wave reaches the maximum of intensity for $\sin^2\left(\frac{1}{2}\delta\right) = 1$, whereas the intensity of this wave equals zero for the directions $\cos\varphi_2 = m\lambda_0/2dn$, which describe the position of the intensity maxima for the transmitted wave.

Adding I_{20}/I_{00} (the relative intensity of the transmitted wave) to

I_{10}/I_{00} (the relative intensity of the reflected wave) we obtain the relation

$$\frac{I_{20}}{I_{00}} + \frac{I_{10}}{I_{00}} = 1. \tag{3.98}$$

The result (3.98) means that the interference patterns for the transmitted and reflected light are complementary to one another. Narrow maxima in the transmitted light have their equivalents in narrow minima of the reflected light.

Naturally, all this remains true under the assumption of the lack of light absorption.

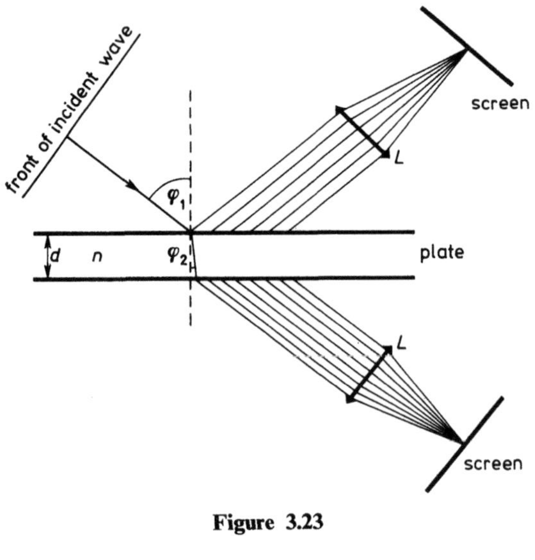

Figure 3.23

The interference pattern obtained while illuminating a plane-parallel plate by a monochromatic plane wave, gives lines of equal inclination located in infinity. For their observation a convergent lens placed at the focal distance from the screen (Fig. 3.23) should be used.

If the plate is illuminated by a bulk source of light, the curves of equal inclination take a ring shape. It results from the condition for the occurrence of maxima:

$$\cos\varphi_2 = \frac{m\lambda_0}{2dn}. \tag{3.99}$$

The less is the angle φ_2, the nearer to the centre of the system of interference rings lies the corresponding maximum.

Differentiating formula (3.99), we get

$$-2dn\sin\varphi\,\mathrm{d}\varphi = \lambda_0\mathrm{d}m. \tag{3.100}$$

For $\mathrm{d}m = 1$, i.e. considering $\mathrm{d}\varphi$ as the angle between two adjacent maxima, we get

$$\mathrm{d}\varphi = -\frac{\lambda_0}{2dn\sin\varphi}. \tag{3.101}$$

As can be seen from the above, the angular distance between the lines of equal inclination is inversely proportional to the plate thickness d.

In order to characterize more closely the shape of the maxima of the intensity curve, we introduce the notion of sharpness F, which is defined as the ratio of the distance between two adjacent interference maxima to the width 2ε of one of them. The latter quantity denotes the distance between two points for which $I_{20} = \frac{1}{2}I_{00}$ (Fig. 3.24). It happens for $\delta = 2m\pi \pm \varepsilon$. For small ε we have

$$\sin\frac{\delta}{2} = \sin\frac{\varepsilon}{2} \approx \frac{\varepsilon}{2}.$$

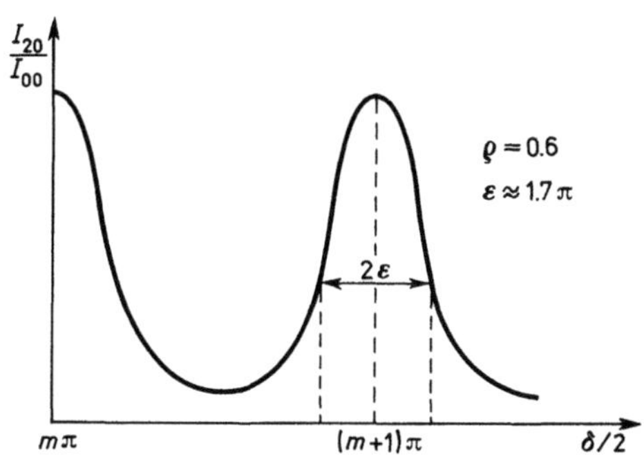

Figure 3.24

Considering all this and equality (3.96), we obtain from formula (3.92) the expression

$$\varepsilon = \frac{1 - \rho^2}{\rho}. \tag{3.102}$$

Hence

$$F = \frac{\pi}{\varepsilon} = \frac{\pi\rho}{1 - \rho^2}. \tag{3.103}$$

The sharpness F characterizes the profile of an interference fringe. It is fully determined by the reflection coefficient. For $\rho \to 1$ (for total reflection) we have $F \to \infty$. This is the case of monochromatic light. If light is quasi-monochromatic, the intensity distribution is the sum of the intensity distributions connected with each monochromatic components. If these monochromatic components cover the region of the length $\Delta\lambda$ converged about the mean value $\bar{\lambda}_0$, then their m-th order maxima are spread out over an area defined by

$$|\Delta\delta| = 2\pi m \frac{\Delta\lambda}{\bar{\lambda}_0}. \tag{3.104}$$

This relation is a consequence of formula (3.89) and the condition for the intensity maximum described by (3.92).

If $\Delta\delta$ is small compared with ε, we can talk about the coincidence of partial patterns for individual monochromatic components of light. As a result, the intensity distribution for quasi-monochromatic light will be the same as for monochromatic light.

3.12. Fabry–Pérot's Interferometer

This instrument consist of two mutually parallel glass plates P_1 and P_2 with plane surfaces. The inner surfaces of these plates are coated with partially transparent films of high reflectivity (Fig. 3.25).

The space between the plane-parallel plates, while illuminating it by monochromatic light of almost perpendicular incidence direc-

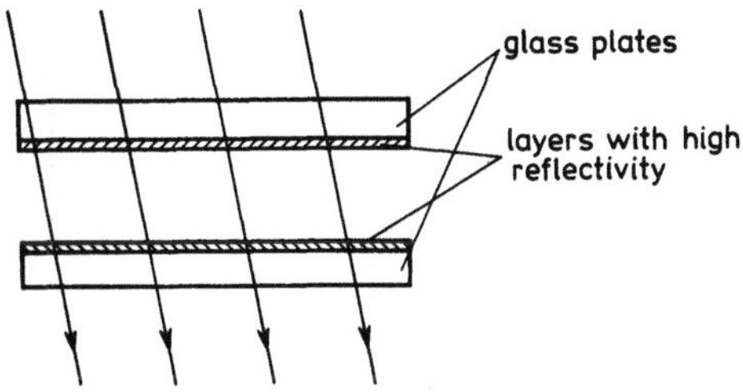

Figure 3.25

tion, produces multistream interference fringes. If the interferometer is illuminated by a bulk source of light, we obtain the family of curves (rings) of equal inclination situated in infinity. In order to observe the rings on the screen, we use the lens L placed in such a way that its focal plane coincides with the screen plane.

For monochromatic light the intensity distribution of the transmitted light $I(\delta)$ is given by formula (3.92).

For light being a superposition of two monochromatic waves $(\lambda_0 \pm \frac{1}{2}\Delta\lambda)$ formulas (3.89) and (3.92) for the distribution of the transmitted light consist of two expressions with argument δ shifted by the angle $\pm\frac{1}{2}\alpha$. This distribution is

$$I(\delta, \alpha) = \frac{I_{00}}{1 + [2\rho/(1-\rho^2)]^2 \sin^2\left[\frac{1}{2}(\delta + \frac{1}{2}\alpha)\right]} + \\ + \frac{I_{00}}{1 + [2\rho/(1-\rho^2)]^2 \sin^2\left[\frac{1}{2}(\delta - \frac{1}{2}\alpha)\right]}. \tag{3.105}$$

The intensity at the point situated in the middle between the maxima for two components is

$$I(2\pi m, \alpha) = \frac{2I_{00}}{1 + [2\rho/(1-\rho^2)]^2 \sin^2\left(\frac{1}{4}\alpha\right)}. \tag{3.106}$$

If the resolution of the adjacent maxima is required, the value of

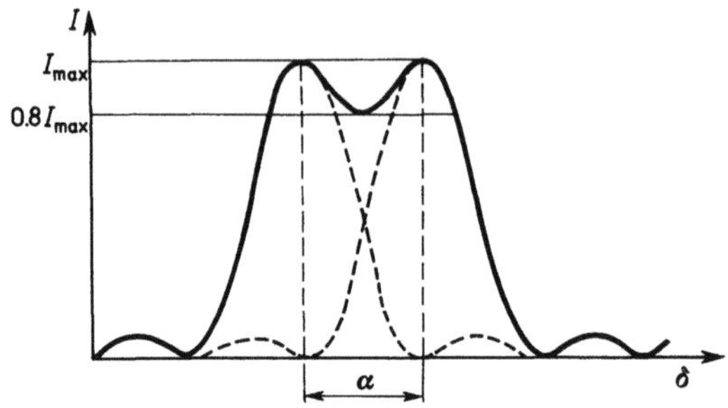

Figure 3.26

the minimum intensity between them should, on the strength of the Rayleigh criterion (Fig. 3.26), satisfy the condition

$$I(2\pi m, \alpha) \leq 0.81\, I_{\max}(\delta, \alpha).$$

For the distribution (3.105) maxima occur for

$$\delta = 2m\pi \mp \frac{\alpha}{2},$$

and are equal

$$I_{\max} = I_{00} + \frac{I_{00}}{1 + [2\rho/(1-\rho^2)]^2 \sin^2\left(\frac{1}{2}\alpha\right)}. \tag{3.107}$$

If two neighbouring maxima are to be distinguished, their angular distance should satisfy the equality

$$\frac{2I_{00}}{1 + [2\rho/(1-\rho^2)]^2 \sin^2\left(\frac{1}{4}\alpha\right)}$$

$$= \left\{ I_{00} + \frac{I_{00}}{1 + [2\rho/(1-\rho^2)]^2 \sin^2\left(\frac{1}{2}\alpha\right)} \right\} \times 0.8. \tag{3.108}$$

Applying the approximation $\sin\alpha \approx \alpha$ (which is true for small angles) and formulas (3.102) and (3.103), we obtain the equation

$$F^4\alpha^4 - 4\pi^2 F^2 \alpha^2 - 2\pi^4 = 0, \tag{3.109}$$

from which we get the required displacement between fringes

$$\alpha = \frac{1.97\pi}{F}. \tag{3.110}$$

This is the limit value of the displacement α, which allows us to state the occurrence of two monochromatic components.

In accordance with formula (3.104) the angular displacement between two extremes of the m-th order for the monochromatic waves $\lambda - \frac{1}{2}\Delta\lambda$ and $\lambda + \frac{1}{2}\Delta\lambda$ is

$$|\Delta\delta| = 2\pi m \frac{\Delta\lambda}{\lambda}. \tag{3.111}$$

Comparing (3.110) with (3.111), we get the equality

$$\frac{\lambda}{\Delta\lambda} = 3.2\, m \frac{\rho}{1-\rho^2}. \tag{3.112}$$

In the case of perpendicular light incidence $(\cos\varphi_1 = 1)$, the condition (3.99) for the occurrence of maxima in the transmitted light takes the form

$$2\frac{1}{\lambda} nd = m. \tag{3.113}$$

Thus finally, we get the following expression for the resolving power of the Fabry–Pérot interferometer

$$\frac{\lambda}{\Delta\lambda} = 3.2 \frac{nd}{\lambda} \frac{2\rho}{1-\rho^2}. \tag{3.114}$$

This resolution is proportional to the sharpness and to optical thickness. The interferometer can be used as an interference filter.

Bibliography

1. Born, M. and Wolf, E.: (1965), 'Principles of Optics', Pergamon Press, Oxford, London.
2. Françon, N.: (1969), 'Holographie', Masson et Cie Editeurs, Paris.
3. Korolev, F.: (1966), 'Theoretical Optics' (in Russian), Vysshaya Shkola, Moskva.
4. Perina, J.: (1973), 'Coherence of Light', Van Nostrand Reinhold Company, New York.

4. Elements of Diffraction Theory

When an electromagnetic wave propagates in a homogeneous medium, the geometrical shape of its wavefront remains unchanged. However, when the wave propagates in an inhomogeneous medium which may contain opaque diaphragms or areas of relatively large refraction index variations, the wavefront is deformed and there is a change in the spatial distribution of wave intensity. The occurring phenomena are known as *diffraction.*

These phenomena are best visible when caused by opaque diaphragms, which serve to cut out parts of the wavefront.

The laws governing the propagation of such a limited wavefront constitute a section of physics called the *theory of diffraction.* Its foundation is *Huygens' principle.* In its original form, as taught in secondary school, it is true only for an unlimited wavefront. Its drawback is the fact that it is a qualitative law, and also that it allows the existence of a returning wave. (Elementary spherical waves from secondary sources create an envelope creating the wavefront in the direction of propagation in the next moment in time, and a secondary wave propagating in the direction of the source as well.) This returning wave is not observed in homogeneous media; it appears only at interfaces between different media, where reflection occurs.

These difficulties are removed by a more profound analysis of wave phenomena, requiring a more precise formulation of Huygens' principle. The diffraction of scalar waves will be discussed in the first place. This theory, within the scalar approximation in optics, gives a reasonably accurate account of the distributions of light intensity in the area behind the diaphragm.

4.1. The Helmholtz–Kirchhoff Integral Theorem

In the case of a monochromatic wave field in the form

$$V(\mathbf{r}, t) = U(\mathbf{r})e^{i\omega t} \tag{4.1}$$

the wave equation becomes Helmholtz's equation

$$\Delta U(\mathbf{r}) + k^2 U(\mathbf{r}) = 0, \quad \text{where} \quad k^2 = \frac{\omega^2}{v^2}. \tag{4.2}$$

For this case, Kirchhoff's integral (see Sec. 1.11) in an area without any sources reduces to the form

$$U(P) = \frac{1}{4\pi}\iint_S \left\{U(Q)\frac{\partial}{\partial n}\frac{e^{ikr}}{r} - \frac{e^{ikr}}{r}\frac{\partial}{\partial n}U(Q)\right\} df \tag{4.3}$$

which is known as the *Helmholtz–Kirchhoff integral theorem.* Its physical sense is as follows:

A monochromatic field at point P may be expressed by the field at points belonging to a closed surface surrounding the said point.

This theorem is a fundamental one in the scalar theory of diffraction.

4.2. Kirchhoff's Theory of Diffraction

Let us consider the diffraction on an aperture in an infinite diaphragm. Suppose that the wave perturbation is incident on the diaphragm from the left side, as shown in Fig. 4.1. We assume that the linear dimensions of the aperture, though large in proportion to the wavelength, are nevertheless small compared to the distances between the points P, P_1 and the screen. To find the perturbation at point P, we apply the Helmholtz–Kirchhoff theorem to the closed surface comprised by the area of the aperture A, the dark side of the screen B, and part of the spherical surface ξ centred at P.

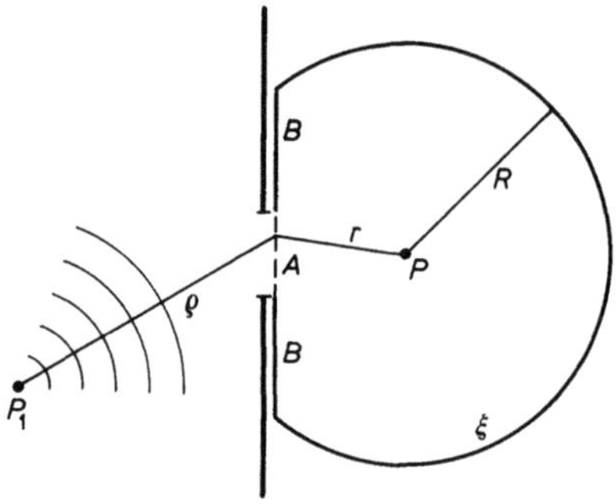

Figure 4.1

The Helmholtz–Kirchhoff theorem may be written in the form

$$U(P) = \frac{1}{4\pi}\left(\iint_B + \iint_\xi + \iint_A\right)\left(U\frac{\partial}{\partial n}\frac{e^{ikr}}{r} - \frac{e^{ikr}}{r}\frac{\partial}{\partial n}U\right)df. \tag{4.4}$$

It is difficult to give the values of U and $\partial U/\partial n$ on A, B and ξ which should be inserted into integral (4.4). This difficulty was overcome by Kirchhoff, who assumed

$$\text{for points } Q \in A \quad \begin{cases} U = U^{(i)} \\ \dfrac{\partial U}{\partial n} = \dfrac{\partial U^{(i)}}{\partial n}, \end{cases}$$

$$\text{for points } Q \in B \quad \begin{cases} U = 0, \\ \dfrac{\partial U}{\partial n} = 0, \end{cases} \tag{4.5}$$

where $U^{(i)}$ is the wave field incident on the diaphragm. These assumptions are known as *Kirchhoff's boundary conditions*, and form the foundations of the Kirchhoff diffraction theory.

There still remains to be estimated the contribution from the spherical part of the surface ξ. We can assume that the radius of this surface $R \to \infty$; the values of U and $\partial U/\partial n$ on ξ will be arbitrarily small, which suggests that the contribution from ξ may be ignored. But as $R \to \infty$, the area ξ also increases so that the conditions $U \to 0$ and $\partial U/\partial n \to 0$ when $R \to \infty$ are not sufficient to make the integral over ξ vanish. The necessary conditions have been formulated more precisely by Sommerfeld and are named after him.

4.3. Sommerfeld's Conditions

According to the Helmholtz–Kirchhoff theorem, the state of the field at a given point P is expressed by the state of the wave field at points on a surface surrounding the given point.

We construct a closed surface S composed of a flat surface S_1 on the diffraction screen and a spherical surface ξ of radius R centred at the given point. Then

$$U(P) = \frac{1}{4\pi} \iint\limits_{S_1 + \xi} \left(\frac{\mathrm{e}^{ikr}}{r} \frac{\partial U}{\partial n} - U \frac{\partial}{\partial n} \frac{\mathrm{e}^{ikr}}{r} \right) \mathrm{d}f. \tag{4.6}$$

As the radius R increases, the contribution from the surface ξ to $U(P)$ should vanish. When can this take place? On the surface ξ: $\mathbf{n} = \mathbf{R}/R$ and $r = R$, and

$$\frac{\partial}{\partial n} \frac{\mathrm{e}^{ikR}}{R} = \left(ik - \frac{1}{R} \right) \frac{\mathrm{e}^{ikR}}{R} \approx ik \frac{\mathrm{e}^{ikR}}{R}. \tag{4.7}$$

The integrand in (4.6) at the points of the surface ξ therefore assumes the form

$$\frac{\mathrm{e}^{ikR}}{R} \left(\frac{\partial U}{\partial n} - ikU \right) R^2 \mathrm{d}\Omega, \tag{4.8}$$

where $\mathrm{d}\Omega$ is the solid angle with point P at the apex and $R^2 \mathrm{d}\Omega$ is the element of the surface ξ.

The integral over the surface ξ will be zero at $R \to \infty$ when

$$\lim_{R \to \infty} R \left(\frac{\partial U}{\partial n} - ikU \right) = 0 \quad \text{(the radiation condition).} \tag{4.9a}$$

Taking the component e^{ikR}/R^2 into account in the expression for $(\partial/\partial n)\,(e^{ikR}/R)$ leads to an additional condition:

$$\lim_{R\to\infty} U = 0 \quad \text{(the condition of finiteness).} \tag{4.9b}$$

4.4. The Young–Rubinowicz Diffraction Model

Within the framework of Kirchhoff's diffraction theory there exists a possibility of two different interpretations of diffraction phenomena. The first model, originally suggested by T. Young and based on the so-called *edge wave*, was developed and supplied with a theoretical foundation by A. Rubinowicz. The second is Fresnel's diffraction model, based on the idea of zones, which was for a long time considered to be the only correct one. We shall first discuss the first model.

Thomas Young considered the phenomenon of wave diffraction to be the result of superposition of two waves at the point of observation: a wave propagating in an unbounded space, and a wave appearing in the presence of the diaphragm, which may be regarded as a wave reflected from the edge of the diaphragm. This model of diffraction was later abandoned by its creator, who accepted the Fresnel zone theory. It reappeared in a slightly modified form a hundred years later, when Adalbert Rubinowicz demonstrated that it is possible to separate Kirchhoff's diffraction field into two components. One component of the diffraction field is thus related to the wave incident on the deflecting diaphragm, while the other represents a wave created by the interaction of the original field with the edge of the diffracting aperture. This interaction is indeed of the nature of reflection.

A. Rubinowicz based the above interpretation of Kirchhoff's diffraction field on the following facts. He noticed that if the integrand in (4.3) is written in the form of a product

$$\mathbf{n}\cdot\mathbf{V}(P, Q), \tag{4.10}$$

where

$$\mathbf{V}(P,Q) = \frac{1}{4\pi}\left[\frac{e^{ikr}}{r}\,\mathrm{grad}_Q U(Q) - U(Q)\,\mathrm{grad}_Q \frac{e^{ikr}}{r}\right],$$

$\mathbf{n}$ is a versor perpendicular to the element of the surface of integration and r is the distance between point Q in the surface of integration and point P, then it can be shown that $\mathrm{div}_Q \mathbf{V}(P,Q) = 0$. In other words, the vector field $\mathbf{V}(P,Q)$ as a function of point Q is a sourceless field and as such may be expressed as the rotation of another vector field, i.e.

$$\mathbf{V} = \mathrm{curl}_Q \mathbf{W}(P,Q). \tag{4.11}$$

This fact allows the diffraction integral (4.4) taken over the deflecting aperture to be written as

$$U_K(P) = \oint_K \mathbf{W}(P,Q)\cdot \mathrm{d}\mathbf{s} + \sum_j \oint_{C_j} \mathbf{W}(P,Q)\cdot \mathrm{d}\mathbf{s}, \tag{4.12}$$

where the first curvilinear integral is taken over the deflecting edge K and the others are taken over curves encircling singularities of the field $\mathbf{W}(P,Q)$ on the original surface of integration. The fact that the field $\mathbf{W}(P,Q)$ indeed has to have singularities is a consequence of the fact that the integral (4.3) taken over a closed surface surrounding point P would in their absence be identically equal to zero, which would contradict the Helmholtz–Kirchhoff integral theorem.

To understand the nature of these singularities let us consider a case where the wave incident on the diaphragm is an isotropic spherical wave. It may then be shown (Rubinowicz A., 1966) that for such a wave the function $\mathbf{W}_0(P,Q)$ (which will be henceforward called a *potential*) takes the form

$$\mathbf{W}_0(P,Q) = \frac{1}{4\pi}\frac{e^{ik(r+\rho)}}{r\rho}\frac{(\mathbf{r}\times\boldsymbol{\rho})}{r\rho + \mathbf{r}\cdot\boldsymbol{\rho}}, \tag{4.13}$$

where $\mathbf{r}$, $\boldsymbol{\rho}$, $r = |\mathbf{r}|$ and $\rho = |\boldsymbol{\rho}|$, are in accordance with the drawing (Fig. 4.2).

It is evident from equation (4.13) that when $\mathbf{r}$ is antiparallel to $\boldsymbol{\rho}$,

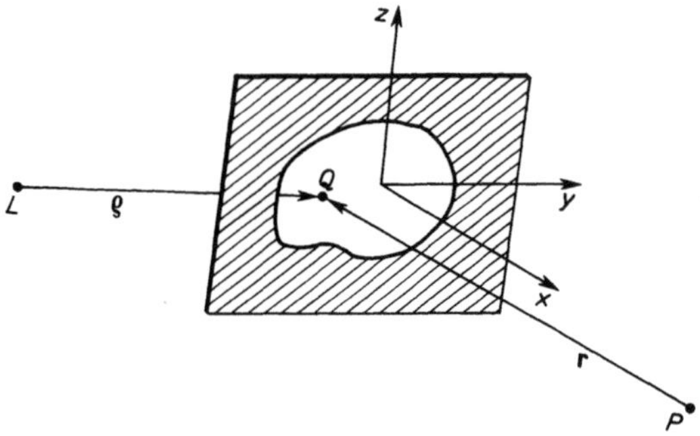

Figure 4.2

the numerator and denominator both vanish. It can however be shown that

$$\frac{|\mathbf{r} \times \boldsymbol{\rho}|}{|r\rho + \mathbf{r}\cdot\boldsymbol{\rho}|} = \frac{\sin\alpha}{1+\cos\alpha} = \tan\frac{\alpha}{2}, \tag{4.14}$$

where α is the angle between the vectors $\mathbf{r}$ and $\boldsymbol{\rho}$. When $\alpha \to \pi$, $\tan\frac{\alpha}{2} \to \infty$ which confirms the existence of a singularity.

To sum up, it may be stated that in the case of a spherical primary wave incident on a diaphragm, the potential $\mathbf{W}(P, Q)$ has a singularity which is located at the intersection of the line joining the point source L with the observation point P, and the surface of integration.

A calculation of the curvilinear integral of $\mathbf{W}(P, Q)$ around its singularity gives at point P the expression

$$U_0 = \oint_{C_j} \mathbf{W}_0(P, Q)\cdot \mathrm{d}\mathbf{s} = \delta \frac{\mathrm{e}^{ikR}}{R}, \tag{4.15}$$

where

$$\delta = \begin{cases} 1 & \text{for points within the cone of illumination,} \\ 0 & \text{for observation points in the area of shadow,} \end{cases}$$

and R is the distance of the point of source L from the observation point P. This describes, for points within the cone of illumination, a wave which would reach the observation point if there were no diaphragm, and is zero for all points in geometrical shadow. This wave is called a *geometric wave.*

The integral over the deflecting edge K

$$\oint_K \mathbf{W}_0(P, Q)\mathbf{ds} = U_D$$

will be called a *diffracted wave.*

Kirchhoff's diffraction field $U_K(P)$ describes a wave motion which is continuous everywhere in space except at the light source. From equation (4.12), it may be presented in the form

$$U_K(P) = U_0(P) + U_D(P).$$

A discontinuity of the first component generates a discontinuity of the second, and a discontinuity of the diffracted wave must be such as to compensate for that of the incident wave.

A calculation of the curvilinear integral taken over the edge of the deflecting aperture makes it possible to determine a number of interesting diffraction phenomena for a large class of apertures in Kirchhoff screens.

4.5. Asymptotic Expressions for the Diffracted Wave

To show the basic characteristics of the edge (diffracted) wave U_D it is sufficient to consider the case of a divergent spherical wave in the form

$$U_0 = \frac{e^{ik\varrho}}{\rho}$$

incident on a Kirchhoff screen with an aperture limited by a spatial curve K. With these assumptions, the diffracted wave U_D will be expressed by the formula

$$U_D = \frac{1}{4\pi}\oint_K \frac{\exp[ik(r+\rho)]}{r\rho}\,\frac{(\mathbf{r}\times\boldsymbol{\rho})}{r\rho + \mathbf{r}\cdot\boldsymbol{\rho}}\,\mathrm{ds}. \tag{4.17}$$

To obtain an asymptotic expression for U_D, the stationary phase method is applied. This method is based on the fact that for large k (i.e. small wavelengths) the value of the expression under the integral sign undergoes rapid oscillations, which causes that the contributions to U_D from various elements of the deflecting edge K add up to zero. The only elements of the deflecting edge which have a non-zero contribution to U_D are those in proximity to the points Q_j defined by the condition

$$\frac{\mathrm{d}(r+\rho)}{\mathrm{d}s} = \mathbf{s}_0 \cdot \mathrm{grad}(r+\rho) = 0, \tag{4.18}$$

where the versor $\mathbf{s}_0$ is tangential to the deflecting edge element in question.

This condition defines areas of relatively stationary phase: $k(r+\rho)$. The points Q_j for which (4.18) is true are called *active points*, and their surroundings (which are elements of the deflecting edge K) having non-zero contributions to U_D are *active areas*. They are placed in such a manner that the sum of the distance between the light source and the active point, and between the active point and the observation point is extremal.

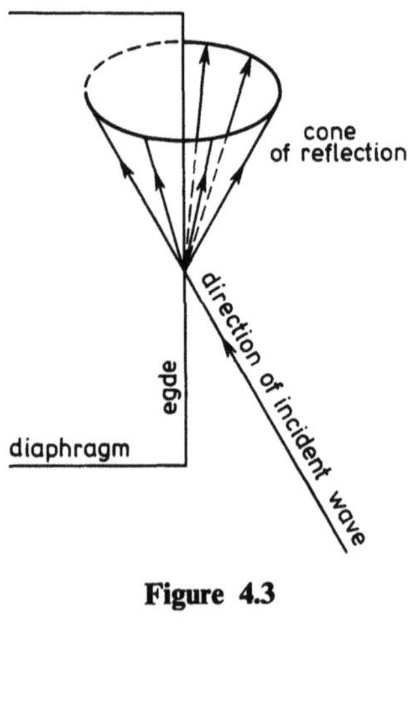

Figure 4.3

The condition (4.18) may also be written in the form

$$\mathbf{s}_0 \cdot \frac{\boldsymbol{\rho}}{\rho} = -\mathbf{s}_0 \cdot \frac{\mathbf{r}}{r}. \tag{4.19}$$

This tells us that a given active surrounding of the aperture edge reflects light in directions which form the so-called *cone of reflection* (Fig. 4.3). The axis of this cone is formed by the active element of the edge. Equation (4.19) is analogous to Fresnel's principle in geometric optics, the only difference being that the incident and reflected rays do not have to lie on one plane.

The diffraction effect occurring at a given observation points is a result of the interference of the incident wave with edge waves. The latter are reflected from points on the aperture edge connected with the observation point by equation (4.18).

Because the diffracted wave has to compensate for the discontinuity of the incident wave (geometrical wave) at the shadow boundary, it is evident that its amplitude must be of the same order of magnitude as that of the incident wave. It will be shown below that the absolute value of the amplitude of the diffracted wave on both sides of the shadow boundary is indeed equal to half the amplitude of the incident wave.

In order to ensure the continuity of the overall wave motion at the shadow boundary, the phases of the incident and edge waves approaching the shadow boundary from the illuminated side must be opposite. Approaching this boundary from the dark side, however, the phases of the two waves are equal.

The above-mentioned properties of the deflected wave are a direct result of the asymptotic expression obtained for it from equation (4.17). To obtain this expression in a form which would hold for observation points near the shadow boundary, where the diffraction effects are most pronounced, it is necessary before applying the stationary phase method to construct a supplementary function

$$\tilde{U} = \frac{\partial}{\partial k}\left[\mathrm{e}^{-ikR} U_D\right], \tag{4.20}$$

where R is the distance of the source L from the observation point P. An explicit form of the new function $\tilde{U}$ is given by the expression

$$\tilde{U} = -\frac{\mathrm{i}}{4\pi}\oint_K \mathrm{e}^{\mathrm{i}k(r+\varrho-R)}\frac{(r+\rho-R)(\mathbf{r}\times\boldsymbol{\rho})\cdot\mathbf{s}_0}{r\rho(r\rho+\mathbf{r}\cdot\boldsymbol{\rho})}\,\mathrm{d}s. \tag{4.21}$$

After applying the stationary phase method to the above expression, the following asymptotic term is obtained:

$$\tilde{U} \sim \frac{\mathrm{i}}{\sqrt{2k\pi}}\sum_j\left\{\mathrm{e}^{\mathrm{i}[k(r+\varrho-R)\pm\pi/4]}\times\right.$$

$$\left.\times\frac{\mathbf{s}_0\cdot(\mathbf{r}\times\boldsymbol{\rho})(r+\rho-R)}{2r\rho(r\rho+\mathbf{r}\cdot\boldsymbol{\rho})\sqrt{|\xi''|}}\int\limits_{s_j-\Delta s}^{s_j+\Delta s}\mathrm{e}^{\mathrm{i}\frac{k}{2}\xi''(s-s_j)^2}\,\mathrm{d}s\right\}_j, \tag{4.22}$$

where

$$\xi = r+\rho-R, \qquad \xi'' = \frac{\mathrm{d}^2\xi}{\mathrm{d}s^2},$$

s_j is the length of the arc corresponding to the j-th active point, $2\Delta s$ is the active area of the edge. After effecting a substitution of variables in equation (4.22) in the form

$$v = \sqrt{\frac{k}{\pi}|\xi''|}\,(s-s_j)$$

and taking into account the fact that for large values of k the integration limits may be set to $\pm\infty$, the following results:

$$\tilde{U} \sim -\frac{\mathrm{i}}{2\sqrt{2k\pi}}\sum_j\left\{\mathrm{e}^{\mathrm{i}[k(r+\varrho-R)\pm\pi/4]}\frac{(\mathbf{r}\times\boldsymbol{\rho})\cdot\mathbf{s}_0(r+\rho-R)}{r\rho(r\rho+\mathbf{r}\cdot\boldsymbol{\rho})\sqrt{|\xi''|}}\right\}_j. \tag{4.23}$$

To obtain the asymptotic form for the deflected wave it is necessary to integrate expression (4.23) with respect to k from $+\infty$ to k, multiply by $\mathrm{e}^{\mathrm{i}kR}$ and introduce a new variable

$$v_1 = \sqrt{\frac{k}{\pi}}(r + \rho - R).$$

The sought-for asymptotic expression for the diffracted wave U_D then results

$$U_D \sim \frac{\mathrm{i}}{2}\sum_j \left\{ \frac{(\mathbf{r} \times \boldsymbol{\rho})\cdot \mathbf{s}_0 \sqrt{r + \rho - R}}{r\rho(r\rho + \mathbf{r}\cdot\boldsymbol{\rho})\sqrt{|\xi''|}} \times \right.$$

$$\left. \times \int_{+\infty}^{\sqrt{2k(r+\varrho-R)/\pi}} \mathrm{e}^{\mathrm{i}\pi v^2/2}\,\mathrm{d}v \cdot \mathrm{e}^{\mathrm{i}[kR \pm \pi/4]} \right\}_j, \tag{4.24}$$

where the sign in front of $\mathrm{i}\pi/4$ in the exponent is a plus for $\xi'' > 0$ and a minus for $\xi'' < 0$.

Expression (4.24) is true for an arbitrarily placed observation point. In the immediate vicinity of the shadow boundary it becomes

$$U_D = -\frac{\mathrm{e}^{\mathrm{i}(kR \pm \pi/4)}}{R\sqrt{2}} \cos\alpha \int_{+\infty}^{\sqrt{2k(r+\varrho-R)/\pi}} \mathrm{e}^{\mathrm{i}\pi v^2/2}\,\mathrm{d}v, \tag{4.25}$$

where $\cos\alpha = \begin{cases} -1 \text{ for observation points in the illuminated region,} \\ +1 \text{ for observation points in the shadow,} \end{cases}$

α is the angle between the vectors $\mathbf{n} \times \boldsymbol{\rho}$ and $\mathbf{r} \times \boldsymbol{\rho}$, where $\mathbf{n}$ is a versor perpendicular to the shadow boundary.

The transition from (4.24) to (4.25) is achieved through the use of the following relations and assumptions:

a. $$\frac{(\mathbf{r} \times \boldsymbol{\rho})\cdot \mathbf{s}_0 \sqrt{r + \rho - R}}{2r\rho(\mathbf{r}\cdot\boldsymbol{\rho} + r\rho)} = \frac{\cos\alpha \sin[(\mathbf{r}, \boldsymbol{\rho})/2]\sin(\boldsymbol{\rho}, \mathbf{s}_0)}{\sqrt{|\xi''|}\sqrt{r\rho(r + \rho + R)}},$$

b. $$\xi'' = \left(\frac{1}{r} + \frac{1}{\rho}\right)\sin^2(\boldsymbol{\rho}, \mathbf{s}_0),$$

c. in the immediate vicinity of the shadow boundary, the only important contribution comes from the active surrounding nearest the observation point.

From (4.25) we may calculate the jump of the deflected wave value at the shadow boundary. It is equal to:

$$U_{D|_1} - U_{D|_2} = -\frac{e^{ikR}}{R}, \tag{4.26}$$

where $U_{D|_1}$ denotes the value of the deflected wave at the shadow boundary from the side of the light, and $U_{D|_2}$ denotes the value of the deflected wave from the side of the shadow.

It is interesting to note that $(r + \rho - R) \to 0$ very close to the shadow boundary and is different from zero in regions far from this boundary. As a result of the properties of Fresnel's integral appearing in (4.25), the amplitude of the diffracted wave at large distances from the shadow boundary decreases rapidly. The presence of $\cos\alpha$ and the jump change of its value for the observation points in vicinity of the shadow boundary show that the diffracted wave undergoes here a phase shift of π.

Experiments confirm these properties of diffracted waves. The existence of wave motion divergent from the deflecting edge is confirmed by the fact that the deflecting edge can be seen to shine when observed from the area of shadow. The eye of the observer, however, must be guarded against receiving the incident wave.

An objective experimental proof of the existence of an edge wave was given in 1912 by Kalashnikov. He showed that a photographic plate placed in the shadow was darkened by a specially situated set of pins. This can only have been caused by a wave from the edge, as is unambiguously shown by the shadows of the pins (Fig. 4.4).

The existence of active areas on the edge of the aperture is in accordance with an experimental fact, namely, that for a given observation point the only areas of the edge which shine are those situated around points satisfying the condition that the sum of their distance from the source and the observation point is extremal.

In the case of a circular aperture and a point light source located on a line perpendicular to the plane of the aperture and running through its centre, two shining regions may be observed on both ends of the diameter connected with extremal values of the sum mentioned above.

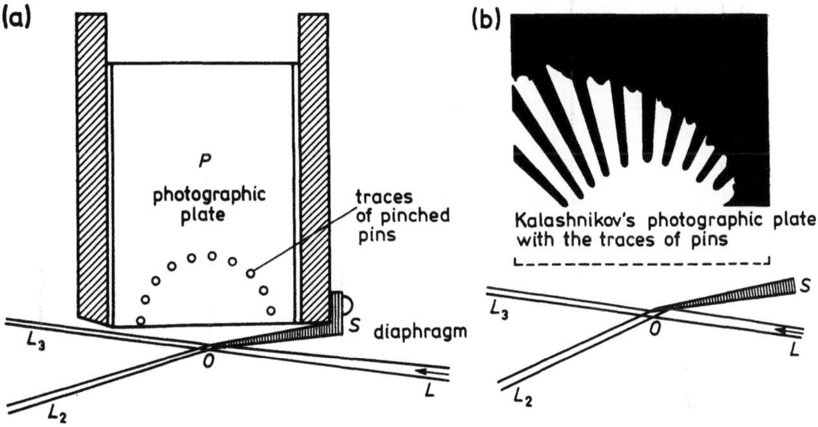

Figure 4.4

A phase difference of π between two diffracted waves from different sides of the shadow boundary may be confirmed by obtaining an image of the aperture edge in the light of these waves only, eliminating the incident wave. The image of the aperture edge thus obtained is a dark line (Banerji S., 1919).

All these experimental facts are explained extensively within the framework of the Young–Rubinowicz diffraction model. They did not have any explanation based on Fresnel's diffraction model.

4.6. The Fresnel–Kirchhoff Diffraction Formula

Further simplification of Kirchhoff's integral giving $U(P_0)$ at a point lying beyond the screen can be carried out in the case where the distance r between a point of the pinhole Q and the observation point P_0 is many times greater than the wavelength, i.e. where $k = 2\pi/\lambda \gg 1/r$. Then

$$\frac{\partial}{\partial n}\frac{e^{ikr}}{r} = \cos(\mathbf{n}, \mathbf{r})\left(ik - \frac{1}{r}\right)\frac{e^{ikr}}{r} \approx ik\cos(\mathbf{n}, \mathbf{r})\frac{e^{ikr}}{r}.$$

Substituting this expression into Kirchhoff's integral we get

$$U(P)=\frac{1}{4\pi}\iint\limits_{S_1}\left\{\frac{e^{ikr}}{r}\left(\frac{\partial U}{\partial n}-ikU\cos(\mathbf{n},\mathbf{r})\right)\right\}df. \tag{4.27}$$

If on the pinhole S_1 falls a spherical wave

$$U(P)=\frac{Ae^{ik\varrho}}{\rho}, \tag{4.28}$$

coming from a point source placed at point P_1, and if $\rho=|P_1Q|$ is the distance between the point source and the current point Q at the pinhole, then by the condition $k \gg 1/\rho$ (the distance between Q and P_1 exceeds by many times the wavelength), we get

$$U(P)=\frac{A}{i\lambda}\iint\limits_{S_1}\frac{e^{ik(r+\varrho)}}{r\rho}\,\frac{\cos(\mathbf{n},\mathbf{r})-\cos(\mathbf{n},\boldsymbol{\rho})}{2}\,df. \tag{4.29}$$

This result is true only in the case where the pinhole is illuminated by a point source of light, and it is called the *Fresnel–Kirchhoff diffraction formula.*

Expression (4.29) is symmetrical in relation to the source point P_1 and to the observation point P_0. This means that if the point source is situated at the observation point, the effect produced at point P_1 will be the same as the one produced at point P_0 in the case where a point source of the same intensity is situated at P_1. This statement is called *Helmholtz's reciprocity theorem.*

4.7. Physical Interpretation of the Fresnel–Kirchhoff Formula

Introducing the notation

$$U(Q)=\frac{Ae^{ik\varrho}}{\rho}\left[\frac{\cos(\mathbf{n},\mathbf{r})-\cos(\mathbf{n},\boldsymbol{\rho})}{2}\right], \tag{4.30}$$

the Fresnel–Kirchhoff diffraction formula may be rewritten in the form

$$U(P) = \left(\frac{-ik}{2\pi}\right) \iint_{S_1} U(Q) \frac{e^{ikr}}{r} \, df. \tag{4.31}$$

On the basis of (4.31) it can be stated that the field at point P is a result of superposition of secondary vibrations coming from point sources of light distributed on the pinhole. The amplitude $U(Q)$ of each secondary wave is proportional to the amplitude of the incident wave $(A e^{ik\varrho}/\rho)$, but differs from it in the following:

(1) the factor $1/\lambda$,

(2) the direction factor $\frac{1}{2}[\cos(\mathbf{n}, \mathbf{r}) - \cos(\mathbf{n}, \boldsymbol{\rho})] \neq 1$,

(3) the phase shift equal to $\frac{1}{2}\pi$.

Assuming in (4.30) that

$$\tfrac{1}{2}[\cos(\mathbf{n}, \mathbf{r}) - \cos(\mathbf{n}, \boldsymbol{\rho})] = 1,$$

we obtain from (4.31) the *Huygens–Fresnel formula*

$$U(P) = \left(\frac{-ik}{2\pi}\right) \iint_{S_1} U(Q) \frac{e^{ikr}}{r} \, df. \tag{4.32}$$

4.8. Fresnel's Zones

From the mathematical point of view it is admissible to deform the integration surface filling the deflecting aperture. Each sheet of surface spanned on the edge of the aperture is fully equivalent, if only the deformation is performed within a space free of sources. This fact allows us to choose as the integration surface the fragment W of the wave front of the spherical wave and the fragment of a conic surface ξ (Fig. 4.5). On the integration surface chosen in this way the following equalities are held:

$$\cos(\mathbf{n}, \boldsymbol{\rho}) = 1 \quad \text{for } Q \in W,$$

$$\cos(\mathbf{n}, \boldsymbol{\rho}) = 0 \quad \text{for } Q \in \xi,$$

$$\cos(\mathbf{n}, \mathbf{r}) = \cos(\pi - \chi) = -\cos\chi \quad \text{for } Q \in W.$$

For large values of $\rho = |P_1 Q|$ the contributions coming from

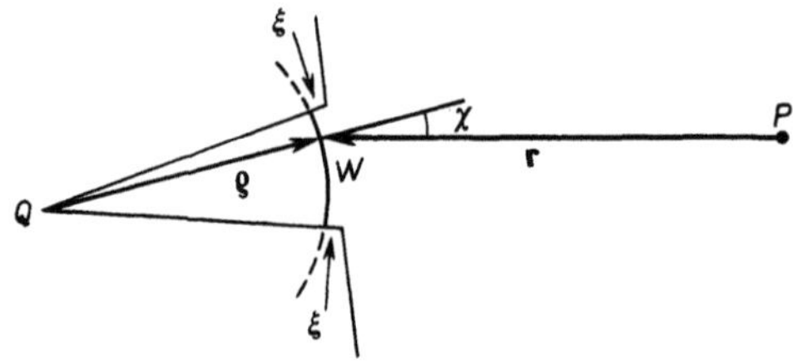

Figure 4.5

ξ may be neglected. All this leads to a new form of the Fresnel–Kirchhoff diffraction formula:

$$U(P) = -\frac{\mathrm{i}A\mathrm{e}^{\mathrm{i}k\varrho}}{2\lambda\rho}\iint_W \frac{\mathrm{e}^{\mathrm{i}kr}}{r}(1+\cos\chi)\,\mathrm{d}f, \tag{4.33}$$

where ρ is the radius of curvature of the spherical wave front. The result can be read as *Fresnel's formulation of Huygens' principle.* The state of the field at point P can be treated as a superposition of the contributions from the element $\mathrm{d}W$ of the wave front given in the form

$$\begin{aligned}\mathrm{d}U(P) &= -\frac{\mathrm{i}A}{2\lambda}\frac{\mathrm{e}^{\mathrm{i}k\varrho}}{\rho}\frac{\mathrm{e}^{\mathrm{i}kr}}{r}(1+\cos\chi)\,\mathrm{d}f\\ &= K(\chi)\frac{A\mathrm{e}^{\mathrm{i}k\varrho}}{\rho}\frac{\mathrm{e}^{\mathrm{i}kr}}{r}\,\mathrm{d}f,\end{aligned} \tag{4.34}$$

where $K(\chi)$ is called the *direction factor*. The classical formulation of Huygens' principle is not sufficient for the quantitive estimation of diffraction effects. This estimation has become possible only after complementing the theory with the construction of zones, which was suggested by Fresnel. The leading idea of introducing this construction is a remark that the secondary partial waves, coming from different points of the wave front, interfere. The construction of Fresnel zones is as follows (Fig. 4.6):

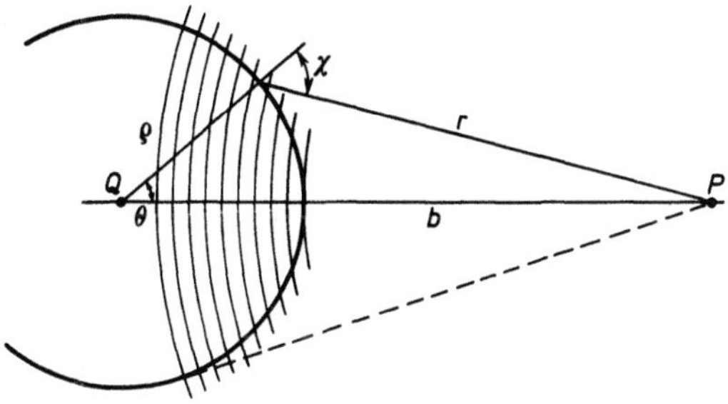

Figure 4.6

From the observation point P we construct spherical surfaces of radii

$$b,\ b+\frac{\lambda}{2},\quad b+\lambda,\quad b+\frac{3}{2}\lambda,\ \dots,\ b+\frac{m}{2}\lambda.$$

The spherical surfaces divide the wave front W into zones z_1, z_2, $z_3, \dots, z_m$. (We assume that $\rho \gg \lambda$ and $b \gg \lambda$.) Then because of a small change of χ,

$$K(\chi) = \frac{-\mathrm{i}}{2\lambda}(1+\cos\chi) \tag{4.35}$$

can be regarded as a constant value within the area of a zone.

From Fig. 4.6 follows the equality:

$$r^2 = \rho^2 + (\rho+b)^2 - 2\rho(\rho+b)\cos\theta,$$

and hence

$$r\mathrm{d}r = \rho(\rho+b)\sin\theta\,\mathrm{d}\theta.$$

An element of the integration surface is expressed by the formula

$$\mathrm{d}f = \rho^2\sin\theta\,\mathrm{d}\theta\,\mathrm{d}\varphi = \frac{\rho}{\rho+b}r\mathrm{d}r\mathrm{d}\varphi,$$

where φ is the azimuth angle. The contribution from the m-th zone to the wave field at point P is given by the expression:

$$U_m(P) = -2\pi \frac{Ae^{ik\varrho}}{\rho + b} K_m \int\limits_{b+(m+1)\lambda/2}^{b+m\lambda/2} e^{ikr}\,dr$$

$$= \frac{2i}{k} K_m \frac{Ae^{ik(\varrho+b)}}{\rho + b} e^{ikm\lambda/2}\left(1 - e^{ik\lambda/2}\right)\pi. \tag{4.36}$$

Taking into account that $k\lambda = 2\pi$, (4.36) simplifies to

$$U_m(P) = 2i\lambda(-1)^{m+1} K_m \frac{Ae^{ik(\varrho+b)}}{\rho + b}. \tag{4.37}$$

The contributions from the successive zones have opposite signs. Summing up all contributions from individual zones we obtain the resultant field at the observation point P

$$U(P) = 2i\lambda \frac{Ae^{i(\varrho+b)k}}{\rho + b} \sum_{l=1}^{n} (-1)^{l+1} K_l. \tag{4.38}$$

The series may be presented in the following way:

$$\sum_{l=1} (-1)^{l+1} K_l = \frac{K_1}{2} + \left(\frac{K_1}{2} - K_2 + \frac{K_3}{2}\right) +$$

$$+ \left(\frac{K_3}{2} - K_4 + \frac{K_5}{2}\right) + \ldots + R_m,$$

where

$$R_m = \begin{cases} +\frac{1}{2}K_m & \text{for odd } m, \\ +\frac{1}{2}K_{m-1} - K_m & \text{for even } m. \end{cases}$$

Assuming

$$K_m = \frac{K_{m-1} - K_{m+1}}{2},$$

we can state that

$$\sum_{l}^{m}(-1)^{l+1}\mathrm{K}_l = \begin{cases} \dfrac{K_1}{2} + \dfrac{K_m}{2} & \text{for odd } m, \\ \dfrac{K_1}{2} - \dfrac{K_m}{2} & \text{for even } m. \end{cases}$$

Thus the resultant field at the point P will be

$$U(P) = \mathrm{i}\lambda(K_1 \pm K_m)\frac{A\mathrm{e}^{\mathrm{i}k(\varrho+b)}}{\rho+b} = \tfrac{1}{2}[U_1(P) + U_m(P)]. \quad (4.39)$$

If $m = \infty$, then

$$U(P) = \tfrac{1}{2}U_1(P) = \mathrm{i}\lambda K_1 \frac{A\mathrm{e}^{\mathrm{i}k(\varrho+b)}}{\rho+b}. \quad (4.40)$$

The total disturbance at the point P equals half the disturbance connected with the first zone. If we have to do with a small pinhole, then $m = 1$ and the state of field disturbance at the point P is

$$U(P) = U_1(P) = 2\mathrm{i}\lambda K_1 \frac{A\mathrm{e}^{\mathrm{i}k(\varrho+b)}}{\rho+b}. \quad (4.41)$$

For the central zone (see (4.34)), for which $\chi = 0$, the direction coefficient $K(0) = -\mathrm{i}/\lambda$. For a zone distant from the centre, for which $\chi = \pi/2$, the value of the direction coefficient is $K(\pi/2) = \mathrm{i}/2\lambda$, and it is different from the zero value taken originally by Fresnel.

4.9. Babinet's Principle

From Kirchhoff's integral we can directly draw conclusions as to the intensity distribution of the deflected light in the case of complementary screens, i.e., such screens that the apertures in one of them correspond exactly to the covered parts in the other, and the other way round.

If $U_1(P)$ and $U_2(P)$ denote the values of disturbances at the point P in the presence of the first or the second screen respectively, and if with the same illuminating source they may be expressed by the integrals taken over the complementary apertures, then

$$U_1(P) + U_2(P) = U(P), \tag{4.42}$$

where $U(P)$ is the state of disturbance at the point P in the absence of any screen. It is known as the *Babinet principle.*

Conclusions from the Babinet principle are as follows:

1. If $U_1(P) = 0$, then $U_2(P) = U$ – at points where in the presence of one of the screens the intensity is zero, the intensity in the presence of the complementary screen is the same as in the absence of any screen.

2. If $U = 0$, then $U_1 = -U_2$ – it says that at points where U is zero, the phases of U_1 and U_2 differ by π and the intensities are equal, i.e., $I_1 = |U_1|^2 = I_2 = |U_2|^2$. If for example a point source is transformed by a defectless lens, the light intensity U in the plane of the image will be zero everywhere except the close neighbourhood of the source image Q.

4.10. The Fraunhofer and Fresnel Diffractions

If the distances from a point source P_1 and a point of observation P to the screen are long in comparison with the linear sizes of the aperture, the factor $[\cos(\mathbf{n}, \boldsymbol{\rho}) - \cos(\mathbf{n}, \mathbf{r})]$ will practically undergo no changes within the aperture. Supposing that the angles $(\mathbf{n}, \boldsymbol{\rho})$ and $\pi - (\mathbf{n}, \mathbf{r})$ are not too large, we can assume that

$$\cos(\mathbf{n}, \boldsymbol{\rho}) - \cos(\mathbf{n}, \mathbf{r}) \approx 2\cos\delta, \tag{4.43}$$

$$\frac{1}{\rho r} = \frac{1}{\rho_0 r_0}, \tag{4.44}$$

where ρ_0 is the distance from the source P_1 to the origin of the coordinate system (Fig. 4.7), r_0 – the distance from the point of observation P to the origin of the coordinate system, r – the distance from the point of observation P to a current point Q of the deflecting aperture, ρ – the distance from the source to the point Q.

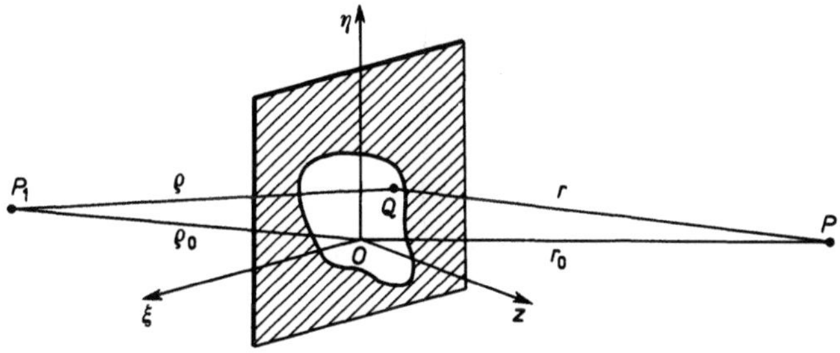

Figure 4.7

On the ground of the above assumptions the diffraction Fresnel–Kirchhoff formula may be written in the form

$$U(P) = -\frac{Ai}{\lambda}\frac{\cos\delta}{r_0\rho_0}\iint_{S_1} e^{ik(r+\varrho)}\,df. \tag{4.45}$$

Let us choose a reference system with the x- and the y-axis in the plane of the deflecting aperture, and with the z-axis pointed towards the half-space containing the point of observation P. Suppose that the points P_1, P and Q have the coordinates

$$P_1 = (x_1, y_1, z_1), \quad P = (x, y, z), \quad Q = (\xi, \eta, 0).$$

Then

$$\rho^2 = (x_1 - \xi)^2 + (y_1 - \eta)^2 + z_1^2,$$
$$r^2 = (x - \xi)^2 + (y - \eta)^2 + z^2,$$
$$\rho_0^2 = x_1^2 + y_1^2 + z_1^2, \quad r_0^2 = x^2 + y^2 + z^2.$$

Hence

$$\begin{aligned}\rho^2 &= \rho_0^2 - 2(x_1\xi + y_1\eta) + \xi^2 + \eta^2,\\ r^2 &= r_0^2 - 2(x\xi + y\eta) + \xi^2 + \eta^2.\end{aligned} \tag{4.46}$$

Assuming that the linear sizes of the aperture are small compared with ρ_0 and r_0, r and ρ can be expanded in a power series with respect to the variables: ξ/ρ_0, η/ρ_0, ξ/r_0, η/r_0. We obtain

$$
\begin{aligned}
r &\approx r_0 - \frac{x\xi + y\eta}{r_0} + \frac{\xi^2 + \eta^2}{2r_0} - \frac{(x\xi + y\eta)^2}{2r_0^3} \cdots, \\
\rho &\approx \rho_0 - \frac{x_1\xi + y\eta}{\rho_0} - \frac{\xi^2 + \eta^2}{2\rho_0} - \frac{(x_1\xi + y_1\eta)^2}{2\rho_0^3} \cdots
\end{aligned}
\tag{4.47}
$$

Substituting these approximations into the Fresnel–Kirchhoff diffraction formula we arrive at the expression

$$
U(P) = -\frac{\mathrm{i}\cos\delta}{\lambda} \frac{A\exp[\mathrm{i}k(r_0 + \rho_0)]}{r_0\rho_0} \iint_{S_1} \exp[\mathrm{i}kf(\xi, \eta)]\,\mathrm{d}\xi\,\mathrm{d}\eta,
\tag{4.48}
$$

where

$$
\begin{aligned}
f(\xi, \eta) = &-\frac{x_1\xi + y_1\eta}{\rho_0} - \frac{x\xi + y\eta}{r_0} + \left(\frac{\xi^2 + \eta^2}{2}\right)\left(\frac{1}{r_0} + \frac{1}{\rho_0}\right) - \\
&- \frac{(x_1\xi + y_1\eta)^2}{2\rho_0^3} - \frac{(x\xi + y\eta)^2}{2r_0^3}.
\end{aligned}
\tag{4.49}
$$

Introducing the notations

$$
l_0 = \frac{-x_1}{\rho_0}, \quad l = \frac{x}{r_0},
$$

$$
m_0 = \frac{-y_1}{\rho_0}, \quad m = \frac{y}{r},
$$

we have

$$
\begin{aligned}
f(\xi, \eta) = (l_0 - l)\xi + (m_0 - m)\eta + \tfrac{1}{2}\left(\frac{1}{\rho_0} + \frac{1}{r_0}\right)(\xi^2 + \eta^2) - \\
- \frac{(l_0\xi + m_0\eta)^2}{2\rho_0} - \frac{(l\xi + m\eta)^2}{2r_0} \cdots
\end{aligned}
\tag{4.49a}
$$

In this way the problem of finding the wave field at the point P has been resolved itself into calculating the integral (4.48).

If now in this integral the approximation

$$f(\xi, \eta) \approx (l_o - l)\xi + (m_o - m)\eta = p\xi + q\xi$$

is used, we speak of the so-called *Fraunhofer diffraction*, and then

$$U(P) = C \iint_{S_1} \exp[-\mathrm{i}k(p\xi + q\eta)]\, \mathrm{d}\xi\, \mathrm{d}\eta. \tag{4.50}$$

If when calculating the integral (4.48) we can not neglect the square terms in (4.49a), we speak of the *Fresnel diffraction.*

4.11. The Regions of Validity for the Fresnel and Fraunhofer Approximations

In the Fraunhofer diffraction only the linear terms of the approximated function $f(\xi, \eta)$ are taken into account. The square terms in the expansion (4.49a) are omitted in this approximation. It is obvious that we can neglect these terms for r_0 and $\rho_0 \to \infty$, but it is not the only case when it can be done. It is enough if the following inequality holds:

$$\tfrac{1}{2}k\left|\left(\frac{1}{r_0} + \frac{1}{\rho_0}\right)(\xi^2 + \eta^2) - \frac{(l_0\xi + m_0\eta)^2}{r_0} - \frac{(l\xi + m\eta)^2}{\rho_0}\right| \ll 2\pi. \tag{4.51}$$

Hence after considering the inequalities

$$(l_0\xi + m_0\eta)^2 \leq (l_0^2 + m_0^2)(\xi^2 + \eta^2),$$
$$(l\xi + m\eta)^2 \leq (l^2 + m^2)(\xi^2 + \eta^2),$$

we obtain the conditions for the distances from the screen to the observation point P and to the source P_1 in relation to the dimensions of the deflecting aperture and the wavelength:

$$\frac{(\xi^2 + \eta^2)_{\max}}{\lambda} \ll |r_0|, \quad \frac{(\xi^2 + \eta^2)_{\max}}{\lambda} \ll |\rho_0| \tag{4.52}$$

and the condition for the position of the source and the observation point with regard to the z-axis:

$$\frac{1}{r_0} + \frac{1}{\rho_0} = 0, \quad l_0^2, m_0^2, l^2, m^2 \ll \frac{(r_0)\lambda}{(\xi^2 + \eta^2)_{\max}}. \tag{4.53}$$

From this it follows that the observation point and the point source should be situated close to the z-axis.

The Fresnel approximation takes into account also the second power terms in the expansion (4.49a). It can be done if in the expansion

$$\begin{aligned}\rho &= \sqrt{z^2 + (x_1 - \xi)^2 + (y_1 - \eta)^2} \\ &\approx z\left[1 + \tfrac{1}{2}\frac{(x_1 - \xi)^2}{z^2} + \tfrac{1}{2}\frac{(y_1 - \eta)^2}{z^2} - \right. \\ &\left. - \tfrac{1}{8}\left(\frac{(x_1 - \xi)^2}{z^2} + \frac{(y_1 - \eta)^2}{z^2}\right)^2\right] + \dots\end{aligned}$$

it is assumed that

$$\tfrac{1}{8}[(x_1 - \xi)^2 + (y_1 - \eta)^2]^2 \ll z^3, \tag{4.54}$$

i.e., the maximum phase change contributed by this term will be much less than radian. The similar condition applies to r.

4.12. The Solution of Diffraction Problem by Means of Decomposition in Plane Waves

Any solution of the wave equation for monochromatic fields (see Sec. 2.4) can be presented as a superposition of plane waves

$$U(x, y, z) = \frac{1}{4\pi^2}\iint\limits_{-\infty}^{+\infty} g(k_1, k_2)\exp\left(+iz\sqrt{k^2 - k_1^2 - k_2^2}\right) \times \\ \times \exp[i(k_1 x + k_2 y)]\,dk_1\,dk_2, \tag{4.55}$$

where $g(k_1, k_2)$ is a complex function describing the amplitude and the phase of a separate plane wave propagating in the direction defined by the set of real variables k_1 and k_2. If $k_1^2 + k_2^2 > k^2$, this expression contains, except ordinary plane waves, also so-called *non-homogeneous plane waves.*

Formula (4.55) allows us to solve the diffraction problem in the case of plane diffracted screens. A plane diffracted screen can always be regarded as the plane $z = 0$, the field at points of the diffracted aperture is the same as the field of the incident wave, while at non-illuminated points of the diffracted screen it equals zero. These assumptions made by Kirchhoff define the field in the plane $z = 0$. Hence on the ground of formula (2.16), $g(k_1, k_2)$ can be determined. When the function $g(k_1, k_2)$ is known, we can determine from (4.55) the field in the plane $z > 0$.

The result obtained is approximate. The approximation consists in replacing a real field in the plane $z = 0$ by a field being in agreement with Kirchhoff's approximation. However, if the distance to the screen is quite large, the result is always consistent with experiment. The precise solution of the problem agrees in this case with Kirchhoff's approximation for sufficiently great distances.

If the screen is opaque, it can be defined by the transmittance function

$$t(x, y) = \begin{cases} 1 & (x, y) \in A, \\ 0 & (x, y) \notin A. \end{cases}$$

The Kirchhoff condition for the field in the plane $z = 0$ may now be written in the form

$$U_t = U_0(x, y, 0)\, t(x, y),$$

where $U_0(x, y, 0)$ is the incident field.

On the ground of formula (2.16) we obtain the expression

$$g_t(u_1, u_2) = \iint_{-\infty}^{+\infty} U_0(x, y, 0) t(x, y) \exp[-i(u_1 x + u_2 y)] \, dx\, dy. \tag{4.56}$$

Regarding (4.56) as a two-dimensional Fourier transform and using its properties, we get

$$g_t(u_1, u_2) = \iint\limits_{-\infty}^{+\infty} g_0(\xi, \eta) T(u_1 - \xi, u_2 - \eta) \mathrm{d}\xi \mathrm{d}\eta, \tag{4.57}$$

where

$$g_0(u_1, u_2) = \iint\limits_{-\infty}^{+\infty} U_0(x, y, 0) \exp[-\mathrm{i}(u_1 x + u_2 y)] \mathrm{d}x \mathrm{d}y,$$

$$T(u_1, u_2) = \iint\limits_{-\infty}^{+\infty} t(x, y) \exp[-\mathrm{i}(u_1 x + u_2 y)] \mathrm{d}x \mathrm{d}y.$$

From the above it can be seen that:

The spectrum of spatial frequencies (angular spectrum) for a wave transmitted through the screen is defined by a convolution of the spectrum of the incident wave and the one characterizing the screen itself.

In the case of a plane wave of unit amplitude falling perpendicularly on the aperture, the results has a particularly simple form. In this case

$$g_0(u_1, u_2) = 4\pi^2 \delta(u_1, u_2), \qquad g_t(u_1, u_2) = T(u_1, u_2)$$

and then the angular spectrum for the transmitted wave can be found as the transform of the aperture transmittance.

To sum up, we can state that introducing an aperture confining spatially the incident wave leads to widening the angular spectrum of disturbance. The smaller the aperture the wider – for a plane incident wave – the angular spectrum of the wave behind the aperture.

Similarly as in the Heisenberg indeterminacy relations, the product of the spatial width of impulse and the width of frequency band remains constant. If an aperture in the screen is small, the output signal will be spatially confined and have a wide band of spatial frequencies. This results in broadening the diffraction pattern, other its properties remaining unchangeable.

4.13. The Fresnel and Fraunhofer Diffractions as Filtration of Spatial Frequencies

The Huygens–Fresnel formula may be written in the form

$$U(x_0, y_0, z) = \iint_A h(x_0, y_0, z; x_1, y_1) U(x_1, y_1) \mathrm{d}x_1 \mathrm{d}y_1, \tag{4.58}$$

where

$$h(x_0, y_0, z; x_1, y_1) = \frac{1}{\mathrm{i}\lambda} \frac{\exp(\mathrm{i}kr_{10})}{2r_{10}} [\cos(\mathbf{n}, \mathbf{r}_{01}) - \cos(\mathbf{n}, \boldsymbol{\rho})].$$

Within the Fresnel approximation we can assume that

(1) $r_{10} \approx z\left[1 + \frac{1}{2}\left(\frac{x_0 - x_1}{z}\right)^2 + \frac{1}{2}\left(\frac{y_0 - y_1}{z}\right)^2\right],$

(2) $\frac{1}{2}[\cos(\mathbf{n}, \mathbf{r}_{01}) - \cos(\mathbf{n}, \boldsymbol{\rho})] \approx 1.$

Then

$$h(x_0, y_0, z, x_1, y_1) \approx \frac{\mathrm{e}^{ikz}}{\lambda \mathrm{i} z} \exp\left\{ik\frac{1}{2z}[(x_0 - x_1)^2 + (y_0 - y_1)^2]\right\}, \tag{4.59}$$

$$U(x_0, y_0, z) = \iint_A U(x_1, y_1) \times$$

$$\times \exp\left\{ik\frac{1}{2z}[(x_0 - x_1)^2 + (y_0 - y_1)^2]\right\} \mathrm{d}x_1 \mathrm{d}y_1 \frac{\mathrm{e}^{ikz}}{\mathrm{i}\lambda z}. \tag{4.60}$$

The expression (4.60) can be written as

$$U(x_0, y_0, z)$$

$$= \frac{\mathrm{e}^{ikz}}{\mathrm{i}\lambda z} \exp\left[\frac{1}{2z}(x_0^2 + y_0^2)\right] \iint_{-\infty}^{+\infty} U(x_1, y_1) \exp\left[ik\frac{1}{2z}(x_1^2 + y_1^2)\right] \times$$

$$\times \exp\left[ik\frac{1}{z}(x_0 x_1 + y_0 y_1)\right] \mathrm{d}x_1 \mathrm{d}y_1. \tag{4.61}$$

Replacing the area A by the whole plane (x_1, y_1) is justified since according to Kirchhoff's assumption we have

$$U(x_1, y_1)\begin{cases} \neq 0 & \text{for} \quad x_1, y_1 \in A, \\ = 0 & \text{for} \quad x_1, y_1 \notin A. \end{cases}$$

As can be seen from formula (4.61), the function $U(x_0, y_0, z)$ is the Fourier transform of the function $U(x_1, y_1)\exp\left[\mathrm{i}\frac{k}{2z}(x_1^2 + y_1^2)\right]$.

Studying the Fresnel diffraction in the plane of spatial frequencies gives us a base for a new physical interpretation. Subjecting the spatially invariable weight function (4.59) to the Fourier transformation, we obtain the transfer function

$$H(u_1, u_2, z) = \exp(ikz)\exp\left[-\mathrm{i}z\frac{1}{2k}(u_1^2 + u_2^2)\right], \tag{4.62}$$

where $u_1 = kx_0/z$, $u_2 = ky_0/z$, which holds by the Fresnel diffraction. The first factor represents a general phase lag, to which every component of the angular spectrum is subjected while propagating between two planes separated by the distance z. The second factor describes phase dispersion dependent in a square way upon the frequency.

From the comparison of formula (4.62) with the expression

$$H_0(u_1, u_2, z) = \exp\left[\mathrm{i}z\sqrt{k^2 - (u_1^2 + u_2^2)}\right] \tag{4.63}$$

appearing in formula (4.55) it follows that the transfer function (4.62) is an approximation of the function (4.63) valid for $u_1^2 + u_2^2 \ll k^2$.

A diffraction pattern for the Fraunhofer approximation is simpler. Requiring in (4.61)

$$z \gg \frac{(x_1^2 + y_1^2)k}{2},$$

we obtain

$$U(x_0, y_0, z) = \frac{\exp(ikz)\exp\left[i\frac{k}{2z}(x_0^2 + y_0^2)\right]}{i\lambda z} \times$$

$$\times \iint_{-\infty}^{+\infty} U(x_1, y_1)\exp[-i(u_1x_1 + u_2y_1)]\,dx_1\,dy_1. \quad (4.64)$$

In the region of the Fraunhofer diffraction the field at the point of observation can be found as a product of the Fourier transform of the function $U(x_1, y_1)$ and the phase function $\exp\left[i\frac{k}{2z}(x_0^2 + y_0^2)\right]$.

There is no transfer function for the Fraunhofer diffraction. However, since the Fraunhofer diffraction is an approximation of the Fresnel diffraction, the function (4.62) may be considered valid also in this case.

4.14. Symmetry Properties in Patterns of the Fraunhofer Diffraction

If a plane wave falls perpendicularly on an aperture in an opaque screen, the wave phase is constant at all points of the aperture, and the following theorem can be formulated:

Each pattern produced in the Fraunhofer diffraction has the centre of symmetry P_0, independently of the existence of this kind of symmetry for the diffracted aperture.

The Fraunhofer diffraction field is of the form

$$U(x_0, y_0, z) = \frac{\exp(ikz)\exp\left[i\frac{k}{2z}(x_0^2 + y_0^2)\right]}{i\lambda z} \times$$

$$\times \iint_A U(x, y)\exp\left[-i\frac{k}{z}(x_0x + y_0y)\right]dx\,dy, \quad (4.65)$$

where (x_0, y_0) and (x, y) are the coordinates of points lying in the planes of the pattern and the diffracted aperture, respectively.

The function $U(x, y)$ defines a wave at the points of the surface of the aperture which coincides with the plane $z = 0$. The coordinate z in formula (4.65) determines therefore the distance between the plane of the image and the diffracted screen.

The intensity $I(x_0, y_0, z)$ is proportional to the time-averaged square of field amplitude, i.e., it is equal to

$$\langle U(x_0, y_0, z)\, U^*(x_0, y_0, z)\rangle. \tag{4.66}$$

Neglecting the constant factors, we obtain

$$I(x_0, y_0, z) = \left\{\iint_A |U(x, y)| \cos\left[\frac{k}{z}(x_0 x + y_0 y)\right] \mathrm{d}x\, \mathrm{d}y\right\}^2 + \left\{\iint_A |U(x, y)| \sin\left[\frac{k}{z}(x_0 x + y_0 y)\right] \mathrm{d}x\, \mathrm{d}y\right\}^2. \tag{4.67}$$

The intensity distribution $I(x_0, y_0, z)$ in the plane of the image will have the centre of symmetry if holds the equality

$$I(x_0, y_0, z) = I(-x_0, -y_0, z). \tag{4.68}$$

From (4.67) it is obvious that (4.68) takes place always, independently of the shape of the aperture in the screen.

Treating the Fraunhofer diffraction formula as a two-dimensional Fourier transform, and introducing the symbols

$$u_1 = \frac{k x_0}{z}, \quad u_2 = \frac{k y_0}{z},$$

the diffraction field $U(x_0, y_0, z) = U(u_1, u_2)$ can be regarded as the transform of the function $U(x, y)$. The module of this transform is an even function of spatial frequencies u_1 and u_2, and its phase an odd function of u_1 and u_2. If $U(x, y)$ is a real and even function, i.e. if

$$U(x, y) = U(-x, -y), \quad U(x, y) = U^*(x, y), \tag{4.69}$$

its Fourier transform with regard to its arguments

$$U(u_1, u_2) = U(-u_1, -u_2), \qquad U(u_1, u_2) = U^*(u_1, u_2) \quad (4.70)$$

is also a real and even function.

In any case the intensity in the plane of the image is an even function of coordinates, i.e., it has the centre of symmetry. The

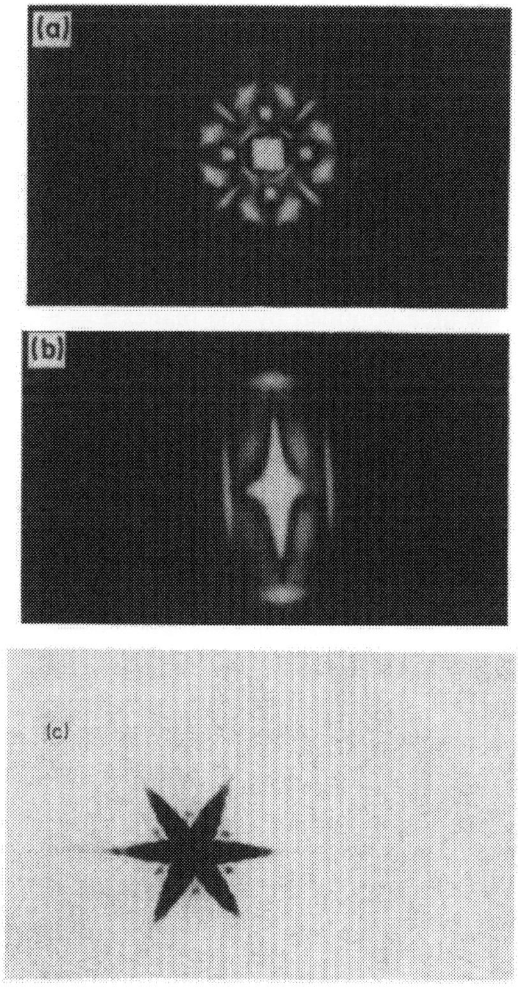

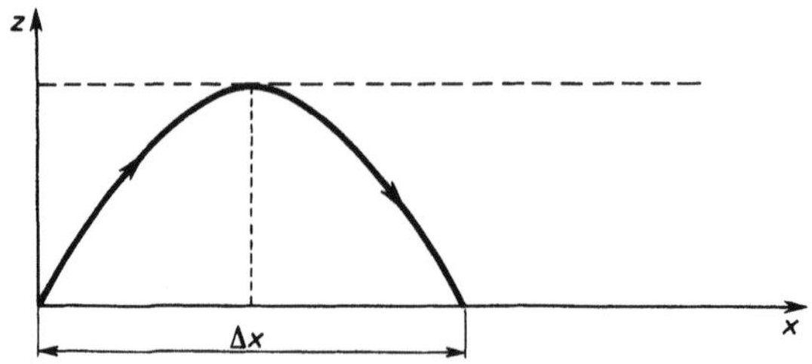

Figure 12.5

the point of emerging of the ray is $(0, z_0)$, while the angle formed by this ray and the x-axis is θ_1^0 and fulfils the condition

$$\cos^2\theta_1^0 \geq \frac{z_s^2}{2z_s^2 - z_0^2}. \tag{12.49}$$

Rays satisfying condition (12.49) in a medium with the refractive index (12.48) have a limited region of propagation. They propagate oscillating in the area confined by the planes $z = \pm z_s$. In this case in a non-homogeneous medium arises a waveguide channel of width $2z_s$. The justification for this is given below.

For a medium with the refractive index (12.48) the equation of ray (12.46) takes the form

$$\frac{dz}{dx} = \frac{\sqrt{\left[2 - \left(\dfrac{z}{z_s}\right)^2\right] - \left[2 - \left(\dfrac{z_0}{z_s}\right)^2\right]\cos^2\theta_1^0}}{\sqrt{2 - \left(\dfrac{z_0}{z_s}\right)^2}\cos\theta_1^0}. \tag{12.50}$$

This equation has the solution of the form

$$z = z_s\sqrt{2 - \left[2 - \left(\frac{z_0}{z_s}\right)^2\right]\cos^2\theta_1^0}\,\sin\left[\frac{x}{z_s\sqrt{2 - \left(\dfrac{z_0}{z_s}\right)^2}\cos\theta_1^0} + \right.$$

$$\left. + \arcsin\frac{\dfrac{z_0}{z_s}}{\sqrt{2 - \left[2 - \left(\dfrac{z_0}{z_s}\right)^2\right]\cos^2\theta_1^0}}\right]. \tag{12.51}$$

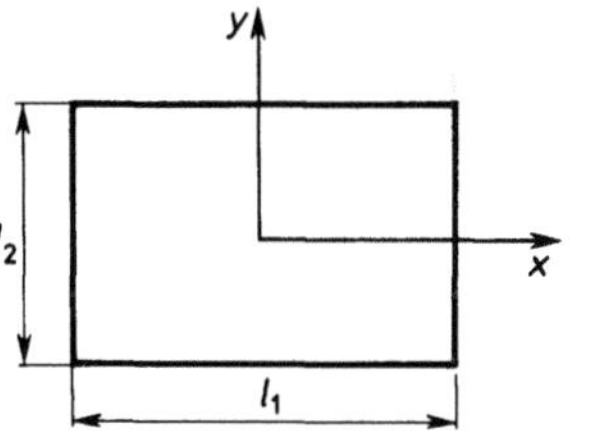

Figure 4.9

unit amplitude falling normally, the field of wave at a point lying very far from the screen can be shown in the form

$$U(x_0, y_0) = \frac{\exp(ikz)\exp\left[i\frac{k}{2z}(x_0^2 + y_0^2)\right]}{i\lambda z} \times$$

$$\times \iint_{-\infty}^{+\infty} t(x_1, y_1)\exp[-i(u_1x_1 + u_2y_1)]\,dx_1dy_1$$

$$= \frac{\exp\left[ikz + i\frac{k}{2z}(x_0^2 + y_0^2)\right]}{i\lambda z} \times$$

$$\times \operatorname{sinc}\left(\frac{l_1x_0}{\lambda z}\right)\operatorname{sinc}\left(\frac{l_2y_0}{\lambda z}\right)l_1l_2, \tag{4.72}$$

since

$$u_1 = \frac{kx_0}{z}, \quad u_2 = \frac{ky_0}{z}, \quad \operatorname{sinc}x = \frac{\sin\pi x}{\pi x}. \tag{4.73}$$

The radiation intensity, which is equal to the square of amplitude module, is

$$I = |U(x_0, y_0)|^2 = \frac{\lambda^2}{\pi^4}\sin^2\left(\frac{kl_1y_0}{2z}\right)\sin^2\left(\frac{kl_2y_0}{2z}\right)\frac{z^2}{x_0^2y_0^2}. \tag{4.74}$$

The plot of this function is shown in Fig. 4.10.

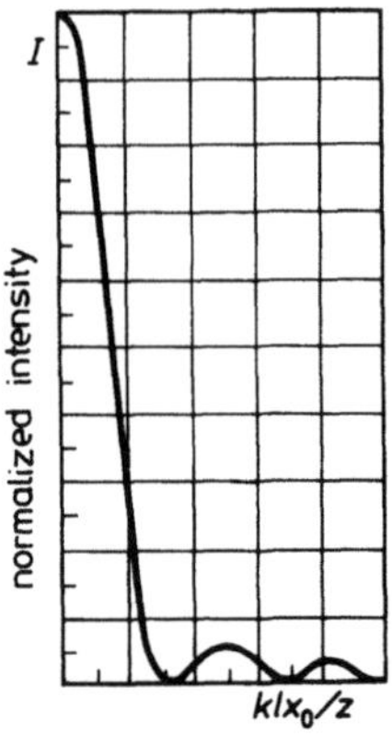

Figure 4.10

4.16. Diffraction Grating

A diffraction grating is a diaphragm characterized by the transmittance function

$$t(x, y) = \sum_{n=0}^{N} \operatorname{rect}\left(\frac{x_1 + n\Delta}{l_1}\right) \operatorname{rect}\left(\frac{y_1}{l_2}\right). \tag{4.75}$$

If the diaphragm is illuminated by an incident plane wave, the Fraunhofer diffraction field has the form

$$\begin{aligned} &U(x_0, y_0, z) \\ &= \frac{e^{ikz}}{i\lambda z} \exp\left[i\frac{k}{2z}(x_0^2 + y_0^2)\right] \iint_{-\infty}^{+\infty} \sum_{n=0}^{N} \operatorname{rect}\left(\frac{x_1 + n\Delta}{l_1}\right) \operatorname{rect}\left(\frac{y_1}{l_2}\right) \times \\ &\quad \times \exp[-i(u_1 x_1 + u_2 y_1)]\, dx_1\, dy_1, \end{aligned} \tag{4.76}$$

where $u_1 = kx_0/z$, $u_2 = ky_0/z$. Having calculated the integral, formula (4.76) can be written as

$$U(x_0, y_0, z)$$
$$= \frac{e^{ikz}}{i\lambda z} \exp\left[i\frac{k}{2z}(x_0^2 + y_0^2)\right] 4 \frac{\sin\left(\frac{kx_0 l_1}{2z}\right)\sin\left(\frac{ky_0 l_2}{2z}\right)}{\frac{kx_0}{2z}l_1 \frac{ky_0}{2z}l_2} \times$$
$$\times \sum_{n=0}^{N} e^{in(x_0 k\Delta/z)} \frac{l_1 l_2}{4} = 4z \frac{e^{ikz}}{i\lambda} \exp\left[i\frac{k}{2z}(x_0^2 + y_0^2)\right] \times$$
$$\times \frac{\sin\left(\frac{kx_0 l_1}{2z}\right)\sin\left(\frac{ky_0 l_2}{2z}\right)\sin\left(N\frac{x_0 k\Delta}{2z}\right)}{k^2 x_0 y_0 \sin\left(\frac{x_0 k\Delta}{2z}\right)} \times$$
$$\times \exp\left\{i\left[N\frac{x_0 k\Delta}{2z}\right]\right\}. \tag{4.77}$$

To obtain formula (4.77), we have used the property of the Fourier transform for a function with shifted argument

$$F\{g(x - a, y - b)\} = G(u_1, u_2)\exp[-i(u_1 a + u_2 b)]$$

and the expression for the sum of N terms of the geometric series

$$\sum_{n=0}^{N} a_1 q^n = \frac{a_1(q^N - 1)}{q - 1}.$$

The light intensity for the field (4.77) is

$$I_t(x_0, y_0)$$
$$= \frac{\lambda^2 z^2}{\pi^4 x_0^2 y_0^2} \sin^2\left(\frac{kx_0 l_1}{2z}\right)\sin^2\left(\frac{ky_0 l_2}{2z}\right) \frac{\sin^2\left(N\frac{x_0 k\Delta}{2z}\right)}{\sin^2\left(\frac{x_0 k\Delta}{2z}\right)}. \tag{4.78}$$

Comparing this result with the intensity distribution given by an individual rectangular aperture, we can state that

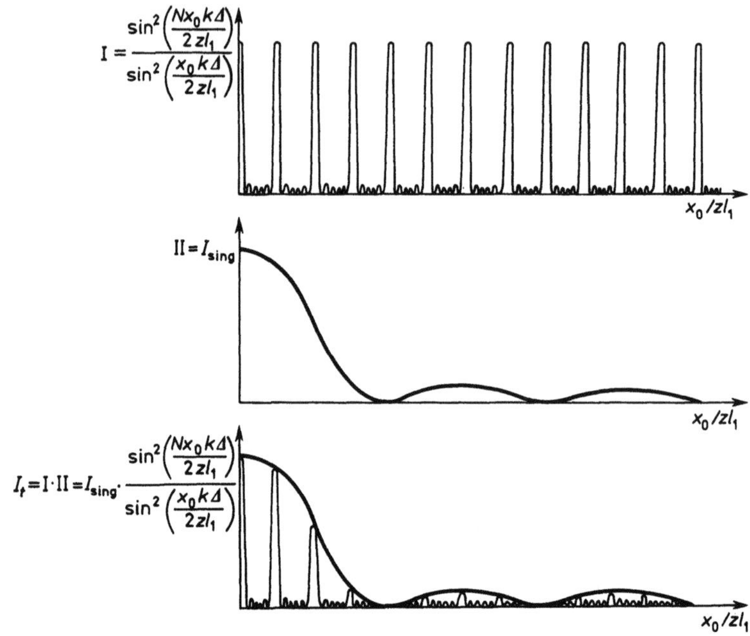

Figure 4.11

$$I_t = I_{\text{sing}} \frac{\sin^2\left(N\dfrac{x_0 k \Delta}{2z}\right)}{\sin^2\left(\dfrac{x_0 k \Delta}{2z}\right)}, \tag{4.79}$$

where I_{sing} is the intensity distribution for a singular rectangular aperture. Therefore the diagram I_t is a product of two curves: I and II (Fig. 4.11).

4.17. The Fourier–Bessel Transformation

A function g has the axial symmetry if it can be written as a function of variable $r = \sqrt{x^2 + y^2}$, i.e., if

$$g(x, y) = g(r). \tag{4.80}$$

Such functions play a big role in numerous optical system.

In the Fourier transform of this kind of function

$$G(u_1, u_2) = \iint\limits_{-\infty}^{+\infty} g(x, y)\exp[-\mathrm{i}(u_1 x + u_2 y)]\,\mathrm{d}x\,\mathrm{d}y \tag{4.81}$$

we can apply the symmetry properties of the function g, and go over to the coordinate system for the polar one. We use here the relations

$$x = r\cos\theta, \qquad y = r\sin\theta$$

and

$$u_1 = u\cos\varphi, \qquad u_2 = u\sin\varphi.$$

This enables us to write the integral (4.81) in the form

$$G(\rho, \varphi) = \int\limits_0^{2\pi} \mathrm{d}\theta \int\limits_0^{\infty} \mathrm{d}r\, rg(r)\mathrm{e}^{-\mathrm{i}ur\cos(\theta-\varphi)}. \tag{4.82}$$

Making use of the fact that

$$J_0(ru) = \frac{1}{2\pi}\int\limits_0^{2\pi} \mathrm{e}^{-\mathrm{i}ur\cos(\theta-\varphi)}\,\mathrm{d}\theta,$$

where J_0 is the Bessel function of the first kind and the zeroth-order, we can write the integral (4.82) as

$$G(u) = 2\pi \int\limits_0^{\infty} rg(r)J_0(ur)\,\mathrm{d}r. \tag{4.83}$$

Thus the Fourier transformation of a function with axial symmetry has also this kind of symmetry, and it can be found from formula (4.83).

This particular form of Fourier transform is known as the *Fourier–Bessel transformation* or the *Hankel transformation of the zeroth-order.*

The transformation inverse to (4.83) is

$$g(r) = 2\pi \int\limits_0^{\infty} G(u)J_0(ru)u\,\mathrm{d}u. \tag{4.84}$$

For a function revealing the axial symmetry there is no difference between simple and reverse transformation. All properties of the Fourier transformations remain true also for the Fourier–Bessel transformations.

4.18. The Fraunhofer Diffraction at a Circular Aperture

We shall consider diffraction at a circular aperture of radius R. In this case the transmittance function is of the form

$$t(r) = \operatorname{circ}\left(\frac{r}{R}\right), \tag{4.85}$$

where the circus function is defined as follows:

$$\operatorname{circ}(r) = \begin{cases} 1 & \text{for} \quad r \leq 1, \\ 0 & \text{for} \quad r > 1. \end{cases}$$

The axial symmetry allows us to write the Fraunhofer integral in the form

$$\varphi(r_0) = \frac{e^{ikz}}{i\lambda z} \exp\left[i\frac{k}{2z}(x_0^2 + y_0^2)\right] \times$$

$$\times\, 2\pi \int_0^\infty \psi(r_1) \operatorname{circ}\left(\frac{r_1}{R}\right) J_0(r_1 u) r_1 \, dr_1, \tag{4.86}$$

where

$$u = \sqrt{\left(\frac{kx_0}{z}\right)^2 + \left(\frac{ky_0}{z}\right)^2} = \frac{r_0}{z} k,$$

and r_0 is the length of the position vector lying in the plane of observation.

If on the diffracting screen with a circular aperture falls a plane wave of unit amplitude ($\psi(r_1) = 1$), the field in the plane of observation will be expressed by the formula

$$\varphi(r_0) = \frac{e^{ikz}}{i\lambda z} \exp\left(i\frac{kr_0^2}{2z}\right) 2\pi R^2 \frac{J_1(Ru)}{Ru}. \tag{4.87}$$

Formula (4.87) was obtained with the use of the relations

$$\int_0^R \operatorname{circ}\left(\frac{r_1}{R}\right) J_0(kr_1) r_1 \mathrm{d}r_1 = \frac{1}{k^2} \int_0^{kR} J_0(\xi)\xi \mathrm{d}\xi,$$

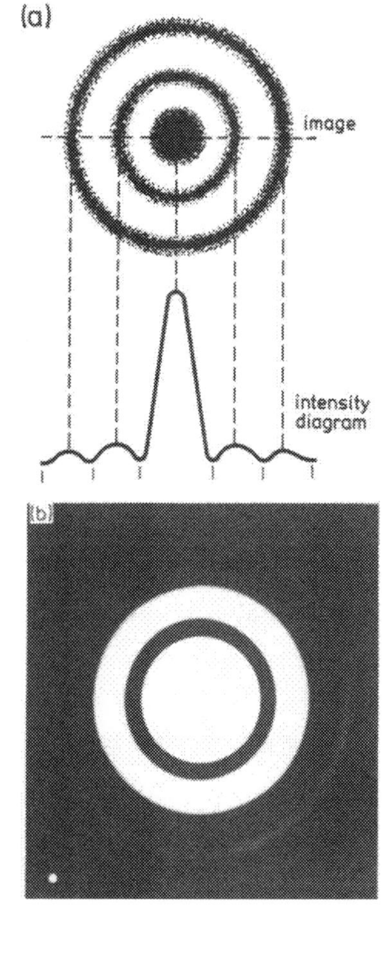

$$\int_0^x \xi J_0(\xi)\mathrm{d}\xi = xJ_1(x),$$

where J_1 is the first-order Bessel function. The intensity distribution in the plane of observation is

$$I(r_0) = R^4 \frac{k^2}{z^2}\left|\frac{J_1(uR)}{uR}\right|^2 = \frac{k^2R^4}{z^2}\left|\frac{J_1(r_0Rk/z)}{r_0Rk/z}\right|^2. \tag{4.88}$$

The diagram of the intensity distribution is the *Airy curve* (Fig. 4.12). The radius of the first dark circle on the diffraction pattern is

$$r_0^{\text{dark}} = 1.22\frac{\lambda z}{2R} = 0.61\frac{z\lambda}{R}.$$

4.19. The Fresnel Diffraction at a Square Aperture

Assume that on a square aperture of side l falls a monochromatic plane wave. The screen is characterized by the transmittance function

$$t(x_1, y_1) = \mathrm{rect}\left(\frac{x_1}{l}\right)\mathrm{rect}\left(\frac{y_1}{l}\right). \tag{4.89}$$

Under these assumptions the Fresnel integral may be written as

$$\begin{aligned} U(x_0, y_0) &= \frac{\mathrm{e}^{\mathrm{i}kz}}{\mathrm{i}\lambda z}\iint_{-l/2}^{l/2} \exp\left\{\mathrm{i}\frac{k}{2z}[(x_1 - x_0)^2 + (y_0 - y_1)^2]\right\}\mathrm{d}x_1\mathrm{d}y_1 \\ &= \frac{\mathrm{e}^{\mathrm{i}kz}}{\mathrm{i}\lambda z}\int_{-l/2}^{l/2} \exp\left[\mathrm{i}\frac{k}{2z}(x_0 - x_1)^2\right]\mathrm{d}x_1 \times \\ &\times \int_{-l/2}^{l/2} \exp\left[\mathrm{i}\frac{k}{2z}(y_0 - y_1)^2\right]\mathrm{d}y_1. \end{aligned} \tag{4.90}$$

Using the following substitution of variables:

$$\sqrt{\frac{\pi}{2}}\xi = \sqrt{\frac{k}{2z}}(x_0 - x_1), \quad \sqrt{\frac{\pi}{2}}\eta = \sqrt{\frac{k}{2z}}(y_0 - y_1),$$

the field behind the screen may be written in the form

$$U(x_0, y_0) = \frac{e^{ikz}}{i\lambda z}\frac{\pi z}{k}\int_{\xi_1}^{\xi_2}\exp\left(i\frac{\pi}{2}\xi^2\right)d\xi\int_{\eta_1}^{\eta_2}\exp\left(i\frac{\pi}{2}\eta^2\right)d\eta, \quad (4.91)$$

where

$$\xi_1 = +\sqrt{\frac{k}{\pi z}}\left(\frac{l}{2} + x_0\right), \quad \eta_1 = +\sqrt{\frac{k}{\pi z}}\left(\frac{l}{2} + y_0\right),$$

$$\xi_2 = -\sqrt{\frac{k}{\pi z}}\left(\frac{l}{2} - x_0\right), \quad \eta_2 = -\sqrt{\frac{k}{\pi z}}\left(\frac{l}{2} - y_0\right).$$

The above integrals may also be written in another form

$$\int_{\xi_1}^{\xi_2}\exp\left(i\frac{\pi}{2}\xi^2\right)d\xi = \int_{\xi_1}^{\xi_2}\cos\frac{\pi}{2}\xi^2 d\xi + i\int_{\xi_1}^{\xi_2}\sin\frac{\pi}{2}\xi^2 d\xi$$
$$= [C(\xi_2) - C(\xi_1)] + i[S(\xi_2) - S(\xi_1)],$$

where

$$C(\alpha) = \int_0^\alpha \cos\frac{\pi t^2}{2}dt, \quad S(\alpha) = \int_0^\alpha \sin\frac{\pi t^2}{2}dt.$$

These are Fresnel's integrals, whose values are tabulated. On account on the above remark the field beyond the screen is

$$U(x_0, y_0) = \frac{e^{ikz}}{2i}\left\{C(\xi_2) - C(\xi_1) + i[S(\xi_2) - S(\xi_1)]\right\} \times$$
$$\times \left\{C(\eta_2) - C(\eta_1) + i[S(\eta_2) - S(\eta_1)\right\}. \quad (4.92)$$

For small values of variable z, which implies a large value of $k/\pi z$, we can calculate the field $U(x_0, y_0)$ taking as the integration limits for ξ_j and η_j the values from the Tables (4.93a) and (4.93b); these values are different for different regions of variability of coordinates x_0 and y_0.

x_0 \ y_0	$< -\frac{l}{2} <$	
ξ_1 \ η_1	$-\infty$	$+\infty$
$C(\xi_1)$ \ $C(\eta_1)$	$-\frac{1}{2}$	$+\frac{1}{2}$
$S(\xi_1)$ \ $S(\eta_1)$	$-\frac{1}{2}$	$+\frac{1}{2}$

(4.93a)

x_0 \ y_0	$< +\frac{l}{2} <$	
ξ_2 \ η_2	$-\infty$	$+\infty$
$C(\xi_2)$ \ $C(\eta_2)$	$-\frac{1}{2}$	$+\frac{1}{2}$
$S(\xi_2)$ \ $S(\eta_2)$	$-\frac{1}{2}$	$+\frac{1}{2}$

(4.93b)

In these tables there are placed also the values of Fresnel's integrals corresponding to the appropriate integration limits. Putting the values (4.93) in the expression (4.92) implies finally

$$U(x_0, y_0) = e^{ikz} \operatorname{rect}\left(\frac{x_0}{l}\right) \operatorname{rect}\left(\frac{y_0}{l}\right), \tag{4.94}$$

which corresponds to geometric optics.

The decomposition of the field $U(x_0, y_0)$ in the limit case of the Fresnel diffraction is a product of the transmittance function for the screen (called also the *pupil function*) and the field value equal to its value in the case of free propagation. The image of the aperture produced in the Fresnel diffraction is identical to that obtained with the use of laws of geometric optics.

4.20. Sinusoidal Diffraction Amplitude Grating

The transmittance function in this case has the form

$$t(x, y) = \left[\frac{1}{2} + \frac{m}{2}\cos(2\pi f_0 x)\right]\mathrm{rect}\left(\frac{x}{l}\right)\mathrm{rect}\left(\frac{y}{l}\right), \tag{4.95}$$

and its diagram is shown in Fig. 4.13.

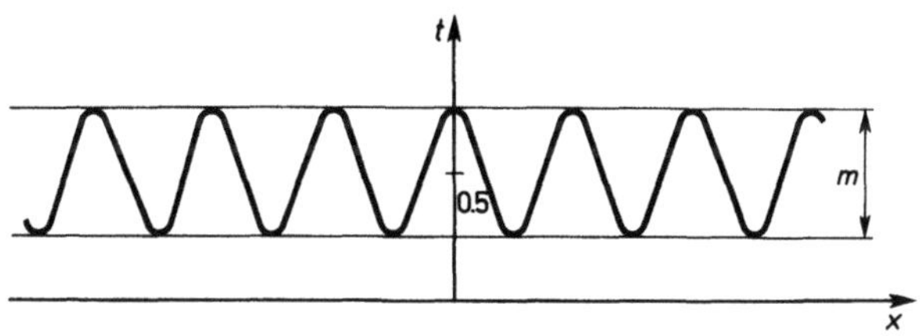

Figure 4.13

The structure of the grating is limited by the square of side l. The parameter m determines the difference between the maximum value and the minimum value of the transmittance, and f_0 stands for grating frequency. If the screen is illuminated by a plane monochromatic wave of unit amplitude falling perpendicularly on the screen, the field distribution on the aperture is described by the transmittance function $t(x, y)$. The Fraunhofer diffraction field is the Fourier transform of the function $t(x, y)$. This transform can be found from the equalities:

$$\iint_{-\infty}^{+\infty} \frac{1}{2} e^{i(k_1 x + k_2 y)}\,dx\,dy = 2\pi^2 \delta(k_1)\delta(k_2),$$

$$\iint_{-\infty}^{+\infty} \frac{m}{4} e^{i2\pi f_0 x} e^{i(k_1 x + k_2 y)} dx dy = m\pi^2 \delta(2\pi f_0 + k_1)\delta(k_2),$$

$$\iint_{-\infty}^{+\infty} \frac{m}{4} e^{-i2\pi f_0 x} e^{i(k_1 x + k_2 y)} dx dy = m\pi^2 \delta(k_1 - 2\pi f_0)\delta(k_2), \quad (4.96)$$

$$\iint_{-\infty}^{+\infty} \operatorname{rect}\left(\frac{x}{l}\right) \operatorname{rect}\left(\frac{y}{l}\right) e^{i(k_1 x + k_2 y)} dx dy$$

$$= l^2 \int_{-1/2}^{1/2} e^{ik_1 l \xi} d\xi \int_{-1/2}^{1/2} e^{ik_2 l \eta} d\eta = 4l^2 \frac{\sin(k_1 l/2)\sin(k_2 l/2)}{(k_1 l/2)(k_2 l/2)}.$$

The equalities given above, the convolution theorem, and the properties of Dirac's function imply

$$\iint_{-\infty}^{+\infty} t(x, y) e^{i(k_1 x + k_2 y)} dx dy = \iint_{-\infty}^{+\infty} [2\pi^2 \delta(k_1 - k_{\mathrm{I}})\delta(k_2 - k_{\mathrm{II}}) +$$

$$+ m\pi^2 \delta(2\pi f_0 + k_1 - k_{\mathrm{I}})\delta(k_2 - k_{\mathrm{II}}) +$$

$$+ m\pi^2 \delta(k_1 - k_{\mathrm{I}} - 2\pi f_0)\delta(k_2 - k_{\mathrm{II}})] \times$$

$$\times 4l^2 \left[\frac{\sin(k_{\mathrm{I}} l/2)}{k_{\mathrm{I}} l/2} \frac{\sin(k_{\mathrm{II}} l/2)}{k_{\mathrm{II}} l/2}\right] dk_{\mathrm{I}} dk_{\mathrm{II}}$$

$$= 4l^2 \left[2\pi^2 \frac{\sin(k_1 l/2)}{k_1 l/2} \frac{\sin(k_2 l/2)}{k_2 l/2} + \right.$$

$$+ \frac{m\pi^2 \sin(2\pi f_0 + k_1) l/2}{(2\pi f_0 + k_1) l/2} \frac{\sin(k_2 l/2)}{k_2 l/2} +$$

$$\left. + \frac{m\pi^2 \sin(k_2 l/2)}{k_2 l/2} \frac{\sin(-2\pi f_0 + k_1) l/2}{(-2\pi f_0 + k_1) l/2} \right]. \quad (4.97)$$

Assuming that (at points of the aperture) $U(x, y) = 1$, and having in view that the role of k_1 and k_2 was taken over by the variables $u_1 = kx_0/z_0$ and $u_2 = ky_0/z_0$, we can write the Fraunhofer integral

$$U(x_0, y_0, z_0) = \frac{e^{ikz_0}}{i\lambda z_0} \exp\left[i\frac{k}{2z_0}(x_0^2 + y_0^2)\right] \times$$

$$\times \int_{-\infty}^{+\infty} t(x, y)U(x, y)\exp[-i(u_1x + u_2y)]\,dx\,dy \quad (4.98)$$

in the form

$$U(x_0, y_0, z_0) = \frac{e^{ikz_0}}{i\lambda z_0} \exp\left[i\frac{k}{2z}(x_0^2 + y_0^2)\right] 8\pi^2 l^2 \frac{\sin(kly_0/2z_0)}{kly_0/2z_0} \times$$

$$\times \left[\frac{\sin(klx_0/2z_0)}{klx_0/2z_0} + \frac{m}{2}\frac{\sin(2\pi f_0 + kx_0/z_0)l/2}{(2\pi f_0 + kx_0/z_0)l/2} + \right.$$

$$\left. + \frac{m}{2}\frac{\sin(kx_0/z_0 - 2\pi f_0)l/2}{(kx_0/z_0 - 2\pi f_0)l/2}\right]. \quad (4.99)$$

The expression obtained presents the diffraction field for a sinusoidal diffraction amplitude grating. Calculating the intensity of the diffraction field we neglect all the mixed products among the components of the sum (4.99):

$$I(x_0, y_0, z_0)$$

$$\approx 64\left(\frac{l^2\pi^2}{z_0\lambda}\right)^2 \left[\frac{\sin^2(klx_0/2z_0)}{(klx_0/2z_0)^2} + \frac{m^2}{4}\frac{\sin^2(2\pi f_0 + kx_0/z_0)l/2}{[(2\pi f_0 + kx_0/z_0)l/2]^2} + \right.$$

$$\left. + \frac{m^2}{4}\frac{\sin^2(kx_0/z_0 - 2\pi f_0)l/2}{[(kx_0/z_0 - 2\pi f_0)l/2]^2}\right] \frac{\sin^2(kly_0/2z_0)}{(kly_0/2z_0)^2}. \quad (4.100)$$

The diagram of intensity (Fig. 4.14) has one central and two side maxima. The central maximum gives the zeroth-order of the diffraction pattern, the side maxima – the first-order of the Fraunhofer diffraction pattern. The distance between the maxima is $\lambda f_0 z_0$.

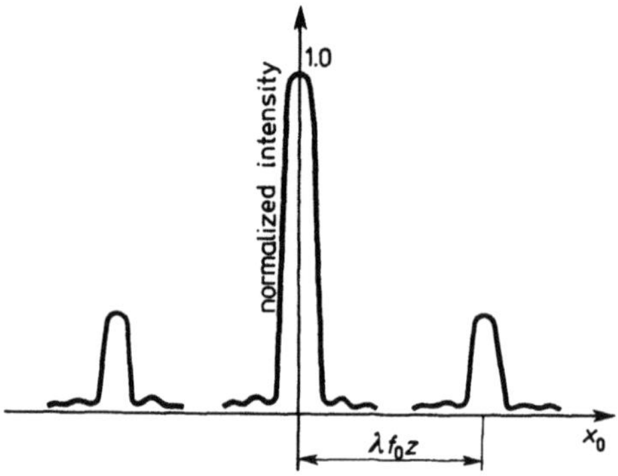

Figure 4.14

4.21. Diffraction at a Plane Screen as a Boundary Problem

The Kirchhoff theory of diffraction is in agreement with experiment but it contains some internal inconsistencies, which make us to seek solutions mathematically more correct. The main difficulty in the theory is that within its frame we should know the boundary conditions for field intensities as well as for their normal derivatives. From the theory of potential it is known that such a formulation of the boundary problem is not correct.

The Kirchhoff integral cannot be regarded as a solution of the boundary problem, because it solves only the so-called *jump problem.*

The application of the Kirchhoff integral gives such solutions that the field in front of and behind the screen (Fig. 4.15) differ by the value of the incident field U_0

$$U_{\mathrm{I}} - U_{\mathrm{II}} = U_0, \qquad \frac{\partial U_{\mathrm{I}}}{\partial n} - \frac{\partial U_{\mathrm{II}}}{\partial n} = \frac{\partial U_0}{\partial n}. \tag{4.101}$$

The diffraction problem can be formulated differently – namely as a boundary problem. The correct formulation of the boundary

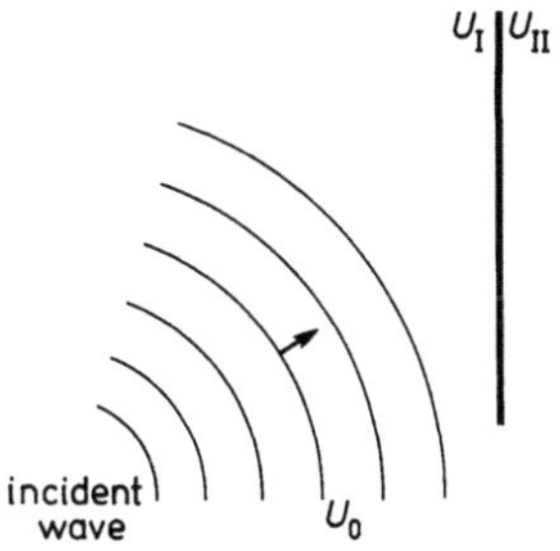

Figure 4.15

conditions for the Helmholz equation requires choosing one of two possibilities:

(1) $U = 0$ – the field vanishes on the screen – the *Dirichlet condition*,

(2) $\partial U/\partial n = 0$ – the normal derivative on the field vanishes on the screen – the *Neumann condition.*

The Rayleigh–Sommerfeld solution is an example of solving the diffraction problem by the boundary conditions formulated above. However, the results have a limited region of application for the plane screens.

Take a plane diffracting screen Σ. If Sommerfeld's conditions are fulfilled, the field at the points behind the screen with a deflecting slit will be expressed by

$$U(P) = \frac{1}{4\pi}\iint_{\Sigma}\left(\frac{\partial U}{\partial n}G - U\frac{\partial G}{\partial n}\right)\mathrm{d}f, \tag{4.102}$$

where G is the Green function. The choice of this function is to some extend optional. We can choose it so that either the function itself equals zero $G|_\Sigma = 0$ or $\partial G/\partial n|_\Sigma = 0$. There is no contradiction between these conditions and the validity of the above formula.

Sommerfeld has shown that the Green function with these properties really exists. He assumed that the function G is generated not only by a point source placed at point P_0 but also by the second source placed at point $\tilde{P}_0$, which is a mirror reflection of the point P_0 and lies on the other side of the screen.

We can distinguished two cases:

1. A source at point P_0 emits a wave of wavelength λ and a source at point $\tilde{P}_0$ – an identical wave but shifted in phase by π. In this case the Green function will take the form

$$G_-(Q) = \frac{e^{ikr_{01}}}{r_{01}} - \frac{e^{ik\tilde{r}_{01}}}{\tilde{r}_{01}}, \tag{4.103}$$

where r_{01} stands for the distance between points P_0 and Q (Fig. 4.16), $\tilde{r}_{01}$ – for the distance between $\tilde{P}_0$ and Q.

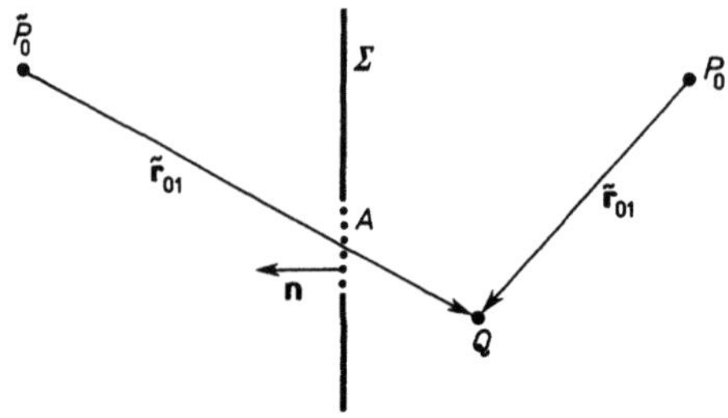

Figure 4.16

For points $Q \in \Sigma$ the following relations are held:

$$r_{01} = \tilde{r}_{01}, \quad \cos(\mathbf{n}, \mathbf{r}_{01}) = -\cos(\mathbf{n}, \mathbf{r}_{01}),$$

and consequently

$$G_-(Q)|_\Sigma = 0.$$

Thus the function G_- equals zero at all points of the plane screen Σ.

Substituting G_- into (4.102), we get

$$U(P_0) = \frac{1}{4\pi} \iint_\Sigma U(Q) \frac{\partial G_-}{\partial n} \, df. \tag{4.104}$$

Now putting in (4.104)

$$U(Q) = U_0 \quad \text{for} \quad Q \in A \text{ (aperture)},$$

$$U(Q) = 0 \quad \text{for} \quad Q \in \text{screen},$$

the state of field at point P_0 will be expressed by the formula

$$U(P_0) = \frac{1}{4\pi} \iint_A U_0(Q) \frac{\partial G_-}{\partial n} \, \mathrm{d}f. \tag{4.105}$$

2. In the case where the two sources emit in phase, the Green function

$$G_+(Q) = \frac{\mathrm{e}^{ikr_{01}}}{r_{01}} + \frac{\mathrm{e}^{ik\tilde{r}_{01}}}{\tilde{r}_{01}} \tag{4.106}$$

for $Q \in \Sigma$ satisfies the condition

$$\left. \frac{\partial G_+(Q)}{\partial n} \right|_{\Sigma} = 0. \tag{4.107}$$

Substituting the Green function (4.106) into formula (4.102), we obtain

$$U(P_0) = -\frac{1}{4\pi} \iint_{\Sigma} G_+(Q) \frac{\partial U}{\partial n} \, \mathrm{d}f. \tag{4.108}$$

If in (4.108) consider the condition

$$\frac{\partial U(Q)}{\partial n} = 0 \quad \text{(on screen)},$$

$$\frac{\partial U(Q)}{\partial n} = \frac{\partial U_0(Q)}{\partial n} \quad \text{(at aperture)},$$

the field at point P_0 will be given by the formula

$$U(P_0) = -\frac{1}{4\pi} \iint_A G_+(Q) \frac{\partial U_0}{\partial n} \, \mathrm{d}f. \tag{4.109}$$

In both cases (1 and 2) in order to know the field at point P_0 it is enough to know either the state of the field or its normal derivative at all points of the aperture and the screen, which remains in agreement with the general theory of differential equations.

4.22. The Rayleigh–Sommerfeld Diffraction Formulas

Considering that for $Q \in \Sigma$ hold the following equalities:

$$r_{01} = \tilde{r}_{01}, \quad \cos(\mathbf{n}, \mathbf{r}_{01}) = -\cos(\mathbf{n}, \tilde{\mathbf{r}}_{01}),$$

then in case 1 we obtain

$$G_-(Q)|_\Sigma = 0,$$

$$\left.\frac{\partial G_-(Q)}{\partial n}\right|_\Sigma = 2\cos(\mathbf{n},\mathbf{r}_{01})\left(ik - \frac{1}{r_{01}}\right)\frac{e^{ikr_{01}}}{r_{01}} \approx 2ki\frac{e^{ikr}}{r_{01}}\cos(\mathbf{n},\mathbf{r}_{01}),$$

while in case 2 we get

$$\left.\frac{\partial G_+(Q)}{\partial n}\right|_\Sigma = 0, \quad G_+(Q)\Big|_\Sigma = 2\frac{\exp(ikr_{01})}{r_{01}}.$$

Suppose that the aperture is illuminated by a spherical wave

$$U(Q) = A\frac{\exp(ik\rho)}{\rho}.$$

Then in the first case we get

$$U(P_0) = -\frac{A}{i\lambda}\iint\limits_A \frac{\exp[ik(r_{01}+\rho)]}{r_{01}\rho}\cos(\mathbf{n},\mathbf{r}_{01})\mathrm{d}f, \qquad (4.110)$$

and in the second

$$U(P_0) = \frac{A}{i\lambda}\iint\limits_A \frac{\exp[ik(r_{01}+\rho)]}{r_{01}\rho}\cos(\mathbf{n},\boldsymbol{\rho})\mathrm{d}f. \qquad (4.111)$$

Formulas (4.110) and (4.111) are known as the *Rayleigh–Sommerfeld diffraction formulas.* The expressions

$$U(P_0) = \frac{1}{4\pi}\iint U_0(Q)\frac{\partial G_-}{\partial n}\mathrm{d}f$$

$$= -\frac{1}{i\lambda}\iint U_0(Q)\frac{\exp(ikr_{01})}{r_{01}}\cos(\mathbf{n},\mathbf{r}_{01})\mathrm{d}f, \qquad (4.112)$$

$$U(P_0) = -\frac{1}{4\pi}\iint G_+ \frac{\partial U_0(Q)}{\partial n}\,\mathrm{d}f$$

$$= \frac{1}{\mathrm{i}\lambda}\iint \frac{\exp(\mathrm{i}kr_{01})}{r_{01}}\frac{\partial U_0(Q)}{\partial n}\,\mathrm{d}f \tag{4.113}$$

represent the superposition principle. The integral (4.112) can be written as

$$U(P_0) = \iint_A h(P_0, Q)\,U(Q)\,\mathrm{d}f, \tag{4.114}$$

where the function $h(P_0, Q)$ defined by

$$h(P_0, Q) = -\frac{1}{\mathrm{i}\lambda}\frac{\exp(\mathrm{i}kr_{01})}{r_{01}}\cos(\mathbf{n}, \mathbf{r}_{01})$$

is the so-called *weight function.*

4.23. Transfer Function for the Fresnel–Huygens Diffraction Formula

As shown in Sec. 4.7, the Huygens–Fresnel formula may be written in the form

$$U(P_0) = -\frac{ik}{2\pi}\iint_A U(Q)\frac{\mathrm{e}^{ikr}}{r}\,\mathrm{d}f. \tag{4.115}$$

Considering the decomposition of a sphere wave in plane waves

$$\frac{\mathrm{e}^{ikr}}{r} = -\frac{\mathrm{i}}{2\pi}\iint_{-\infty}^{+\infty} \frac{\exp\left[\mathrm{i}\left(u_1 x + u_2 y + z\sqrt{k^2 - u_1^2 - u_2^2}\right)\right]}{\sqrt{k^2 - u_1^2 - u_2^2}}\,\mathrm{d}u_1\mathrm{d}u_2, \tag{4.116}$$

the expression

$$\frac{\mathrm{i}\exp\left[\mathrm{i}z\sqrt{k^2 - (u_1^2 + u_2^2)}\right]}{\sqrt{k^2 - (u_1^2 + u_2^2)}}$$

may be treated as a Fourier transform of a sphere wave. If $g(u_1, u_2)$ denotes the appropriate transform of the function $U(Q)$, then because of a convolution character of the function (4.115), we can write

$$U(P_0) = \frac{k}{2\pi} \iint\limits_{-\infty}^{+\infty} g(u_1, u_2) \frac{\exp\left(iz\sqrt{k^2 - u_1^2 - u_2^2}\right)}{\sqrt{k^2 - u_1^2 - u_2^2}} \times$$
$$\times \exp\left[i(u_1 x + u_2 y)\right] du_1 du_2. \tag{4.117}$$

The Rayleigh–Sommerfeld solution gives for the state of the field at points of the plane $z > 0$ (in the case of a plane screen placed in the plane $z = 0$) the expression

$$U(P_0) = \frac{1}{2\pi} \iint\limits_{-\infty}^{+\infty} U(x, y, 0) \frac{\partial}{\partial z} \frac{e^{ikr}}{r} dx dy, \tag{4.118}$$

since in this case the normal to the surface of the screen has the direction and sense of the z-axis.

Decomposing a wave in plane waves and having in view the properties of convolution, we can write

$$U(P_0) = \frac{1}{2\pi} \iint\limits_{-\infty}^{+\infty} g(u_1, u_2) \exp\left(iz\sqrt{k^2 - u_1^2 - u_2^2}\right) \times$$
$$\times \exp\left[i(xu_1 + yu_2)\right] du_1 du_2. \tag{4.119}$$

The function $\exp\left(iz\sqrt{k^2 - u_1^2 - u_2^2}\right)$ is the Fourier transform of the expression $(\partial/\partial z)(e^{ikr}/r)$, and $g(u_1, u_2)$ – the transform of the wave function $U(x, y, 0)$:

$$\exp\left[iz\sqrt{k^2 - (u_1^2 + u_2^2)}\right]$$
$$= \iint\limits_{-\infty}^{+\infty} \left[\frac{\partial}{\partial z}\left(\frac{e^{ikr}}{r}\right)\right] \exp\left[-i(u_1 x + u_2 y)\right] dx dy, \tag{4.120}$$

$$g(u_1, u_2) = \iint\limits_{-\infty}^{+\infty} U(x, y, 0)\exp[-\mathrm{i}(u_1 x + u_2 y)]\,\mathrm{d}x\mathrm{d}y. \quad (4.121)$$

Formula (4.117) takes the form (4.119) if

$$(u_1^2 + u_2^2) \ll k^2. \quad (4.122)$$

From the considerations given above we can draw the conclusion:

For the validity of the Huygens–Fresnel formula the spatial spectrum of the wave $U(x, y, 0)$ is required to be converged in the region (4.122).

4.24. The Transition of a Plane Monochromatic Wave through a Prism and the Simplest Focusing System

Let us consider a situation where a plane wave is incident perpendicularly on a face of a prism with vertical angle ε and made of material with the refractive index n. According to the refraction law, the wave undergoes both a change of the propagation direction and the prism phase change

$$\Delta = \frac{2\pi}{\lambda_0}(n + 1)\varepsilon x. \quad (4.123)$$

This expression is obtained from the analysis of the situation presented in Fig. 4.17 under the assumption of small value of the

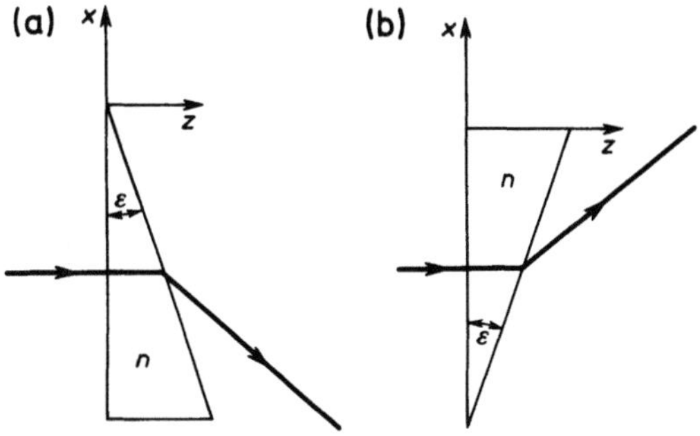

Figure 4.17

vertical angle ε and regarding the fact that the paths covered by the propagating wave along z-direction are different in glass and in air in dependence on the x-th coordinate.

From a mathematical point of view the passage of a plane wave $u \sim \exp\left[i(2\pi/\lambda_0)z\right]$ through a prism is equivalent to multiplication of the formula defining its original form by factor $\exp(-i\Delta)$ in case a (Fig. 4.17) or by factor $\exp(i\Delta)$ – in case b. It is justified, since in the first case the x-th component of the wave vector is negative, in the second case – positive. The simplest focusing system is a sphere lens formed by two spherical caps of curvature radii R_1 and R_2. The base area of these caps is much larger than the square of wavelength of radiation passing through the lens. Since the permittivity of the material of the lens is different from the permittivity of air, the phase velocity of the wave also changes with change of medium.

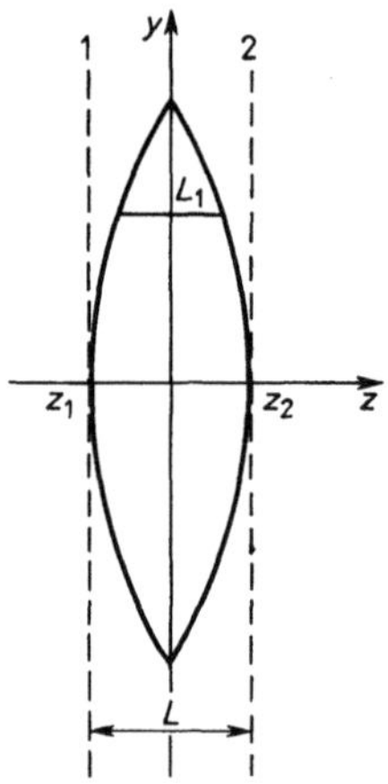

Figure 4.18

Let us consider a situation where on a lens placed in the xy-plane of the Cartesian coordinate system falls a wave described by a function $g_1(x, y)$ and propagating along the z-axis (Fig. 4.18). Since the thickness of lens is rather small, we assume that a ray which crosses the input plane 1 at an arbitrary point crosses the output plane 2 at the point of the same coordinates x, y.

In the paraxial approximation the coefficient of amplitude

transmittance of a wave can be treated as a constant quantity, and only the phase change between planes 1 and 2 is taken into account. This enables us to write the relation between the wave entering the lens $g_1(x, y)$ and the wave leaving the lens $g_2(x, y)$ as

$$g_2(x, y) = g_1(x, y)\mathrm{e}^{ikn(x,y)L}, \tag{4.124}$$

where L is the distance between planes 1 and 2, and $n(x, y)$ – the refractive index of a medium confined between these planes. For an arbitrary point of coordinates (x, y, z) lying in the area between plane 1 $(z = z_1)$ and 2 $(z = z_2)$, we can write

$$n(x, y)L = L_2(x, y) + nL_1(x, y),$$

where

$$L_1(x, y) + L_2(x, y) = |z_1| + |z_2| = L$$

is the distance between planes 1 and 2, $L_1(x, y)$ – the thickness of the lens at any point, $|z_1|$ and $|z_2|$ – the distances from the plane $z = 0$ in which the centre of the lens lies to the input and output places respectively.

In view of this the expression (4.124) may be written as

$$g_2(x, y) = g_1(x, y)\mathrm{e}^{ikL}\mathrm{e}^{ik(n-1)L_1(x,y)}. \tag{4.125}$$

According to Fig. 4.19 we can write the equality

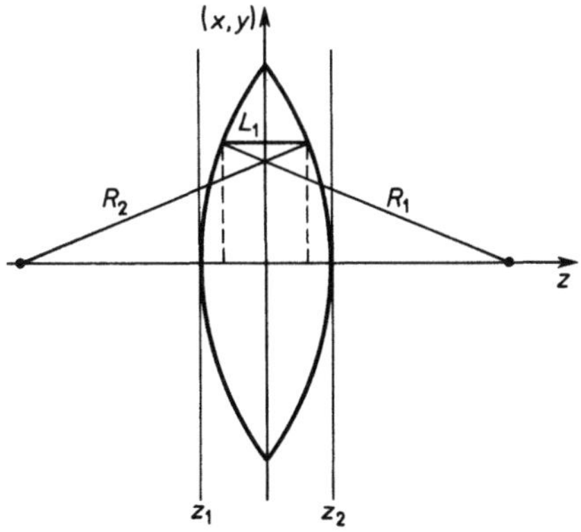

Figure 4.19

$$L_1(x, y) = |z_1| - \left(R_1 - \sqrt{R_1^2 - x^2 - y^2}\right) + |z_2| - \\ - \left(R_2 - \sqrt{R_2^2 - x^2 - y^2}\right), \qquad (4.126)$$

which is replaced by the approximate expression

$$L_1(x, y) = |z_1| + |z_2| - (x^2 + y^2)\left(\frac{1}{2R_1} + \frac{1}{2R_2}\right). \qquad (4.127)$$

Replacing in (4.125) $L_1(x, y)$ by (4.127), we are finally arriving at an explicit relationship between the entering and leaving waves

$$g_2(x, y) = g_1(x, y)\exp\left[-ik\left(\frac{x^2 + y^2}{2f} - nL\right)\right], \qquad (4.128)$$

where

$$\frac{1}{f} = (n - 1)\left(\frac{1}{R_1} + \frac{1}{R_2}\right)$$

is the power of the lens and f – its focal length.

In order to give the physical interpretation of the transformation (4.128) performed by the lens on an incident wave front, we compare (4.128) with an appropriate approximation for a spherical wave propagating from the origin of the coordinate system. In the paraxial approximation (i.e. for $|z| = f \gg x$ and y) this wave can be presented as

$$\frac{\exp\left(ik\sqrt{x^2 + y^2 + z^2}\right)}{\sqrt{x^2 + y^2 + z^2}} \cong \frac{1}{f}\mathrm{e}^{ikf}\exp\left[ik\frac{(x^2 + y^2)}{2f}\right]. \qquad (4.129)$$

The first exponential factor on the right-hand side of (4.129) describes a constant phase growth, the second a phase change of the spherical wave in the paraxial approximation.

Comparing (4.129) with (4.128), we can state that if a plane wave (i.e. $g_1(x, y) = \mathrm{e}^{ikz}$) is incident on a convergent lens ($f > 0$), behind the lens this wave is spherical converging to the point of intersection of the focal plane with the optical axis of the lens. On the other hand, in the case of divergent lens ($f < 0$) the exponential factor in

(4.128) will describe a divergent spherical wave coming from the point on the left side of the lens at a distance equal to f.

Applying the Fresnel–Kirchhoff diffraction formula (4.31), in which the paraxial approximation has been taken into account, we obtain the wave field in the focal plane of the lens

$$g(x_f, y_f) = \frac{\mathrm{e}^{\mathrm{i}kf}}{\mathrm{i}\lambda f} \exp\left[\mathrm{i}\frac{k}{2f}(x_f^2 + y_f^2)\right] \times$$
$$\times \iint \mathrm{d}x\mathrm{d}y g_2(x, y) \exp\left[\mathrm{i}\frac{k}{2f}(x^2 + y^2)\right] \times$$
$$\times \exp\left[-\mathrm{i}\frac{k}{f}(x_f x + y_f y)\right], \tag{4.130}$$

where x_f, y_f denote the coordinates in the focal plane. Combining (4.128) with (4.130), we can rewrite (4.130) as

$$g(x_f, y_f) = \frac{1}{\mathrm{i}\lambda f} \exp\left[\mathrm{i}\frac{k}{2f}(x_f^2 + y_f^2)\right] \exp[\mathrm{i}(kf + nL)] \times$$
$$\times \iint_{-\infty}^{+\infty} g_1(x, y) \exp[-\mathrm{i}(k_1 x + k_2 y)] \mathrm{d}x\mathrm{d}y, \tag{4.131}$$

where

$$k_1 = k\frac{x_f}{f}, \quad k_2 = k\frac{y_f}{f}.$$

Therefore the wave field in the focal plane of the lens is, with the accuracy of the phase factor, the Fourier transform of the wave field incident on the lens; the variables k_1 and k_2 being spatial frequencies here.

4.25. The Bragg Diffraction

Periodic acoustic waves produce in a medium in which they propagate a periodic change of its refractive index. A phase grating is generated; its refractive index is

$$n = n_0 + \eta \cos(\boldsymbol{\beta} \cdot \mathbf{r} - \Omega t), \tag{4.132}$$

where n_0 – the refractive index of free medium in the absence of forced mechanical disturbances of any kind, $\boldsymbol{\beta}$ – the wave vector of acoustic wave, η and Ωt are the amplitude and phase of periodic non-homogeneities of the refractive index.

The refractive index n is sinusoidally modulated in space and has a constant value in planes perpendicular to the vector β.

Two planes characterized by the same value of the refractive index are separated by the distance

$$D = \frac{2\pi}{|\boldsymbol{\beta}|}.$$

Let us consider a plane electromagnetic wave falling askew on the border of a transparent medium in which acoustic waves have been generated.

The area of occurrence of the acoustic waves in the medium has a shape of a layer of thickness l. This situation is illustrated in Fig. 4.20. Within the time of passage of light through the medium the strain distribution in the column of acoustic wave can be regarded as constant. It is because of the difference in propagation velocities in the medium for light and the acoustic wave. These velocities differ by six orders.

A light wave propagating in a medium with non-homogeneous refractive index undergoes the diffraction phenomenon. For any

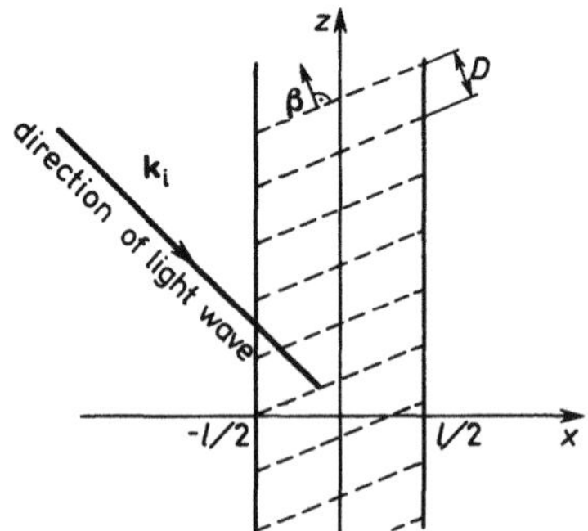

Figure 4.20

instant of time a diffraction pattern is connected with a temporary value of the acoustic field. Thus the dependence upon time for the acoustic field implies the dependence upon time for the diffraction pattern. The latter will, practically without delay, follow the changes of the acoustic field.

A light wave in the medium described by formula (4.132) satisfies the equation

$$0 = \left(\frac{\partial^2}{\partial x^2} + \frac{\partial^2}{\partial z^2}\right)\mathbf{E} + k_0^2 n_0^2\left[1 + \frac{2\eta}{n_0}\cos(\boldsymbol{\beta}\cdot\mathbf{r} - \Omega t)\right]\mathbf{E}, \quad (4.133)$$

which remains valid by the assumptions

$$\eta \ll n_0, \qquad \Omega < \omega,$$

where $\omega = k_0 c$. Then (4.132) yields

$$n^2 \cong n_0^2 + 2n_0\eta\cos(\boldsymbol{\beta}\cdot\mathbf{r} - \Omega t). \quad (4.134)$$

Our purpose is to find the solution of (4.133) in a form of the sum of the primary field $\mathbf{E}_0$ (incident wave) and the scattered wave $\mathbf{E}^S$:

$$\mathbf{E} = \mathbf{E}_0 + \mathbf{E}^S. \quad (4.135)$$

We assume that the primary field will be

$$\mathbf{E}_0 = \mathbf{A}\exp(-\mathbf{i}\mathbf{k}_i\cdot\mathbf{r}), \quad (4.135a)$$

and the field of the scattered wave is supposed to be a superposition of plane waves:

$$\mathbf{E}^S = \iiint \mathbf{B}(\mathbf{k}_s)\exp(-\mathbf{i}\mathbf{k}_s\cdot\mathbf{r})\,\mathrm{d}k_{sx}\mathrm{d}k_{sy}\mathrm{d}k_{sz}. \quad (4.135b)$$

Substituting the postulated form of the field (4.135) into equation (4.133) leads to the equality:

$$\begin{aligned} &2n_0\eta k_0^2\mathbf{A}\exp(-\mathbf{i}\mathbf{k}_i\cdot\mathbf{r})\cos(\boldsymbol{\beta}\cdot\mathbf{r} - \Omega t) + \\ &+ \iiint(n_0^2k_0^2 - k_s^2)\mathbf{B}(\mathbf{k}_s)\exp(-\mathbf{i}\mathbf{k}_s\cdot\mathbf{r})\mathrm{d}k_s + \\ &+ \int 2n_0\eta k_0^2\cos(\boldsymbol{\beta}\cdot\mathbf{r} - \Omega t)\mathbf{B}(\mathbf{k}_s)\exp(-\mathbf{i}\mathbf{k}_s\cdot\mathbf{r})\mathrm{d}\mathbf{k}_s = 0, \quad (4.136) \end{aligned}$$

where $k_0 = \omega\sqrt{\mu_0\varepsilon_0}$, $k_i = k_0 n_0$. Considering small values of $2\eta n_0$ and $\mathbf{B}(\mathbf{k}_s)$ their products, in the first approximation, can be disregarded, which simplifies equality (4.136) to

$$2n_0\eta k_0^2\mathbf{A}\exp(-\mathrm{i}\mathbf{k}_i\cdot\mathbf{r})\cos(\boldsymbol{\beta}\cdot\mathbf{r} - \Omega t) + \\ + \iiint(n_0^2k_0^2 - k_s^2)\mathbf{B}(\mathbf{k}_s)\exp(-\mathrm{i}\mathbf{k}_s\cdot\mathbf{r})\mathrm{d}k_{sx}\mathrm{d}k_{sy}\mathrm{d}k_{sz} = 0. \quad (4.137)$$

The function $\mathbf{B}(\mathbf{k}_s)$ in equation (4.137) is not known. We can find it in the following way: equation (4.137) is multiplied by the factor $\exp(\mathrm{i}\mathbf{k}_s'\cdot\mathbf{r})$ and then integrated over the whole space. This operation and the equality

$$\iiint\exp[\mathrm{i}(\mathbf{k}_s' - \mathbf{k}_s)\cdot\mathbf{r}]\,\mathrm{d}^3\mathbf{r} = (2\pi)^3\delta(\mathbf{k}_s' - \mathbf{k}_s), \quad (4.138)$$

lead to the result:

$$\mathbf{B}(\mathbf{k}_s) = \frac{-\mathbf{A}k_0^2n_0\eta}{(2\pi)^3(n_0^2k_0^2 - k_s^2)}\iiint_V\{\exp[\mathrm{i}(\boldsymbol{\beta}\cdot\mathbf{r} - \mathbf{k}_i\cdot\mathbf{r} + \mathbf{k}_s\cdot\mathbf{r} - \Omega t)] + \\ + \exp[-\mathrm{i}(\boldsymbol{\beta}\cdot\mathbf{r} + \mathbf{k}_i\cdot\mathbf{r} - \mathbf{k}_s\cdot\mathbf{r} - \Omega t)]\}\,\mathrm{d}^3\mathbf{r}. \quad (4.139)$$

In the integral (4.139) V is the region of space where $\eta\cos(\boldsymbol{\beta}\cdot\mathbf{r} - \Omega t) \neq 0$,

The integrands are strongly oscillating, and of their values decide evidently only these subregions of space where hold the equalities

$$\mathbf{k}_s = \mathbf{k}_i \pm \boldsymbol{\beta}. \quad (4.140)$$

It is the so-called *Bragg condition.*

More precise analysis of the case where the region of interaction with an elastic wave is $\left(|x| < \frac{l}{2}, -\infty < y, z < +\infty\right)$ leads to the result

$$B(\mathbf{k}_s) = \frac{\eta n_0 k_0^2 A}{\pi(k_s^2 - n_0^2k_0^2)}\Big[\delta(k_{sz} + \beta_z - k_{iz}) \times \\ \times \delta(k_{sy} + \beta_y - k_{iy})\frac{\sin[(k_{sx} + \beta_x - k_{ix})l/2]}{k_{sx} + \beta_x - k_{ix}}\mathrm{e}^{-\mathrm{i}\Omega t} + \\ + \delta(k_{sz} - \beta_z - k_{iz})\delta(k_{sy} - \beta_y - k_{iy})\frac{\sin[(k_{sx} - \beta_x - k_{ix})l/2]}{k_{sx} - \beta_x - k_{ix}}\mathrm{e}^{\mathrm{i}\Omega t}\Big].$$

Applying this explicit form of $B(\mathbf{k}_s)$ in formula (4.135b), we obtain for the scattered field the expression

$$E^S = \frac{n_0 \eta k_0^2 A}{\pi} \Bigg\{ \exp(-\mathrm{i}\Omega t) \exp[-\mathrm{i}(k_{iz} - \beta_z)z - \mathrm{i}(k_{iy} - \beta_y)y] \times$$

$$\times \int_{-\infty}^{+\infty} \frac{\exp(-\mathrm{i}k_{sx}x)\sin[(k_{sx} + \beta_x - k_{ix})l/2]}{\{[(k_{iz} - \beta_z)^2 + (k_{iy} - \beta_y)^2] + k_{sx}^2 - n_0^2 k_0^2\}} \times$$

$$\times \frac{1}{k_{sx} + \beta_x - k_{ix}} \mathrm{d}k_{sx} +$$

$$+ \exp(\mathrm{i}\Omega t) \exp[-\mathrm{i}(\beta_z + \mathrm{k}_{iz})z - \mathrm{i}(\beta_y + \mathrm{k}_{iy})y] \times$$

$$\times \int_{-\infty}^{+\infty} \frac{\exp(-\mathrm{i}k_{sx}x)\sin[(k_{sx} - \beta_x - k_{ix})l/2]}{\{[(k_{iz} + \beta_z)^2 + (k_{iy} + \beta_y)^2] + k_{sx}^2 - n_0^2 k_0^2\}} \times$$

$$\times \frac{1}{k_{sx} - \beta_x - k_{ix}} \mathrm{d}k_{sx} \Bigg\}. \tag{4.141}$$

Replacing the integrals in (4.141) by contour integrals and using the Cauchy residue theorem give

$$E^S = 2\eta n_0 k_0^2 A \times$$

$$\times \left[\frac{\exp\{-\mathrm{i}\Omega t - \mathrm{i}[(k_{iz} - \beta_z)z + (k_{iy} - \beta_y)y - k_{sx}x]\}}{\mathrm{i}k_{sx}} \times \right.$$

$$\times \frac{\sin[(k_{sx} + \beta_x - k_{ix})l/2]}{k_{sx} + \beta_x - k_{ix}} +$$

$$+ \frac{\exp\{\mathrm{i}\Omega t - \mathrm{i}[(k_{iz} + \beta_z)z + (k_{iy} + \beta_y)y + k_{sx}x]\}}{\mathrm{i}k_{sx}} \times$$

$$\times \frac{\sin[(k_{sx} - \beta_x - k_{ix})l/2]}{k_{sx} - \beta_x - k_{ix}}, \tag{4.142}$$

where $k_{sx} = \sqrt{n_0^2 k_0^2 - k_{sz}^2 - k_{sy}^2}$, $k_{sz} = k_{iz} \mp \beta$, $k_{sy} = k_{iy} \mp \beta$. The result written above is the required solution for the field of the scattered wave in a medium with periodic properties. In other words, it is the solution of the Bragg diffraction problem in the first approximation of perturbation theory.

From the solution obtained we conclude that the scattered wave reaches its maximum if $\mathbf{k}_s$ fulfils Bragg's condition. This maximum

$$E_{\max}^S = \frac{2\mathrm{i} n_0 k_0^2 A}{k_{sx}} \eta \exp\{\mp \mathrm{i}\Omega t - \mathrm{i}[(k_{iz} \mp \beta_z)z + + (k_{iy} \mp \beta_y)y + k_{sz}x]\} \tag{4.143}$$

is proportional to A and η, and to k_{sx}^{-1}. The latter quantity in general does not fulfil Bragg's condition unless we have $l \gg \lambda$.

Bibliography

1. Banerji, S.: (1919), *Phil. Mag.* **6**, 37, 112.
2. Born, M. and Wolf, E.: (1965), 'Principles of Optics', Pergamon Press, Oxford, London.
3. Goodman, J.: (1968), 'Introduction to Fourier Optics', McGraw-Hill, New York.
4. Hecht, E.: (1979), *Amer. J. Phys.* **40**, 571.
5. Hönl, N.; Maue, A.W. and Westpfahl, K.: (1961), 'Theorie der Beugung, Handbuch der Physik', Bd. XXV/1, Springer-Verlag, Heidelberg.
6. Kalashnikov, A.: (1912), *Zhurn. Rusk. Fiz. Khim. Obshch. Fiz. Otdel.* **44**, 137.
7. Rubinowicz, A.: (1966), 'Die Beugungswelle in der Kirchhoffschen Theorie der Beugung', Springer-Verlag, Berlin, PWN, Warszawa.

5. Diffraction of Electromagnetic Waves

5.1. Introductory Remarks

The scalar description of light diffraction is not satisfactory for the study of polarization effects of a diffraction field. A more complete electromagnetic description should be given.

There are two possible approaches to the problem of diffraction of electromagnetic waves. The first, simpler, it is Kirchhoff's diffraction theory for electromagnetic waves basing on the Larmor–Lorentz principle, which is a vector equivalent of the scalar Huygens–Kirchhoff principle. It allows us to express the field of a monochromatic wave at the point P through the values of fields $\mathbf{E}$ and $\mathbf{H}$ at all points Q of the surface enclosing the point in question.

In order to use the theory we have to define a closed surface S (it is done in a way similar to that in the scalar case (see Sec. 4.2)) and to know the field at all its points. The latter problem we solve, assuming that the electromagnetic field $\mathbf{E}$, $\mathbf{H}$ at the points of the deflecting hole is equal to one that would exist at these points of space in the absence of the opaque screen.

Therefore at the points of the aperture the following equalities hold:

$$\mathbf{n} \times \mathbf{E} = \mathbf{n} \times \mathbf{E}_0, \qquad \mathbf{n} \times \mathbf{H} = \mathbf{n} \times \mathbf{H}_0, \tag{5.1}$$

where $\mathbf{E}_0$ and $\mathbf{H}_0$ are wave vectors for an undisturbed field, $\mathbf{n}$ is a versor normal to the element of the surface. The Kirchhoff electromagnetic diffractional screen is defined by equating to zero the tangent component both of the electric and the magnetic field at all points of the screen

$$\mathbf{n} \times \mathbf{E} = 0, \qquad \mathbf{n} \times \mathbf{H} = 0. \tag{5.2}$$

The scalar analogs of Sommerfeld's conditions, which ensure vanishing of shares coming from the hemisphere surface, are:

1°

$$|\mathbf{E}| < \frac{M}{R^{1+\alpha}}, \qquad |\mathbf{H}| < \frac{M}{R^{1+\alpha}}, \tag{5.3a}$$

where R is the distance from the point under consideration; constants α and M hold the conditions $\alpha > 0$ and $M < \infty$. (It means that the field in infinity decreases faster than R^{-1}. The physical meaning is that all field sources have to be situated within finite distances.)

2° In the case of media without dumping some additional conditions should be imposed on the fields

$$\begin{aligned} &\lim_{R\to\infty} \left[\left(\operatorname{curl} \mathbf{E} \times \frac{\mathbf{R}}{R} \right) - ik\mathbf{E} \right] \cdot \mathbf{R} = 0, \\ &\lim_{R\to\infty} \left[\left(\operatorname{curl} \mathbf{H} \times \frac{\mathbf{R}}{R} \right) - ik\mathbf{H} \right] \cdot \mathbf{R} = 0. \end{aligned} \tag{5.3b}$$

If conditions (5.3) are satisfied, and the properties of the aperture and the screen are defined by (5.1) and (5.2), the surface integral reduces to the integral over the surface of the deflecting hole.

The results of *electromagnetic Kirchhoff's theory of diffraction* are valid only for wavelengths small compared with the linear size of the aperture and only at points lying far from the screen. It is connected with the fact that the Kirchhoff electromagnetic screen is not a physical screen.

Another approach to the problem of diffraction of electromagnetic waves is finding vector solutions of Maxwell's equations satisfying the specified boundary conditions.

From the point of view of the problems occurring in the theory of electromagnetic fields, the problems of diffraction belong to a class of external problems. Within them we should determine the electromagnetic field in an unlimited space outside a certain surface

Σ, on which the tangent component of the electric field E_t and the magnetic field H_t are supposed to be known. In this outer space there may occur some areas where field sources in a form of currents can exist.

To obtain the univocal solutions of Maxwell's equations it is necessary to demand that not only the boundary conditions but also additional conditions related to the field behaviour in infinity (conditions (5.3a) and (5.3b)) should be satisfied. If the surface confining the region where the diffraction takes place has sharp edges, then except the conditions formulated above the so-called *edge condition* should additionally be considered.

Its physical essence is as follows: the edge of an diffraction object does not emit energy. In mathematical formulation the edge condition can be written as

$$\lim_{\varrho \to 0} \rho \mathbf{n} \cdot (\mathbf{E} \times \mathbf{H}) = 0, \tag{5.4}$$

where ρ stands for the radius of the surface of an annular cylinder surrounding the edge, and $\mathbf{n} = \boldsymbol{\rho}/\rho$ is a versor normal to the elements of the cylindrical surface.

The diffraction of electromagnetic waves thought of as a boundary problem is a complex mathematical theory calling for very advanced mathematical methods and for some simplified assumptions as to the boundary conditions imposed on the solutions required. Usually we assume that diffractional screens are characterized by vanishing of the tangent component of the electric field: $\mathbf{n} \times \mathbf{E} = \mathbf{0}$ (ideally conducting screen) or by vanishing of the tangent component of the magnetic field: $\mathbf{n} \times \mathbf{H} = \mathbf{0}$. In the most general case the diffractional screens satisfy the *Leontowicz condition*

$$(\mathbf{n} \times \mathbf{E}) + \sqrt{\frac{\mu}{\varepsilon}}\, \mathbf{n} \times (\mathbf{n} \times \mathbf{H}) = \mathbf{0}.$$

5.2. The Green Tensors

For Maxwell's equations, similarly as it was for Helmholtz's equation, we can formulate a mathematical form of a principle which allows us to express the field of a monochromatic wave at point P through the values of the fields $\mathbf{E}$ and $\mathbf{H}$ at all points Q of the surface surrounding the point in question.

The principle is known as the *Larmor–Lorentz principle*. For its formulation it is necessary to know the so-called *Green tensors* $\mathfrak{G}_{el}(P,Q)$ and $\mathfrak{G}_{mag}(P,Q)$ being the functions of two points: the point of integration Q and the point of observation P.

The physical sense of the Green tensor is as follows: the l-th column of the tensor $\mathfrak{G}_{el}$ ($\mathfrak{G}_{mag}$) represents, with the accuracy to dimensional coefficients, an electromagnetic field activated by a radiating electric (magnetic) dipole orientated along the l-th axis of the Cartesian coordinate system under consideration.

The tensor functions discussed above are the solutions of the equations

$$\begin{aligned}&\operatorname{curl}\operatorname{curl}\mathbf{G}^l_{el}-k_0^2\varepsilon_r\mu_r\mathbf{G}^l_{el}=\delta(P,Q)\mathbf{l},\\&\operatorname{curl}\operatorname{curl}\mathbf{G}^l_{mag}-k_0^2\mu_r\varepsilon_r\mathbf{G}^l_{mag}=\delta(P,Q)\mathbf{l},\end{aligned}\tag{5.5}$$

where the index l denotes an appropriate column of the tensor, versor $\mathbf{l}$ has the direction of an appropriate axis of a chosen coordinate system, $\delta(P,Q)\mathbf{l}$ is the unit point moment of the electric (magnetic) dipole; ε_r (μ_r) is the relative permittivity (permeability) of the medium, $k_0^2=\omega^2\varepsilon_0\mu_0$.

In the case of an isotropic medium the tensors $\mathfrak{G}_{el}(P,Q)$ and $\mathfrak{G}_{mag}(P,Q)$ are connected through

$$\mathfrak{G}_{el}=\mathfrak{G}_{mag}\tag{5.6}$$

(this has been proved in the work of Hönl, Maue and Westpfahl (1961)).

An explicit form of $\mathfrak{G}_{el}(P,Q)$ and $\mathfrak{G}_{mag}(P,Q)$ is obtained in the following way.

Having applied the operator div to equation (5.5), we obtain the relation

$$-k_0^2\varepsilon_r\mu_r\operatorname{div}\mathbf{G}^l_{el}=\mathbf{l}\cdot\nabla\delta(P,Q),\tag{5.7}$$

hence, on applying to the both sides the operator ∇, we get

$$-k_0^2\varepsilon_r\mu_r \operatorname{grad}\operatorname{div}\mathbf{G}_{\mathrm{el}}^l = (\mathbf{I}\cdot\nabla)\nabla\delta(P,Q). \tag{5.8}$$

If note that (5.5) can be written in the form

$$\operatorname{grad}\operatorname{div}\mathbf{G}_{\mathrm{el}}^l - \Delta\mathbf{G}_{\mathrm{el}}^l - k_0^2\varepsilon_r\mathbf{G}_{\mathrm{el}}^l = \delta(P,Q)\mathbf{I},$$

where Δ is Laplacian, then, on the strength of the above, we obtain the equation

$$(\Delta + k_0^2\varepsilon_r\mu_r)\mathbf{G}_{\mathrm{el}}^l = -\left[\mathbf{I} + \frac{1}{k_0^2\varepsilon_r\mu_r}(\mathbf{I}\cdot\nabla)\nabla\right]\delta(P,Q). \tag{5.9}$$

If this equation is to be satisfied, it is enough to assume that

$$\mathbf{G}_{\mathrm{el}}^l(P,Q) = \left[\mathbf{I} + \frac{1}{k_0^2\varepsilon_r\mu_r}\nabla(\mathbf{I}\cdot\nabla)\right]g(P,Q). \tag{5.10}$$

The scalar function $g(P,Q)$ is a solution of the equation

$$(\Delta + k_0^2\varepsilon_r\mu_r)g(P,Q) = -\delta(P,Q). \tag{5.11}$$

It has the form

$$g(P,Q) = \frac{\exp(ik|\mathbf{r}_P - \mathbf{r}_Q|)}{4\pi|\mathbf{r}_P - \mathbf{r}_Q|}, \tag{5.12}$$

i.e., it depends only on the distance between two points P and Q, since the vector $\mathbf{r}_P$ determines the position of the point P, and the vector $\mathbf{r}_Q$ is the position vector for the point Q in a chosen coordinate system.

The solution of equation (5.11) is sought in the form of a superposition of plane waves of unknown amplitudes $g(k_1, k_2)$. The function $\delta(P,Q)$ can be presented in a similar form. Substituting these expressions into (5.11) and performing the operations, we obtain the relation which enables us to calculate the unknown amplitudes of plane waves. The solution of (5.11) can be written as

$$g(P,Q) = -\frac{1}{8\pi^3}\iiint\limits_{-\infty}^{+\infty}\frac{\exp(i\mathbf{k}\cdot[\mathbf{r}_P - \mathbf{r}_Q])}{-\mathbf{k}\cdot\mathbf{k} + k_0^2\varepsilon_r\mu_r}\,dk_1dk_2dk_3. \tag{5.13}$$

Calculating this integral, we obtain function (5.12). Thus the explicit form of the tensor $\mathfrak{G}_{el}(P,Q)$ can be expressed by the formula

$$\mathfrak{G}_{el}(P,Q) = -\frac{1}{4\pi}\left\{\left[\frac{1}{k_0^2\varepsilon_r\mu_r}\nabla\nabla + \mathfrak{J}\right]\frac{\exp(ik|\mathbf{r}_P - \mathbf{r}_Q|)}{|\mathbf{r}_P - \mathbf{r}_Q|}\right\}, \quad (5.14)$$

where $\nabla\nabla$ is a dyadic product of the nabla operators. It is an operator of a tensor type with matrix

$$\nabla\nabla = \begin{bmatrix} \frac{\partial^2}{\partial x^2} & \frac{\partial^2}{\partial x \partial y} & \frac{\partial^2}{\partial x \partial z} \\ \frac{\partial^2}{\partial x \partial y} & \frac{\partial^2}{\partial y^2} & \frac{\partial^2}{\partial y \partial z} \\ \frac{\partial^2}{\partial x \partial z} & \frac{\partial^2}{\partial y \partial z} & \frac{\partial^2}{\partial z^2} \end{bmatrix},$$

while $\mathfrak{J}$ is the unit tensor, represented by a matrix having non-zero elements only on its diagonal. These elements equal unity.

5.3. The Larmor–Lorentz Principle

The Larmor–Lorentz principle, being a vector analog of the Huygens–Kirchhoff principle, says that a field at each inner point is unambigously determined by the values of the tangent components for the electric and magnetic field on the surface confining the area in question.

In order to obtain the explicit form of the Larmor–Lorentz principle, we consider a monochromatic electromagnetic wave whose vectors **E**, **H** satisfy the equations

$$\begin{aligned} &\operatorname{curl}\operatorname{curl}\mathbf{E}(Q) - k_0^2\varepsilon_r\mu_r\mathbf{E}(Q) = \mathbf{0}, \\ &\operatorname{curl}\operatorname{curl}\mathbf{H}(Q) - k_0^2\mu_r\varepsilon_r\mathbf{H}(Q) = \mathbf{0}. \end{aligned} \quad (5.15)$$

Multiplying scalarly the first of equations (5.5) by **E** and the first of (5.15) by $\mathbf{G}_{el}^l$, and subtracting the results of these two operations we arrive at the identity

$$\mathbf{E}(Q)\cdot\operatorname{curl}\operatorname{curl}\mathbf{G}^{l}_{\mathrm{el}}(P,Q)-\mathbf{G}_{\mathrm{el}}(P,Q)\cdot\operatorname{curl}\operatorname{curl}\mathbf{E}(Q)$$
$$=\delta(P,Q)\mathbf{E}(Q)\cdot\mathbf{l}. \quad (5.16)$$

Integrating (5.16) over the area Ω, being the subarea where the functions in question are defined, and containing inside point P, we obtain, on applying the Gauss law, the required form of the Larmor–Lorentz principle

$$E_l(P)=-\iint\limits_S \mathbf{n}\cdot(\mathbf{E}\times\operatorname{curl}\mathbf{G}^{l}_{\mathrm{el}}-\mathbf{G}^{l}_{\mathrm{el}}\times\operatorname{curl}\mathbf{E})\,\mathrm{d}f. \quad (5.17)$$

Integral (5.17) is performed over the surface S confining the area Ω.

Formula (5.17) can be written in a slightly different form

$$E_l(P)=-\iint\limits_S [(\mathbf{n}\times\mathbf{E})\cdot\operatorname{curl}\mathbf{G}^{l}_{\mathrm{el}}+(\mathbf{n}\times\operatorname{curl}\mathbf{E})\cdot\mathbf{G}^{l}_{\mathrm{el}}]\,\mathrm{d}f. \quad (5.18)$$

The last expression has a clear physical intepretation. The quantity $\mathbf{n}\times\mathbf{E}$ represents the tangent component of the electric field at the points of the integration surface S. It is proportional to the surface density of the magnetic current. The quantity

$$\mathbf{n}\times\operatorname{curl}\mathbf{E}=-\mathrm{i}\omega\mu(\mathbf{n}\times\mathbf{H})$$

is proportional to the tangent component of the magnetic field at the points of the surface S. It can be related to the surface density of electric current. Considering this and using the identity

$$\operatorname{curl}\mathbf{E}^{l}_{\mathrm{el}}=-\mathrm{i}\omega\mu\mathbf{H} \quad (5.19)$$

the first term of the integrand (5.18) can be interpreted as a share to the l-th component of the electric field at point P, coming from the surface magnetic current existing in the vicinity of integration point Q. The second term of the integrand may be interpreted as a share to the l-th component of the electric field at point P, connected with the surface electric current existing in the vicinity of point Q mentioned above.

If, in an analogous way, we handle the second equations of (5.5) and (5.15), we obtain for the field $\mathbf{H}$ at point P the expression

$$H_l(P) = -\iint_S [(\mathbf{n}\times\mathbf{H})\cdot\operatorname{curl}\mathbf{G}^l_{\text{mag}} + (\mathbf{n}\times\operatorname{curl}\mathbf{H})\cdot\mathbf{G}^l_{\text{mag}}]\,\mathrm{d}f. \tag{5.20}$$

Expressions (5.18) and (5.20) make up the mathematical form of the Larmor–Lorentz principle. The explicit form of the tensors $\mathfrak{G}_{\text{el}}(P,Q)$ and $\mathfrak{G}_{\text{mag}}(P,Q)$ are given by formulas (5.14) and (5.6).

5.4. Kottler's Form of the Larmor–Lorentz Principle

The explicit form of the Green tensors (5.14) and (5.6) enables us to transform the Larmor–Lorentz principle for an electric field to the form

$$\begin{aligned} E_l(P) = {} & \frac{1}{4\pi}\iint_S \left[\frac{\mathrm{e}^{ikR}}{R}\frac{\partial}{\partial n}E_l(Q) - E_l(Q)\frac{\partial}{\partial n}\frac{\mathrm{e}^{ikR}}{R}\right]\mathrm{d}f + \\ & + \frac{1}{4\pi}\frac{\mathrm{i}}{\omega\varepsilon}\iint_S \mathbf{n}\cdot\operatorname{curl}\left[\mathbf{H}(Q)(\mathbf{l}\cdot\nabla)\frac{\mathrm{e}^{ikR}}{R}\right]\mathrm{d}f - \\ & - \frac{1}{4\pi}\iint_S \mathbf{n}\cdot\operatorname{curl}\left[(\mathbf{E}\times\mathbf{l})\frac{\mathrm{e}^{ikR}}{R}\right]\mathrm{d}f, \end{aligned} \tag{5.21}$$

where $R = |\mathbf{r}_P - \mathbf{r}_Q|$, and for a magnetic field to the form

$$\begin{aligned} H_l(P) = {} & -\frac{1}{4\pi}\iint_S \left[H_l(Q)\frac{\partial}{\partial n}\left(\frac{\mathrm{e}^{ikR}}{R}\right) - \frac{\mathrm{e}^{ikR}}{R}\frac{\partial}{\partial n}H_l(Q)\right]\mathrm{d}f + \\ & + \frac{1}{4\pi}\iint_S \mathbf{n}\cdot\operatorname{curl}\left(\mathbf{l}\times\mathbf{H}(Q)\frac{\mathrm{e}^{ikR}}{R}\right)\mathrm{d}f - \\ & - \frac{\mathrm{i}}{4\pi\omega\mu}\iint_S \mathbf{n}\cdot\operatorname{curl}\left[\mathbf{E}(Q)(\mathbf{l}\cdot\nabla)\frac{\mathrm{e}^{ikR}}{R}\right]\mathrm{d}f. \end{aligned} \tag{5.22}$$

If on the closed integration surface S there are no lines on which the fields **E** and **H** undergo stepwise changes, formulations (5.18) and (5.20) mentioned above reduce in the case of the Larmor–Lorentz principle to the expression

$$E_l^0(P) = -\frac{1}{4\pi}\iint_S \mathrm{d}f\left[E_l(Q)\frac{\partial}{\partial n}\frac{\mathrm{e}^{ikR}}{R} - \frac{\mathrm{e}^{ikR}}{R}\frac{\partial}{\partial n}E_l(Q)\right] \tag{5.23}$$

for the electric field, and to the expression

$$H_l^0(P) = -\frac{1}{4\pi}\iint_F \mathrm{d}f\left[H_l(Q)\frac{\partial}{\partial n}\frac{\mathrm{e}^{ikR}}{R} - \frac{\mathrm{e}^{ikR}}{R}\frac{\partial}{\partial n}H_l(Q)\right] \tag{5.24}$$

for the magnetic field.

However, if on the integration surface there is a line of noncontinuity (K) of the fields **E** and **H**, as it is in the case of the Kirchhoff theory of diffraction of electromagnetic fields, then the field at the point of observation P is expressed by the relations:

$$E_l(P) = E_l^0(P) + \frac{\mathrm{i}}{4\pi\omega\varepsilon}\oint_{(K)} \mathrm{d}\mathbf{s}\cdot\mathbf{H}\frac{\partial}{\partial x_l}\frac{\mathrm{e}^{ikR}}{R} - -\frac{1}{4\pi}\oint_{(K)}\frac{\mathrm{e}^{ikR}}{R}(\mathbf{l}\times\mathbf{E})\cdot\mathrm{d}\mathbf{s}, \tag{5.25}$$

$$H_l(P) = H_l^0(P) + \frac{1}{4\pi}\oint_{(K)}\frac{\mathrm{e}^{ikR}}{R}(\mathbf{l}\times\mathbf{H})\cdot\mathrm{d}\mathbf{s} - -\frac{\mathrm{i}}{4\pi\omega\mu}\oint_{(K)} \mathrm{d}\mathbf{s}\cdot\mathbf{E}\frac{\partial}{\partial x_l}\frac{\mathrm{e}^{ikR}}{R}. \tag{5.26}$$

If we are interested in finding the diffraction field in the electromagnetic case within the Kirchhoff diffraction theory, the use of the Larmor–Lorentz principle requires to define a closed surface S, which is defined in much the same way as in the scalar case (see Sec. 4.2). The scalar Sommerfeld conditions are now replaced by conditions (5.3), and the properties of the diffractional screen and the aperture are defined by (5.1) and (5.2).

The surface integral reduces to the integral over the deflecting aperture, and the edge integral is taken over the edge of the aperture, since it is where, considering the conditions (5.1) and (5.2), the noncontinuity line for the fields **E** and **H** occurs.

Comparing expressions (5.25) and (5.26) for the components of the vector fields being the results of diffraction with appropriate expressions given by the scalar approximation (see Sec. 4.2), we can note an essential difference. Applying the scalar approximation to each component of vector field individually, we obtain as a result a diffraction field which is obviously a solution of the wave equation but is not a solution of Maxwell's equations. The scalar approximation gives also no information about possible changes of field polarization, which can appear in diffraction phenomena. Within the scalar approximation the polarization state for the diffraction field is the same as for the incident field. The effects of the changes of field polarization caused by diffraction can be determined only within the vector electromagnetic theory of diffraction.

The Kirchhoff theory is here the first approximation, more precise results can be obtained by solving the problem of diffraction of electromagnetic waves as a boundary problem. The simplest, from mathematical point of view, is the case of ideally conducting plane screens.

5.5. Diffraction of a Plane Electromagnetic Wave at an Ideally Conducting Half-Plane

A. *Formulation of the Problem*

Considering the diffraction of a plane electromagnetic wave at a screen being a half-plane, we can restrict ourselves to the case of functional dependencies on two spatial variables, for example x and

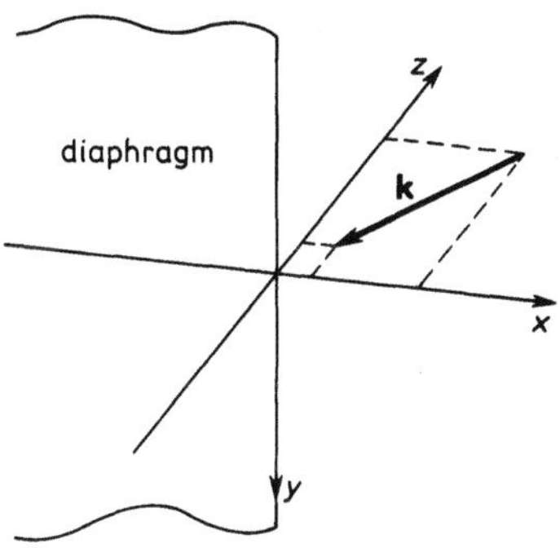

Figure 5.1

z. It corresponds to a situation where the edge of the screen coincides with the y-axis, and the direction of the incident wave remains perpendicular to the edge (Fig. 5.1).

In this case the electromagnetic field is decomposited in two: *TE* and *TM* field. *TE* field is a field for which

$$E_x = E_z = H_y = 0, \qquad H_x = \frac{1}{i\omega\mu_0}\frac{\partial E_y}{\partial z}, \qquad H_z = -\frac{1}{i\omega\mu_0}\frac{\partial E_y}{\partial x}, \tag{5.27}$$

and the component E_y satisfies the equation

$$\left(\frac{\partial^2}{\partial x^2} + \frac{\partial^2}{\partial y^2}\right)E_y + k^2 E_y = 0. \tag{5.28}$$

TM field has the components

$$E_y = H_x = H_z = 0, \qquad E_x = -\frac{1}{i\omega\varepsilon_0}\frac{\partial H_y}{\partial z}, \qquad E_z = \frac{1}{i\omega\varepsilon_0}\frac{\partial H_y}{\partial x} \tag{5.29}$$

and the component H_y which satisfies the equation

$$\left(\frac{\partial^2}{\partial x^2} + \frac{\partial^2}{\partial z^2}\right)H_y + k^2 H_y = 0, \tag{5.30}$$

where $k^2 = \omega^2\varepsilon_0\mu_0$.

As can be easily noted from the above, the *TE* field is determined by one component of the electric field E_y, while the component of the magnetic field H_y plays the same role in the case of the *TM* field. Such decomposition of the total field allows us to handle separately with the diffraction of a plane electromagnetic wave of type *TE* on a perfectly conducting half-plane given by the equations: $z = 0$ and $x \leq 0$.

Let a wave incident on the diffractional screen be a wave of type *TE* and of the form

$$E_y^{(i)} = E_0 \exp[-ik(x\cos\theta + z\sin\theta)], \tag{5.31}$$

where θ is the angle which the wave vector **k** makes with the *x*-axis (Fig. 5.2)

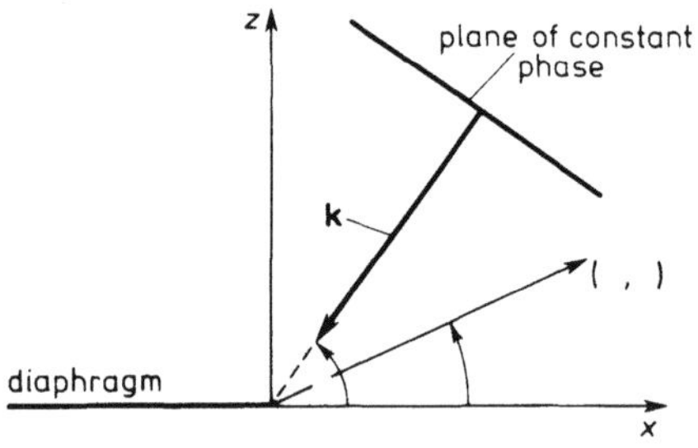

Figure 5.2

Introducing the polar coordinates (ρ, φ)

$$x = \rho\cos\varphi, \qquad z = \rho\sin\varphi,$$

the wave can be presented as

$$E_y^{(i)} = E_0 \exp[-ikr\cos(\theta - \varphi)]. \tag{5.31a}$$

The total field at any point of the space will be

$$E_y(\rho, \varphi, \theta) = E_y^{(i)}(\rho, \varphi, \theta) + E_y^S(\rho, \varphi, \theta),$$

where E_y^S is the secondary field. The total field is required to satisfy:

(1) the equation

$$\left(\frac{\partial^2}{\partial x^2} + \frac{\partial^2}{\partial z^2} + k^2\right) E_y = 0; \tag{5.32}$$

(2) the boundary conditions

$$E_y(\rho, \pi, \theta) = E_y(\rho, -\pi, \theta) = 0$$

(for an ideally conducting plane all tangent components of the electric field are equal to zero at all points of the plane);

(3) the edge condition

$$\lim_{\varrho \to 0} E_y(\rho, \varphi, \theta) < \infty,$$

(4) the conditions in infinity (which in this case reduce to Sommerfeld's conditions (see Sec. 4.3)).

B. Solving the Diffraction Problem for a Plane Electromagnetic Wave TE at a Perfectly Conducting Half-Plane

In order to solve the problem we use the Sommerfeld method and postulate the solution in the form

$$E_y = E_y(\rho, \varphi, \theta) - E_y(\rho, \varphi, 4\pi - \theta), \tag{5.33}$$

where $E_y(\rho, \varphi, \theta)$ should satisfy, in addition to the conditions 1–4 discussed above, the following conditions.

(a) It should be a regular and univocal function in a two-sheet Riemann space. (The latter is built in much the same way as two-sheet Riemann surfaces known in the theory of function of complex variable, the only difference is that a branching point is replaced by a branching line (the diaphragm edge). Crossing one sheet of the Riemann space and entering the other is achieved by crossing the half-plane of the screen. One of the sheet of the Riemann space is physical, the other is non-physical. It is of interest to know how light reflects from a two-sheet space of this kind. We can assume that a wave falling on the screen passes to the non-physical sheet, and what is a reflected wave comes from the very sheet and enters the physical space. The reflected wave is a wave having its source in the non-physical space.)

(b) $E_y(\rho, \varphi, \theta)$ – as a function of the angle φ – should have a period 4π (twice circling of the edge leads again to the same point of the Riemann space).

Fulfilment of the boundary conditions by the postulated solution (5.33) leads to the equations:

$$\begin{aligned} &E_y(\rho, \pi, \theta) - E_y(\rho, \pi, 4\pi - \theta) = 0, \\ &E_y(\rho, -\pi, \theta) - E_y(\rho, -\pi, 4\pi - \theta) = 0. \end{aligned} \tag{5.34}$$

To find the explicit form of the function $E_y(\rho, \varphi, \theta)$, we use the plane of complex variable α. In this plane we take as function (5.31a) the Cauchy integral of the form

$$E_y = \frac{E_0}{2\pi} \oint_A \frac{\exp[-ik\rho\cos(\varphi - \alpha)]}{1 - \exp[i(\theta - \alpha)/2]} \, d\alpha, \tag{5.35}$$

where the path of integration A is chosen in such a way that it encircles the pole of the function $\{1 - \exp[i(\theta - \alpha)/2]\}^{-1}$.

The situation in question is illustrated in Fig. 5.3, in which points where the function $\cos(\varphi - \alpha)$ equals zero

$$\alpha_m = \varphi - m\pi + i\pi\left(m + \tfrac{1}{2}\right) \quad (m = 0, \pm 1, \ldots)$$

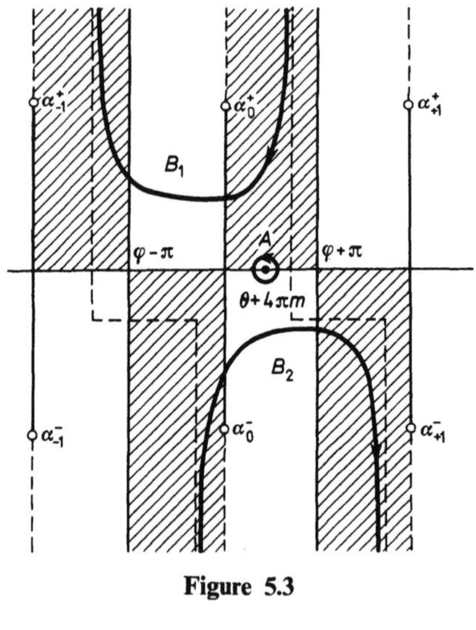

Figure 5.3

and the pole of the integrand function

$$\alpha_l = \theta + 4\pi l \qquad (l = 0, \pm 1, \ldots)$$

has been marked.

The points α_m are the points of branching of a multisheet plane α. The vertical lines drawn from these points to ∞ are the lines of branching. The integral curves B_j are led in such a way that the imaginary part of the function $\cos(\varphi - \alpha)$ is negative. It ensures the convergence of the integral for $\rho \to \infty$.

Transformating the path of integration A into the curve $B = B_1 + B_2$, we choose this curve in such a way that the integral remains still convergent. Replacing the primary path of integration A by B is justified, if the curve B becomes completed to a closed curve (in. Fig. 5.3 the completed parts are marked by dashed lines). It is not, however, an essential problem. The fact of importance is that the integral

$$E_y(\rho, \varphi, \theta) = \frac{E_0}{2\pi} \int_B \frac{\exp[-ik\rho\cos(\varphi - \alpha)]}{1 - \exp[i(\theta - \alpha)/2]} \, d\alpha \tag{5.36}$$

is the solution of equation (5.32) and as is shown below, it satisfies the conditions imposed on this solution. It is the sought explicit form of the function $E_y(\rho, \varphi, \theta)$.

Curve B can be deformed into a curve $C = C_1 + C_2$ (Fig. 5.4) with simultaneous separation of the residuum of the function under the sign of integration. The residuum may appear only if α_0 lies between $\varphi - \pi$ and $\varphi + \pi$. On the ground of the above remarks the function $E_y(\rho, \varphi, \theta)$ may be written as

$$E_y(\rho, \varphi, \theta) = \delta(\varphi) E_0 \exp[-ik\rho\cos(\varphi - \theta)] + \\ + \frac{E_0}{2\pi} \int_C \frac{\exp[-ik\rho\cos(\varphi - \alpha)]}{1 - \exp[i(\theta - \alpha)/2]} \, d\alpha, \tag{5.37}$$

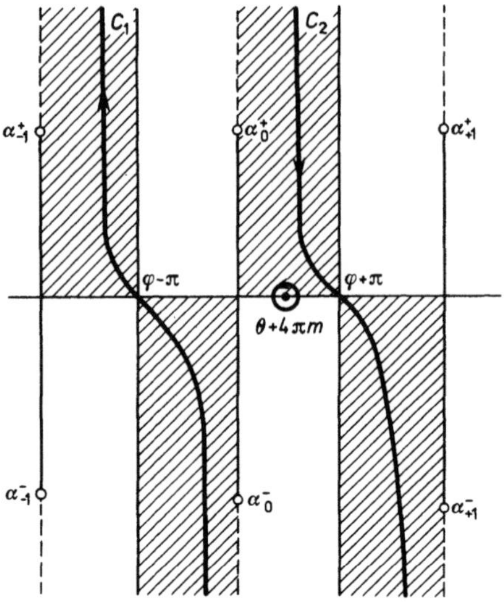

Figure 5.4

where

$$\delta(\varphi) = \begin{cases} 1 & \text{for} \quad -\pi + \theta \leq \varphi \leq \pi + \theta, \\ 0 & \text{for other values of the angle } \varphi. \end{cases}$$

From this we conclude that the first term of (5.37), connected with the incident wave, appears both in the physical and in the non-physical part of the space, which is in agreement with the model discussed above. The region $-\pi + \theta \leq \varphi \leq \pi$ lies in the physical sheet, while the region $\pi \leq \varphi \leq \pi + \theta$ – in the non-physical sheet.

Knowing the explicit form of the function $E_y(\rho, \varphi, \theta)$, we can easily find $E_y(\rho, \varphi, 4\pi - \theta)$, since it differs from (5.37) only in the value of the parameter θ. Thus

$$E_y(\rho, \varphi, 4\pi - \theta) = \delta_1(\varphi) E_0 \exp[-ik\rho\cos(\varphi + \theta)] + \\ + \frac{E_0}{2\pi} \int_C \frac{\exp[-ik\rho\cos(\varphi - \alpha)]}{1 - \exp[-i(\theta + \alpha)/2]} \mathrm{d}f, \quad (5.38)$$

where

$$\delta_1(\varphi) = \begin{cases} 1 & \text{for} \quad -\theta + \pi < \varphi < 3\pi - \theta, \\ 0 & \text{for other values of the angle } \varphi. \end{cases}$$

The first term of (5.38) represents a wave reflected from the screen. It is non-zero in the region $(-\theta + \pi \le \varphi \le \pi)$ of the physical space and in the region $(\pi < \varphi < 3\pi - \theta)$ of the non-physical one.

Each of the integrals (5.37) and (5.38) can be replaced by the sum of two integrals taken over the curves C_1 and C_2 (Fig. 5.3):

$$\int\limits_{C_j} \left[\frac{1}{1 - \exp[\mathrm{i}(\pm\theta - \alpha)/2]} - \tfrac{1}{2} \right] \exp[-\mathrm{i}k\rho\cos(\varphi - \alpha)]\,\mathrm{d}\alpha, \tag{5.39}$$

where $j = 1, 2$, and the signs $\pm$ standing by the angle θ refer to the integral obtained from (5.37) or (5.38) respectively.

Substitution of each of the integrals (5.37) and (5.38), taken over the total curve $C = C_1 + C_2$, by two separate integrals, taken over C_1 and C_2, is not univocal. The choice of a constant of separation is done for the convenience in calculating, as it allows us to use the identity:

$$\left[\frac{1}{1 - \exp[\mathrm{i}(\pm\theta - \alpha)/2]} - \tfrac{1}{2} \right] = + \frac{\mathrm{i}}{2} \cot\left(\frac{\alpha \mp \theta}{4} \right).$$

Using it the integrals (5.39) may be written as

$$\int\limits_{C_j} \frac{\mathrm{i}}{2} \cot\left(\frac{\alpha \mp \theta}{4} \right) \exp[-\mathrm{i}k\rho\cos(\varphi - \alpha)]\,\mathrm{d}\alpha \quad (j = 1, 2). \tag{5.40}$$

To get the expressions for the diffraction field which could be interpreted physically we use asymptotic extensions of the integral occurring in formula (5.40). These extensions are obtained by the use of the saddle point method. The essence of this method consists in a fact that for large values of the parameter k the value of the integral (5.40) is mainly defined by this fragment of the path C_j on which the values of the expression

$$|\exp[-ik\rho\cos(\varphi-\alpha)]| = \exp\{k\mathrm{Re}[-i\rho\cos(\varphi-\alpha)]\} \quad (5.41)$$

are large in comparison with the values on other fragments of the path C_j. The faster the value $\mathrm{Re}[i\rho\cos(\varphi-\alpha)]$ decreases the easier the value of the integral can be evaluated.

With the use of the Cauchy theorem we can always deform the curve C_j into the curve of the fastest decrease. Considering that $\alpha = \alpha_r + i\alpha_u$ and regarding $u = \mathrm{Re}[i\rho\cos(\varphi-\alpha)]$ as a surface in the space (α_r, α_u, u), we can give the following geometrical interpretation.

Since the function u is a harmonic function (i.e. satisfying the equation $\Delta u = 0$), the surface in question $u = u(\alpha_r, \alpha_u)$ cannot have extreme values at points where

$$\frac{\partial}{\partial\alpha}[i\rho\cos(\varphi-\alpha)] = 0, \quad (5.42a)$$

which takes place for $\alpha_0^{\pm} = \varphi \pm \pi$. Extreme values can appear only at the saddle points of the surface.

Because the function $i\rho\cos(\varphi-\delta)$ is analytical, the lines of the fastest decrease of the value $\mathrm{Re}[i\rho\cos(\varphi-\alpha)]$ coincide with the lines

$$\mathrm{Im}[-i\rho\cos(\varphi-\alpha)] = \text{const.} \quad (5.42b)$$

The path of the fastest decrease $\bar{C}_j$ should contain points $\alpha_0^{(n)}$ for which (5.41) reaches the largest values on the curves. Points lying on the line (5.42a) for which (5.41) has the largest values are saddle points.

We can now find asymptotic expressions for integrals (5.40). Introducing the parameter

$$\tau^2 = -\rho + \rho\cos(\varphi-\alpha) \quad (5.43)$$

integrals (5.40) can be presented as

$$\exp(-ik\rho)\int_{\bar{C}_j}\cot\left(\frac{\alpha\mp\theta}{4}\right)\frac{i}{\rho\sin(\varphi-\alpha)}\tau\exp(-ik\tau^2)d\tau. \quad (5.44)$$

To get the first term of the asymptotic extension of these integrals, it is necessary to know the values of the expression

$$\mathrm{i}\cot\left(\frac{\alpha \mp \theta}{4}\right)\frac{\tau}{\rho\sin(\varphi-\alpha)} \tag{5.45}$$

at the saddle point. To achieve this, we expand expression (5.43) in a Taylor series in the vicinity of the saddle points, which yields

$$\tau \approx \mathrm{i}(\alpha - \alpha_0^{\pm})\sqrt{\frac{\rho}{2}}.$$

Likewise, expanding the expression $\rho\sin(\varphi - \alpha)$, we obtain

$$\rho\sin(\varphi - \alpha) \approx \rho\,\frac{\alpha - \alpha_0^{\pm}}{1!}.$$

Thus (5.45) takes at the saddle points the values

$$\cot\frac{\exp(+\mathrm{i}\pi/2)}{\sqrt{2\rho}}\left(\frac{\varphi \pm \pi \mp \theta}{4}\right). \tag{5.46}$$

Hence the first term of asymptotic expansion for integrals (5.44) takes the form

$$\frac{\exp(\mathrm{i}\pi/2 - \mathrm{i}k\rho)}{\sqrt{2k\rho}}\cot\left(\frac{\varphi \pm \pi \mp \theta}{4}\right)\sqrt{\frac{\pi}{2}}\exp\left(-\mathrm{i}\frac{\pi}{4}\right). \tag{5.47}$$

The last two factors in (5.47) yield from the equality

$$\int_0^\infty \exp(-\mathrm{i}u^2)\,\mathrm{d}u = \frac{\sqrt{\pi}}{2}\exp\left(-\mathrm{i}\frac{\pi}{4}\right).$$

It is clear that (5.47) is valid everywhere where holds the inequality $|\cot((\varphi \pm \pi \mp \theta)/4)| < \infty$. This condition is fulfilled for $\varphi \neq \pi \pm \theta$, i.e., everywhere outside the limits of the shadow of the incident wave and the reflected wave.

Bibliography

1. Kleine, M. (ed.): (1950), 'The Theory of Electromagnetic Waves', Dover Publications Inc, New York.
2. Rubinowicz, A.: (1975), 'Selected Papers', PWN, Warszawa.
3. Hönl, H., Maue, A.W. and Westpfahl, K.: (1961), 'Theorie der Beugung, Handbuch der Physik', Bd XXV/1, Springer-Verlag, Berlin.

6. Optics of Crystals

When studying electromagnetic waves in isotropic media the vectors **E** and **D** were parallel. Now we will deal with the optics of anisotropic media, to which belong most crystals (natural anisotropy) as well as transparent isotropic media subject either to external tensions or to strong external electric fields (forced anisotropy).

We will consider the cases where a medium can be treated as nonmagnetic, homogeneous and transparent. According to this assumption, the relations between the intensities of fields **E** and **H** and the corresponding induction vectors **D** and **B** can be written with the use of the summation convention as

$$D_i = \varepsilon_{ik} E_k, \qquad B_i = \mu_0 H_i.$$

In addition, we assume that the components of the dielectric tensor ε_{ik} are real, and its main values positive. Besides, we take under consideration the case of monochromatic waves only, confining ourselves to the problem of the dependence of the phase velocity of a wave on the wave polarization and on the propagation direction, and to the phenomena connected with it.

6.1. Electric Permittivity in an Anisotropic Medium

In an electrically anisotropic medium the relation between vector **D** and field intensity **E** is determined by the relation

$$\mathbf{D} = \varepsilon_0 \varepsilon \mathbf{E}, \tag{6.1}$$

where ε_0 is the electric permittivity in vacuum, ε the relative electric permittivity tensor.

In a non-conducting medium this tensor is symmetric, and in a Cartesian system of coordinates whose axes coincide with its principal axes this tensor becomes a diagonal tensor of the form

$$\varepsilon = \begin{bmatrix} \varepsilon_1 & 0 & 0 \\ 0 & \varepsilon_2 & 0 \\ 0 & 0 & \varepsilon_3 \end{bmatrix}. \tag{6.2}$$

With this tensor we can associate a characteristic tensor surface

$$\varepsilon_1 x^2 + \varepsilon_2 y^2 + \varepsilon_3 z^2 = \text{const}, \tag{6.3}$$

which is known as the *Fresnel ellipsoid.* Three principal axes can be distinguished in it. These axes are characteristic because only for them the vector **D** coincides with the direction of the vector **E**. (If $\mathbf{E} = [E, 0, 0]$, then, because of the diagonal character of the tensor, we have the vector $\mathbf{D} = [D, 0, 0]$, which means that these two vectors are parallel. The same applies to the other axes.)

Sometimes it is more convenient to use a tensor inverse to (6.2). It is defined by

$$\varepsilon^{-1} = \begin{bmatrix} \dfrac{1}{\varepsilon_1} & 0 & 0 \\ 0 & \dfrac{1}{\varepsilon_2} & 0 \\ 0 & 0 & \dfrac{1}{\varepsilon_3} \end{bmatrix}. \tag{6.4}$$

For this tensor there exists a tensor surface of the form

$$\frac{x^2}{n_1^2} + \frac{y^2}{n_2^2} + \frac{z^2}{n_3^2} = \text{const}, \tag{6.5}$$

which is called the *ellipsoid of indices.* The indices are defined by relations

$$n_i = \sqrt{\mu_r \varepsilon_i}.$$

In optics we do not deal with magnetic anisotropy, since its influence on optical phenomena is insignificant, because the magnetization cannot follow fast variations of optical vibrations. That is why the relative permeability for crystals μ_r will be equated to unity.

6.2. The Structure of a Plane Monochromatic Wave in an Anisotropic Medium

Maxwell's equations for a medium postulated above take the form

$$-\mu_0 \frac{\partial \mathbf{H}}{\partial t} = \operatorname{curl}\mathbf{E}, \qquad \frac{\partial \mathbf{D}}{\partial t} = \operatorname{curl}\mathbf{H},$$
$$\operatorname{div}\mathbf{D} = 0, \qquad \operatorname{div}\mathbf{H} = 0 \tag{6.6}$$

(a non-conducting medium: $\mathbf{j} = \mathbf{0}$, without free uncompensated electric charges $\rho = 0$).

From equations (6.6) we can draw a few interesting conclusions for the case of plane monochromatic waves, i.e. for waves

$$\mathbf{E} = \mathbf{E}_0 \exp\left[\mathrm{i}(\mathbf{k}\cdot\mathbf{r} - \omega t)\right]. \tag{6.7}$$

Taking into account relations (6.1) and (6.7), the equation $\operatorname{div}\mathbf{D} = 0$ leads to the dependence

$$\mathbf{k}\cdot\mathbf{D} = 0, \tag{6.8}$$

which proves that the vector **D** is perpendicular to the direction of the wave vector determining the propagation of the surface of constant phase. From first two equations (6.6), having substituted (6.7), we get the equalities

$$+\mu_0 \mathrm{i}\omega\mathbf{H} = \mathrm{i}(\mathbf{k}\times\mathbf{E}), \qquad -\mathrm{i}\omega\mathbf{D} = \mathrm{i}(\mathbf{k}\times\mathbf{H}), \tag{6.9}$$

from which we conclude that the vector **D** is perpendicular to **H**, and **H** is perpendicular to **k** and **E**. The physical meaning of equalities (6.8) and (6.9) is shown in Fig. 6.1, from which it can be seen that:

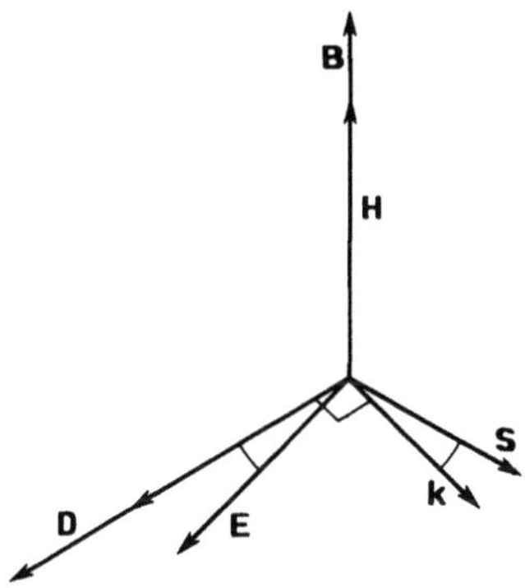

Figure 6.1

(1) vectors **D**, **E**, **k** and $\mathbf{S} = \mathbf{E} \times \mathbf{H}$ (Poynting's vector) are coplanar,

(2) vector **S** determining the direction of energy flow does not coincide with the propagation direction of the surface of constant phase.

Eliminating the field **H** from (6.9), we arrive at the relation

$$\omega^2 \mu_0 \mathbf{D} = k^2 \mathbf{E} - \mathbf{k}(\mathbf{k} \cdot \mathbf{E}), \tag{6.10}$$

which, after considering the equalities

$$k^2 = \frac{\omega^2}{c^2} n^2, \quad \mathbf{k} = k\mathbf{s}, \quad |\mathbf{s}| = 1,$$

can be written in the form

$$\varepsilon_i E_i = n^2 [E_i - s_i(\mathbf{s} \cdot \mathbf{E})] \quad (i = 1, 2, 3). \tag{6.11}$$

Relations (6.11) make up the system of three homogeneous equations with respect to unknown quantities E_i. Non-zero solutions of the system occur only when the determinant of the system equals zero. This leads to the equation

$$\frac{s_1^2}{n^2 - \varepsilon_1} + \frac{s_2^2}{n^2 - \varepsilon_2} + \frac{s_3^2}{n^2 - \varepsilon_3} = \frac{1}{n^2}, \tag{6.12}$$

which on considering the following relations:

$$s_1^2 + s_2^2 + s_3^2 = 1,$$

$$v_i = \frac{c}{\sqrt{\varepsilon_i}} \qquad (i = 1, 2, 3), \tag{6.13}$$

$$v_f = \frac{c}{n},$$

gives the *Fresnel equation* for the velocity of phase propagation of electromagnetic waves in an anisotropic medium

$$\frac{s_1^2}{v^2 - v_1^2} + \frac{s_2^2}{v^2 - v_2^2} + \frac{s_3^2}{v^2 - v_3^2} = 0. \tag{6.14}$$

Equation (6.14) is a square equation with respect to v_f^2. For each value of v_f determined from (6.14) and for a fixed direction $\mathbf{s} = [s_1, s_2, s_3]$, we can find from (6.11) the corresponding components E_i $(i = 1, 2, 3)$.

To different values $v_f^{(k)}$ correspond different vectors $\mathbf{E}^{(k)}$, to which in turn we can assign vectors $\mathbf{D}^{(k)} = \varepsilon \cdot \mathbf{E}^{(k)}$. These vectors are mutually perpendicular, i.e.

$$\mathbf{D}^{(k_1)} \cdot \mathbf{D}^{(k_2)} = 0. \tag{6.15}$$

It can be easily proved by taking equation (6.11) into consideration. From this equation we obtain

$$E_i^{(k)} = \frac{v_i^2}{v_i^2 - (v_f^{(k)})^2} s_i(\mathbf{s} \cdot \mathbf{E}^{(k)}), \tag{6.16}$$

and respectively (for a diagonal form of the tensor)

$$D_i^{(k)} = \frac{\mu_0^{-1}}{v_i^2 - (v_f^{(k)})^2} s_i(\mathbf{s} \cdot \mathbf{E}^{(k)}). \tag{6.17}$$

Substituting (6.17) into (6.15) we an write

$$\sum_i D_i^{(k_1)} D_i^{(k_2)} = \sum_i \frac{s_i^2 (\mathbf{s} \cdot \mathbf{E}^{(k_1)})(\mathbf{s} \cdot \mathbf{E}^{(k_2)})\, \mu_0^{-2}}{\{v_i^2 - (v_f^{(k_1)})^2\}\{v_i^2 - (v_f^{(k_2)})^2\}}$$

$$= (\mathbf{s} \cdot \mathbf{E}^{(k_1)})(\mathbf{s} \cdot \mathbf{E}^{(k_2)}) \frac{\mu_0^{-2}}{(v_f^{(k_1)})^2 - (v_f^{(k_2)})^2} \times$$

$$\times \sum_i \left[\frac{s_i^2}{v_i^2 - (v_f^{(k_1)})^2} - \frac{s_i^2}{v_i^2 - (v^{(k_2)})^2}\right] = 0, \tag{6.18}$$

where the last equality is a consequence of equation (6.14). Thus the validity of (6.15) has been proved, and consequently the fact that vectors **D** connected with different values v_f are mutually perpendicular, which corresponds to two different states of wave polarization.

6.3. Geometrical Construction for Determining the Velocity of Propagation and the Direction of Vibrations of the Electric Vector

By means of the ellipsoid (6.5) we can find geometrically phase velocities and two directions of vibrations of the vector **D** related to a given propagation direction **s**. To this end we draw a plane containing the origin of the coordinate system and perpendicular to **s**. The intersection curve of this plane with the ellipsoid of indices is an ellipse. The lengths of semi-axes for this ellipse are proportional to the inverse $(1/v_f)$ of the phase velocity, while the directions of axes determine the directions of vibrations of the vector **D** (Fig. 6.2). To prove this let us consider the equations defining the ellipse:

$$\begin{aligned} &xs_1 + ys_2 + zs_3 = 0, \\ &\frac{x^2}{\varepsilon_1} + \frac{y^2}{\varepsilon_2} + \frac{z^2}{\varepsilon_3} = 1. \end{aligned} \tag{6.19}$$

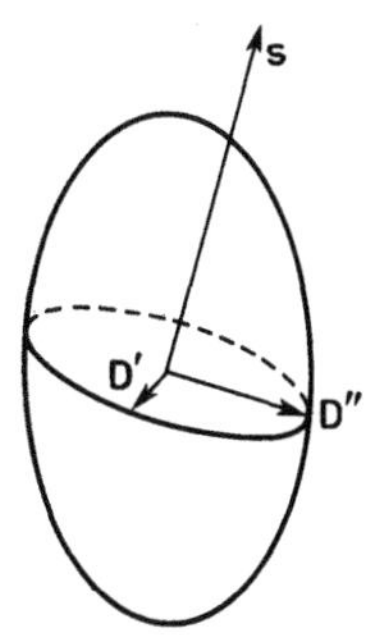

Figure 6.2

Since the principal axes of an ellipse are its extreme chords, we can find them from the condition for the extreme of the quantity

$$r^2 = x^2 + y^2 + z^2, \tag{6.20}$$

under additional conditions (6.19). We solve this problem by means of Lagrange's method of unspecified multipliers. We introduce two multipliers $2\lambda_1$ and λ_2 and with their use we construct the function

$$F = x^2 + y^2 + z^2 + 2\lambda_1(xs_1 + ys_2 + zs_3) + \\ + \lambda_2\left(\frac{x^2}{\varepsilon_1} + \frac{y^2}{\varepsilon_2} + \frac{z^2}{\varepsilon_3} - 1\right).$$

In this way our problem has been resolved into finding the extreme of the function $F(x, y, z)$ without any additional conditions. The necessary conditions for the extremum of F (i.e. vanishing of its derivatives $\partial F/\partial x = \partial F/\partial y = \partial F/\partial z = 0$) give the relations

$$\begin{aligned} x + \lambda_1 s_1 + \lambda_2 \frac{x}{\varepsilon_1} &= 0, \\ y + \lambda_1 s_2 + \lambda_2 \frac{y}{\varepsilon_2} &= 0, \\ z + \lambda_1 s_3 + \lambda_2 \frac{z}{\varepsilon_3} &= 0. \end{aligned} \tag{6.21}$$

Multiplying (6.21) by x, y, z respectively and summing up these equations, we obtain, in view of (6.19), the equality

$$r^2 + \lambda_2 = 0. \tag{6.22}$$

The same operations done on (6.21), the only difference being the replacing of x, y, z by s_1, s_2, s_3, yield the relation

$$\lambda_1 + \lambda_2\left(\frac{xs_1}{\varepsilon_1} + \frac{ys_2}{\varepsilon_2} + \frac{zs_3}{\varepsilon_3}\right) = 0. \tag{6.23}$$

Substituting λ_1 and λ_2 determined by (6.22) and (6.23) into (6.21) gives

$$x\left(1-\frac{r^2}{\varepsilon_1}\right)+s_1r^2\left(\frac{xs_1}{\varepsilon_1}+\frac{ys_2}{\varepsilon_2}+\frac{zs_3}{\varepsilon_3}\right)=0,$$

$$y\left(1-\frac{r^2}{\varepsilon_2}\right)+s_2r^2\left(\frac{xs_1}{\varepsilon_1}+\frac{ys_2}{\varepsilon_2}+\frac{zs_3}{\varepsilon_3}\right)=0, \tag{6.24}$$

$$z\left(1-\frac{r^2}{\varepsilon_3}\right)+s_3r^2\left(\frac{xs_1}{\varepsilon_1}+\frac{ys_2}{\varepsilon_2}+\frac{zs_3}{\varepsilon_3}\right)=0.$$

For a given $\mathbf{s} = [s_1, s_2, s_3]$ these are three homogeneous equations for the variables x, y, z.

Equations (6.24) differ only in notation from equations (6.11). In fact, if we make in (6.24) the substitutions

$$x \to \frac{D_1}{\sqrt{w}}, \quad \frac{x}{\varepsilon_1} \to \frac{E_1}{\sqrt{w}},$$

$$y \to \frac{D_2}{\sqrt{w}}, \quad \frac{y}{\varepsilon_2} \to \frac{E_2}{\sqrt{w}},$$

etc. (where $w = \mathbf{E}\cdot\mathbf{D}$), and

$$r^2 \to \frac{\mathbf{D}\cdot\mathbf{D}}{w} = \frac{\mathbf{D}\cdot\mathbf{D}}{\mathbf{D}\cdot\mathbf{E}} = n^2,$$

where $n^2 = \varepsilon$, we obtain the system of equations

$$D_1 = n^2[E_1 - s_1(\mathbf{E}\cdot\mathbf{s})],$$

$$D_2 = n^2[E_2 - s_2(\mathbf{E}\cdot\mathbf{s})], \tag{6.25}$$

$$D_3 = n^2[E_3 - s_3(\mathbf{E}\cdot\mathbf{s})],$$

which is in agreement with (6.11). Therefore, it can be stated that the roots of the equation equating the determinant of the system (6.24) (r being the lengths of semi-axes of the ellipse) to zero and the roots of system (6.25) are proportional. Besides, two possible directions of the vector $\mathbf{D}$: $\mathbf{D}'$ and $\mathbf{D}''$, coincide with the axes of the ellipse (6.19). The axes of an ellipse are perpendicular to each other, therefore it can be stated that:

The directions of vibrations of two vectors $\mathbf{D}'$ *and* $\mathbf{D}''$ *corresponding to a given propagation direction* $\mathbf{s}$ *are perpendicular to each other.*

In the case of a uniaxial crystal, when $\varepsilon_1 = \varepsilon_2 = \varepsilon^{(o)}$ and $\varepsilon_3 = \varepsilon^{(e)}$, the ellipsoid (6.19) becomes an ellipsoid of revolution with a distinguished axis called the *optical axis.* Let us take a plane in which lies the vector of propagation $\mathbf{s}$ and the optical axis. This plane is called the *principal plane.* An electromagnetic wave of the vector $\mathbf{D}'$ perpendicular to this plane will propagate with a constant velocity independent of the angle δ made by the vector $\mathbf{s}$ and the optical axis, while the wave of the vector $\mathbf{D}''$ lying in the principal plane will change its velocity of propagation with the change of the angle δ.

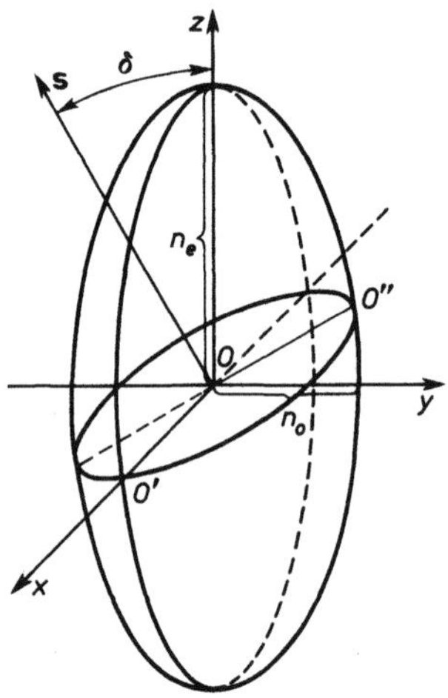

Figure 6.3

This is due to the fact that (Fig. 6.3) the segment OO' perpendicular to the principal plane and joining the centre of the ellipsoid with its surface points lying also on the ellipse (6.19) remains constant independently of the angle δ. It is equal to one of the semi-axes of the ellipsoid in question. On the other hand, the semi-axis of the ellipse OO'' lying in the principal plane will depend on the angle δ. For $\delta = 0$ the ellipse under consideration degenerates into a circle, which means that the velocity of propagation is independent of the direction of vibrations of the vector **D**. Dependence of the segments OO' and OO'' on the angle δ is given by

$$OO' = n_{o'} \qquad OO'' = \frac{1}{\sqrt{\dfrac{\sin^2\delta}{n_e^2} + \dfrac{\cos^2\delta}{n_o^2}}}. \tag{6.26}$$

In the case of a crystal for which the tensor of electric permittivity (6.2) has three different diagonal elements $(\varepsilon_1 \neq \varepsilon_2 \neq \varepsilon_3)$ in the system of crystal axes, the ellipsoid of indices will not show rotational symmetry. However, an arbitrary ellipsoid has always two circular sections C_1 and C_2, being a result of intersecting the ellipsoid with the planes passing through its centre. The versors $\mathbf{s}_1$ and $\mathbf{s}_2$ perpendicular to these planes are coplanar with the longest and shortest principal axes of the ellipsoid. These two directions are called the *optical axes*, as there is only one velocity of propagation along them, which is independent of the direction of the vector **D**.

6.4. Uniaxial Crystals

If the tensor of electric permittivity of a crystal has the form

$$\varepsilon = \varepsilon_0 \begin{bmatrix} \varepsilon^{(o)} & 0 & 0 \\ 0 & \varepsilon^{(o)} & 0 \\ 0 & 0 & \varepsilon^{(e)} \end{bmatrix},$$

the ellipsoid of indices (6.5) is an ellipsoid of revolution. In this case there is only one direction of wave propagation along which the velocity of propagation has a constant value independent of the orientation of the vector **D**. This direction is called the optical axis of a crystal, and crystals with such a property are called *uniaxial crystals*.

To uniaxial crystals belong crystals crystallizing in trigonal, tetragonal and hexagonal systems. The optical axis of such crystals coincides with the axis of symmetry of third, fourth or sixth order respectively; it is at the same time one of the principal axes of the tensor ε. The main value of the tensor related with this axis is denoted by $\varepsilon^{(e)}$. The directions of the other two principal axes are arbitrary, and the appropriate main values of the tensor are identical. They are denoted by $\varepsilon^{(o)}$.

Putting in Fresnel's equation (6.14)

$$\varepsilon_1 = \varepsilon_2 = \varepsilon^{(o)}, \qquad \varepsilon_3 = \varepsilon^{(e)},$$

we obtain the equation

$$(s_1^2 + s_2^2)(v_f^2 - v_o^2)(v_f^2 - v_e^2) + s_3^2(v_f^2 - v_o^2)^2 = 0, \tag{6.27}$$

where

$$v_o = \frac{c}{\sqrt{\varepsilon^{(o)}}}, \qquad v_e = \frac{c}{\sqrt{\varepsilon^{(e)}}}.$$

Let δ denote the angle which the vector $\mathbf{s}$ determining the direction of phase propagation of a plane wave forms with the z-axis. Then

$$s_1^2 + s_2^2 = \sin^2\delta, \qquad s_3^2 = \cos^2\delta, \tag{6.28}$$

and (6.27) becomes

$$(v_f^2 - v_o^2)(v_f^2 - v_e^2)\sin^2\delta + (v_f^2 - v_o^2)\cos^2\delta = 0. \tag{6.29}$$

The roots of this equation are given by

$$v_f'^2 = v_o^2, \qquad v_f''^2 = v_o^2\cos^2\delta + v_e^2\sin^2\delta. \tag{6.30}$$

Let us introduce the term of the *surface of phase velocity* as a locus of points obtained in this way that along every direction of the wave vector $\mathbf{s}$ two segments of lengths equal to v_f' and v_f'', respectively, are laid off. Now we may give the following interpretation of (6.30).

For uniaxial crystals there are two surfaces of wave velocity: one is a sphere of radius $v_f' = v_o$, the other is an ellipsoid. The first corresponds to an ordinary wave with the propagation velocity independent of the direction of propagation, the second to an extraordinary wave with propagation velocity dependent on the angle between the direction of the wave vector and the optical axis.

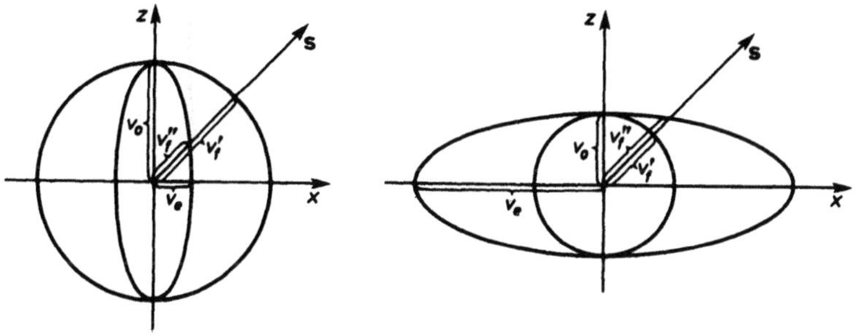

Figure 6.4

These two velocities are equal only if $\delta = 0$, i.e. if the wave vector coincides with the optical axis (Fig. 6.4).

If $v_o > v_e$, the *ordinary* wave travels faster than the *extraordinary* one; a crystal of this kind is called a *positive uniaxial crystal.*

If $v_o < v_e$, the ordinary wave travels more slowly than the extraordinary one, in this case we speak of a *negative uniaxial crystal.*

6.5. Biaxial Crystals

If the tensor of electric permittivity of a crystal has in the system of its axes the form

$$\varepsilon = \begin{bmatrix} \varepsilon_1 & 0 & 0 \\ 0 & \varepsilon_2 & 0 \\ 0 & 0 & \varepsilon_3 \end{bmatrix}, \tag{6.31}$$

i.e. if there are three different elements on its diagonal, then the ellipsoid of indices (6.5) is not a revolution surface. However, since an ellipsoid has two circular sections passing through its centre, we can distinguish two directions along which the velocity of propagation of the wave is independent of the direction of vibrations of the vector **D**.

These two directions are called the optical axes of a crystal, and crystals with such properties are called *biaxial crystals.* To these crystals are numbered crystals crystalizing in the orthorhombic, monoclinic and triclinic systems. In media of this kind the ordinary and the extraordinary wave cannot be distinguished – both waves behave as the extraordinary wave in a uniaxial crystal.

The optical properties of biaxial crystals result from Fresnel's equation (6.14). This equation assigns to each direction of propagation of a plane wave $\mathbf{s} = [s_1, s_2, s_3]$ two different phase velocities v'_f and v''_f and determines the surface of phase velocity. We can write it in the form

$$s_1^2(v_f^2 - v_2^2)(v_f^2 - v_3^2) + s_2^2(v_f^2 - v_1^2)(v_f^2 - v_3^2) + \\ + s_3^2(v_f^2 - v_1^2)(v_f^2 - v_2^2) = 0, \qquad (6.32)$$

from which it is evident that it is a fourth degree equation with regard to the variable v_f. To study the surface of phase velocity (6.32) it is necessary to examine in succession the shape of the intersections of the surface with the planes: $s_1 = 0$, $s_2 = 0$, and $s_3 = 0$. In the first case (for $s_1 = 0$) we obtain

$$(v_f^2 - v_1^2)[s_2^2(v_f^2 - v_3^2) + s_3^2(v_f^2 - v_2^2)] = 0, \qquad (6.33)$$

which implies two possibilities

$$v_f^2 = v_1^2, \quad v_f^2 = s_2^2 v_3^2 + s_3^2 v_2^2. \qquad (6.34)$$

In the second case (for $s_2 = 0$) we obtain

$$v_f^2 = v_2^2, \quad v_f^2 = s_1^2 v_3^2 + s_3 v_1^2. \qquad (6.35)$$

In the third case (for $s_3 = 0$) we get

$$v_f^2 = v_3^2, \quad v_f^2 = s_1^2 v_2^2 + s_2^2 v_1^2. \qquad (6.36)$$

In all three cases of intersecting the surface of phase velocity by appropriate planes we obtain a circle and an oval. The difference between these cases lies merely in the relative positions of the two curves. Thus if we put $\varepsilon_1 < \varepsilon_2 < \varepsilon_3$ or, which is fully equivalent, $v_1 > v_2 > v_3$ (in view of the equality $v_i = 1/\sqrt{\mu\varepsilon_i\varepsilon_0}$), the relative positions of the curves will be such as in Fig. 6.5. In order to draw the figure we have used the following variables:

$$x = v_f s_1, \quad y = v_f s_2, \quad z = v_f s_3. \qquad (6.37)$$

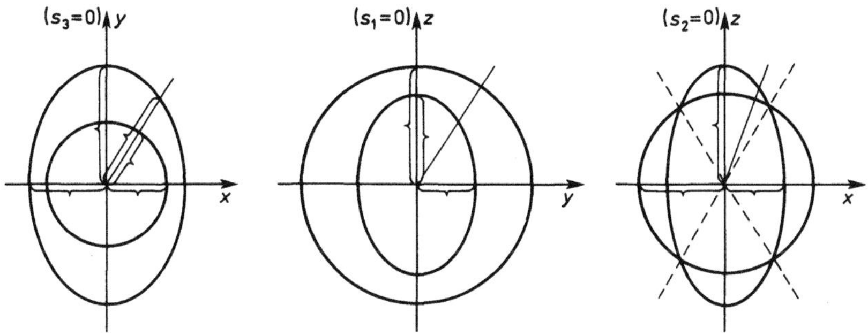

Figure 6.5

For $s_1 = 0$ (remembering that $s_1^2 + s_2^2 + s_3^2 = 1$) we obtain from (6.37)

$$v_f^2 = y^2 + z^2,$$

which permits to present (6.34) in the form

$$y^2 + z^2 = v_1^2, \quad y^2 v_3^2 + z^2 v_2^2 = (y^2 + z^2)^2. \tag{6.38}$$

Likewise, (6.35) and (6.36) assume the form

$$x^2 + z^2 = v_2^2, \quad x^2 v_3^2 + z^2 v_1^2 = (x^2 + z^2)^2 \tag{6.39}$$

and

$$x^2 + y^2 = v_3^2, \quad x^2 v_2^2 + y^2 v_1^2 = (x^2 + y^2)^2 \tag{6.40}$$

respectively.

Thus the intersections of the surface of phase velocities with the planes $x = 0$, $y = 0$, and $z = 0$ produce the curves (6.38), (6.39), and (6.40) respectively. The surface of phase velocity itself has four points of intersections, each of them lying in a different quadrant of the xz-plane. Using the system of equations (6.39), we obtain the coordinates of the point of intersection

$$x = \pm v_2 \sqrt{\frac{v_2^2 - v_1^2}{v_3^2 - v_1^2}}, \quad z = \pm v_2 \sqrt{\frac{v_3^2 - v_2^2}{v_3^2 - v_1^2}}. \tag{6.41}$$

By the definition of the variables x and z formulas (6.37) and (6.41) allow us to determine two directions of phase propagation

$[s_1^{(1)}, 0, s_3^{(1)}]$ and $[s_1^{(2)}, 0, s_3^{(2)}]$ for which the phase velocity is constant, independently of the direction of vibrations of the vector **D**. This confirms the existence of two optical axes. Each of them is orientated at the angle β with respect to the z-direction. β can be defined in the following way: putting $s_1 = \sin\beta$, $s_3 = \cos\beta$ and considering (6.37) and (6.41), we obtain

$$\tan\beta = \pm\sqrt{\frac{v_1^2 - v_2^2}{v_2^2 - v_3^2}} = \frac{s_1}{s_3}. \tag{6.42}$$

From (6.42) it can be seen that the optical axes are symmetrical in relation to the z-axis.

The value of phase velocity for the propagation direction **s** which makes with the optical axes the angles δ_1 and δ_2 is expressed by the formula

$$v_f^2 = \tfrac{1}{2}\left[v_1^2 + v_3^2 + (v_1^2 - v_3^2)\cos(\delta_1 \pm \delta_2)\right]. \tag{6.43}$$

Although v_2 does not appear in formula (6.43) explicitly, it is involved implicitly in the defining relations for the angles δ_1 and δ_2. Formula (6.43) in the case of uniaxial crystal $v_1 = v_2$ implies expressions (6.30). The justification for (6.43) is as follows: in view of the condition

$$s_1^2 + s_2^2 + s_3^2 = 1,$$

equation (6.32) may be written in the form

$$\begin{aligned} v_f^4 - v_f^2\left[s_1^2(v_2^2 - v_1^2) + s_3^2(v_2^2 - v_3^2) + v_1^2 + v_3^2\right] + \\ + s_1^2 v_3^2(v_2^2 - v_1^2) + s_3^2 v_1^2(v_2^2 - v_3^2) + v_1^2 v_3^2 = 0. \end{aligned} \tag{6.44}$$

With the direction of propagation **s** being fixed, (6.44) is a square equation with respect to v_f^2. The form of its solutions is

$$v_f^2 = \tfrac{1}{2}\left\{\left[s_1^2(v_2^2 - v_1^2) + s_3^2(v_2^2 - v_3^2) + v_1^2 + v_3^2\right] \pm \sqrt{\Delta}\right\}, \tag{6.45a}$$

where

$$\begin{aligned} \Delta = \left[s_1^2(v_2^2 - v_1^2) + s_3^2(v_2^2 - v_3^2) + v_1^2 + v_3^2\right]^2 - \\ - 4\left[s_1^2 v_3^2(v_2^2 - v_1^2) + s_3^2 v_1^2(v_2^2 - v_3^2) + v_1^2 v_3^2\right]. \end{aligned} \tag{6.45b}$$

Since the directional cosines of the optical axes are $\pm\sin\beta$, 0, $\cos\beta$, and the angles δ_1 and δ_2 formed by the vector $\mathbf{s}$ with the optical axes are given by the equations

$$\begin{aligned}\cos\delta_1 &= s_1\sin\beta + s_3\cos\beta,\\ \cos\delta_2 &= -s_1\sin\beta + s_3\cos\beta,\end{aligned} \tag{6.46}$$

we can, considering (6.42), write the equalities

$$\begin{aligned}s_1^2 &= \frac{1}{4}\frac{v_1^2 - v_3^2}{v_1^2 - v_2^2}(\cos\delta_1 - \cos\delta_2)^2,\\ s_3^2 &= \frac{1}{4}\frac{v_1^2 - v_3^2}{v_2^2 - v_3^2}(\cos\delta_1 + \cos\delta_2)^2.\end{aligned} \tag{6.47}$$

Taking (6.47) into account in formulas (6.45), we can write

$$\begin{aligned}s_1^2(v_2^2 - v_1^2) + s_3^2(v_2^2 - v_3^2) + (v_1^2 + v_3^2)\\ = (v_1^2 - v_3^2)\cos\delta_1\cos\delta_2 + (v_1^2 + v_3^2)\end{aligned}$$

and

$$\Delta = (v_1^2 - v_3^2)^2\sin^2\delta_1\sin^2\delta_2.$$

These two equalities when applied in (6.45) give immediately the required result (6.43).

6.6. Refraction in Crystals

If a plane wave falls from the vacuum upon a plane surface Σ of an anisotropic medium, there occurs the phenomenon of refraction and reflection of the wave.

Let $\mathbf{s}$ be a versor normal to the wave front of the incident wave, and $\mathbf{s}'$ a versor of the refracted wave. It turns out that there are two possible values of the versor $\mathbf{s}'$. The field vectors of the incident field and of the refracted field are functions of

$$t - \frac{\mathbf{r}\cdot\mathbf{s}}{c}, \quad t - \frac{\mathbf{r}\cdot\mathbf{s}'}{v_f},$$

respectively, where v_f is the velocity of propagation of a wave in crystal.

The condition of field continuity at the interface requires that for any point $\mathbf{r}$ lying on the plane Σ and for any instant of time t the following relation must hold:

$$t - \frac{\mathbf{r} \cdot \mathbf{s}}{c} = t - \frac{\mathbf{r} \cdot \mathbf{s}'}{v_f}, \tag{6.48}$$

which implies

$$\mathbf{r} \cdot \left(\frac{\mathbf{s}'}{v_f} - \frac{\mathbf{s}}{c} \right) = 0. \tag{6.49}$$

Condition (6.49) means that the vector $\mathbf{s}'/v' - \mathbf{s}/c$ is perpendicular to Σ.

The versor of the refracted wave $\mathbf{s}'$ may be determined as follows: from a certain point O on Σ as origin we lay off (Fig. 6.6) in all directions $\mathbf{s}'$ vectors, the length of each of them being $1/v'_f$, where v'_f is the phase velocity corresponding to $\mathbf{s}'$ in accordance with the Fresnel equation (6.14). The locus of the end-points is a two-sheet surface of wave vector. The vector $\mathbf{s}/c$ is plotted in the same manner.

Since the vectors $\mathbf{s}'/v'_f$ must be such that $\mathbf{s}'/v'_f - \mathbf{s}/c$ is perpendicular to Σ, its end-point Q' must be on the normal to the surface Σ which passes through the end-point P of the vector $\mathbf{s}/c$. In general, the normal to Σ cuts the two-sheet surface discussed above in four points. Two of them lie at the same side of the boundary of the crystal (Q' and Q'' in Fig. 6.6). Hence, there are two possible

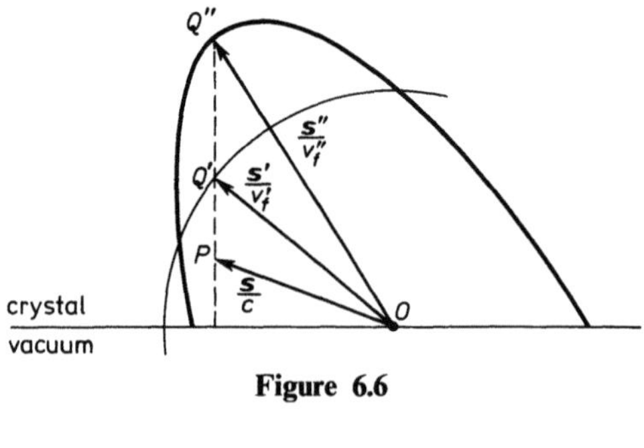

Figure 6.6

directions of phase propagation, so that in general each incident wave gives rise to two refracted waves. This phenomenon is called *birefringence.*

Taking as the boundary plane Σ the plane $z = 0$, we can write equality (6.49) as

$$\frac{xs_x + ys_y}{c} = \frac{xs'_x + ys'_y}{v'_f} = \frac{xs''_x + ys''_y}{v''_f}. \tag{6.50}$$

It must be fulfilled for all values x and y. It can take place only if

$$\frac{s''_x}{s_x} = \frac{s''_y}{s_y}, \quad \frac{s'_x}{s_x} = \frac{s'_y}{s_y}, \tag{6.51}$$

i.e., if the wave vectors lie in the plane of incidence.

If O_i, O_t, O''_t denote the angles formed, respectively, by the incident and the two refracted waves with the normal to Σ, we obtain from (6.51)

$$\frac{\sin O_i}{\sin O'_t} = \frac{c}{v'_f}, \quad \frac{\sin O_i}{\sin O''_t} = \frac{c}{v''_f}. \tag{6.52}$$

Thus each of the refracted waves obeys the law of refraction as it was in the case of isotropic media. However, it is more complicated to determine the direction of wave propagation in a crystal, since the phase velocity depends on the angle of refraction $v_f = v_f(O_t)$.

In a uniaxial crystal one of the sheets of the surface of phase velocity is a sphere, thus in this case the velocity of the refracted wave is independent of the angle of refraction. This wave is called the *ordinary wave.* With this wave is associated the vector **D** perpendicular to the principal plane of the crystal (determined by the direction of propagation and the optical axis). The second sheet of the two-sheet surface of phase velocity in a uniaxial crystal is an ellipsoid. It corresponds to a wave related with the vector **D** lying in the principal plane of the crystal; the velocity of propagation for this wave is not constant and depends on the angle between the optical axis and the direction of propagation.

In both cases if the anisotropy axis is orientated arbitrarily with respect to the boundary surface and for normal incidence $(O_i = 0$

and $O_t' = O_t'' = 0$), the versors of phase propagation **s**$'$ and **s**$''$ in the crystal will coincide, remaining perpendicular to Σ. The both waves will differ only in phase velocity.

It will be different with the trajectory of the rays (Poynting's vectors) which do not coincide, as it was shown in Sec. 6.2, with the direction of phase propagation. A beam of rays of monochromatic light falling on the boundary surface of a crystal is divided into two beams differently deflected with respect to the direction of the incident beam, and not always lying in the same plane as the incident beam and the normal to the boundary surface.

If a beam of parallel rays falls perpendicularly on the surface of a uniaxial crystal, the optical axis not being perpendicular to the wave, then one of the beams passing through the crystal does not change its direction with respect to the direction of the incident beam, the second undergoes deflection (Fig. 6.7). After emerging from the crystal both the beams are parallel to each other. When the crystal is rotated by the angle 2π, the undeflected beam remains unchanged, while the deflected one draws a conical surface within the crystal.

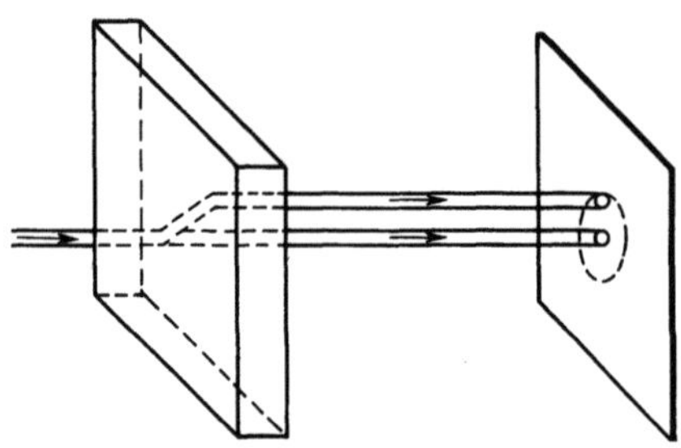

Figure 6.7

6.7. The Velocity of Propagation of Energy of an Electromagnetic Wave in an Anisotropic Medium

As it has been stated in Sec. 6.2, the direction of phase propagation in an anisotropic medium is not always the same as the direction of energy propagation. The same concerns the value of the velocity. The velocity of energy propagation, the so-called *velocity of ray* v_r, is defined by the relation

$$v_r = \frac{|\mathbf{S}|}{w}, \tag{6.53}$$

i.e. by the ratio of the value of Poynting's vector $|\mathbf{S}|$ to the volume density of energy.

In the case of a plane monochromatic wave the density of the field energy is expressed by

$$w = \frac{1}{v_f}\mathbf{s}\cdot(\mathbf{E}\times\mathbf{H}), \tag{6.54}$$

where v_f is the phase velocity, $\mathbf{s}$ a versor perpendicular to the plane of constant phase.

From these two relations it follows that

$$v_f = v_r \cos\alpha, \tag{6.55}$$

where α stands for the angle between the vectors $\mathbf{D}$ and $\mathbf{E}$ (see Fig. 6.1).

The ray velocity, likewise the phase velocity, is a function of direction, and for each direction of a ray

$$\mathbf{t} = \frac{\mathbf{E}\times\mathbf{H}}{|\mathbf{E}\times\mathbf{H}|} \tag{6.56}$$

it may assume two values. We can find these permissible values of the velocity of energy propagation for a fixed direction $\mathbf{t}$ from the equation

$$\frac{t_1^2}{\dfrac{1}{v_r^2}-\mu\varepsilon_1}+\frac{t_2^2}{\dfrac{1}{v_r^2}-\mu\varepsilon_2}+\frac{t_3^2}{\dfrac{1}{v_r^2}-\mu\varepsilon_3}=0, \tag{6.57}$$

which is derived analogically to the Fresnel equation for the phase velocity in an anisotropic medium.

Knowing the permissible values of the ray velocity v_r for all directions $\mathbf{t}$, we can construct the so-called *surface of ray velocity.*

To this end we lay off from a fixed point O inside the crystal in all directions $\mathbf{t}$ the positions vectors $\mathbf{r} = v_r\mathbf{t}$. Their ends form the sought surface of ray velocity, which is an analog of the surface of phase velocity introduced earlier (see Secs. 6.4 and 6.5) and specified by vectors $\mathbf{r} = v_f\mathbf{s}$.

It may be shown that a plane W' tangent to the surface of ray velocity is always perpendicular to a appropriate wave versor or, in other words, the surface of ray velocity is the envelope of planes drawn through the points on the surface of phase velocity, at right angles to the appropriate radius vectors from the origin to these points (Fig. 6.8).

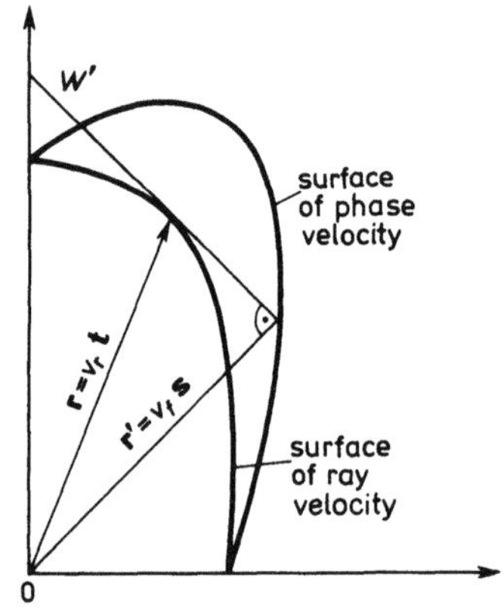

Figure 6.8

The prove of this thesis is based on the equality:

$$\frac{1}{\mu v_r^2}\mathbf{E} = \mathbf{D} - (\mathbf{t}\cdot\mathbf{D})\mathbf{t}, \tag{6.58}$$

which is obtained in the following way. We introduce the vector

$$\mathbf{D}_\perp = \mathbf{D} - (\mathbf{t}\cdot\mathbf{D})\mathbf{t}, \tag{6.59}$$

perpendicular to $\mathbf{t}$ and lying in the plane of the vectors $\mathbf{t}$ and $\mathbf{D}$. This vector is parallel to $\mathbf{E}$ (since $\mathbf{E}$ is perpendicular to $\mathbf{t}$ and lies in a plane specified by the vectors $\mathbf{t}$ and $\mathbf{D}$). Thus it may be expressed (using (6.11) and (6.55)) as

$$\mathbf{D}_\perp = \left(\mathbf{D}\cdot\frac{\mathbf{E}}{|\mathbf{E}|}\right)\frac{\mathbf{E}}{|\mathbf{E}|} = \frac{1}{\mu v_r^2}\mathbf{E} \tag{6.60}$$

which, together with (6.59), confirms the equality (6.58).

Having the equality (6.58) at our disposal, we can start proving the theorem formulated above. We begin multiplying both sides of (6.58) by v_r^2 and using the notation $\mathbf{r} = v_r\mathbf{t}$, which gives

$$\frac{1}{\mu}\mathbf{E} = r^2\mathbf{D} - \mathbf{r}(\mathbf{D}\cdot\mathbf{r}). \tag{6.61}$$

A small variation $\delta\mathbf{E}$ produces, according to the definition (6.56), a change in the versor $\mathbf{t}$, and this consequently permits, on the basis of (6.47), to go over from one point of the surface of ray velocity to the other. The radius vector $\delta\mathbf{r}$ will be tangent to the surface of ray velocity. If now additionally denote by $\delta\mathbf{D}$ the change of vector $\mathbf{D}$ corresponding to the change $\delta\mathbf{E}$, we obtain from (6.61)

$$\begin{aligned}\frac{1}{\mu}\delta\mathbf{E} = 2(\mathbf{r}\cdot\delta\mathbf{r})\mathbf{D} + r^2\delta\mathbf{D} - \delta\mathbf{r}(\mathbf{r}\cdot\mathbf{D}) - \\ - \mathbf{r}(\delta\mathbf{D}\cdot\mathbf{r}) - \mathbf{r}(\mathbf{D}\cdot\delta\mathbf{r}).\end{aligned} \tag{6.62}$$

Multiplying scalarly (6.62) by $\mathbf{D}$, and taking into account the equality

$$\mathbf{D}\cdot\delta\mathbf{E} = \mathbf{E}\cdot\varepsilon\cdot\delta\mathbf{E} = \mathbf{E}\cdot\delta\mathbf{D}$$

and (6.61), we obtain

$$\delta\mathbf{r}\cdot[(\mathbf{D}\times\mathbf{r})\times\mathbf{D}] = 0, \tag{6.63}$$

or finally

$$\delta\mathbf{r}\cdot\mathbf{s} = 0, \tag{6.64}$$

since the vector $(\mathbf{D}\times\mathbf{r})\times\mathbf{D}$ is parallel to the versor $\mathbf{s}$. Formula (6.64) shows that for small changes of $\mathbf{E}$ and $\mathbf{D}$ the variation of vector $\mathbf{r}$ is perpendicular to the wave versor $\mathbf{s}$. Thus the formulated thesis is confirmed. The relation between the surface of phase velocity and the surface of ray velocity is shown in Fig. 6.8. The content of the last theorem may be interpreted in physical terms. To this end we should consider not a single wave but a group formed by plane waves with the same frequency and with slightly different directions of propagation. The wave versors fill the solid angle about a "mean wave versor" $\mathbf{s}_0$.

Suppose that at the instant of time $t = 0$ all waves are in phase at a point 0 and only the amplitudes of those waves whose versors are close to $\mathbf{s}_0$ are significant. After a time unit the wave front of each of these plane waves will propagate by the distance $v_f\mathbf{s}$ and will reach the position W'. Each plane wave front W' is perpendicular to the appropriate vector $\mathbf{r} = v_f\mathbf{s}$. The amplitude of the wave group is the largest in the region where the plane waves amplify mutually, i.e. where the planes W' intersect each other, and this happens on the surface of ray of radius vector $\mathbf{r} = v_r\mathbf{t}$.

6.8. Huygens' Construction

In accordance with the results in Sec. 6.7, from a point source in a uniaxial crystal propagate waves with two velocities of energy propagation. One of the two velocities is independent of the direction of propagation, and the ray surface corresponding to it is a sphere. The other depends on the angle between the direction of propagation and the optical axis. The surface for this ray velocity is an ellipsoid of revolution whose axis is the optical axis of the crystal. These two surfaces of velocity are tangent with each other at the points of their intersection with the optical axis, because in this direction the disturbances of both kinds propagate with the same velocity. This situation is presented in Fig. 6.9.

As there are two different velocities of propagation, after time τ from the excitation at point O, there will appear in the crystal two surfaces: one of them circular, the second elliptical. If point O is one of numerous points of a plane wave propagating in the crystal, then drawing (according to the Huygens formula) the phase envelopes for elementary waves coming from various points of the primary wave we obtain the successive positions of the fronts of the propagating disturbances.

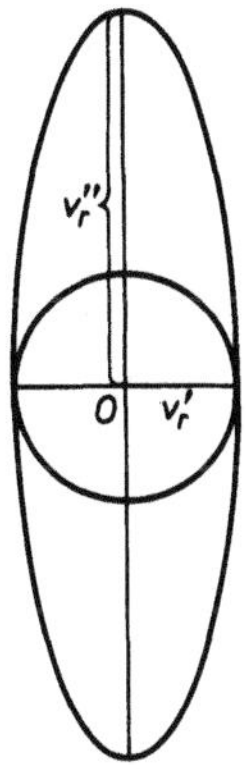

Figure 6.9

The birefringence in crystal can be explained as follows: let the optical axis of the crystal lie in the plane of incidence and make the angle β with the boundary surface AB, and let a plane wave fall from the outside upon the boundary surface perpendicularly to it (see Fig. 6.10). The envelopes of the elementary waves of both kinds will be parallel to the phase surface of the incident wave and to each other. Joining the point of incidence S with the points C and D at which the envelopes S_o and S_e osculate with the appropriate surfaces of ray velocity produced by the elementary waves, we determine the directions of the refracted rays. In the case under consideration only a ray corresponding to the spherical surface of ray velocity remains perpendicular to the front of the refracted wave S_o. This ray is not subject to refraction and it obeys the ordinary refraction law – it is the ordinary ray. The extraordinary ray is

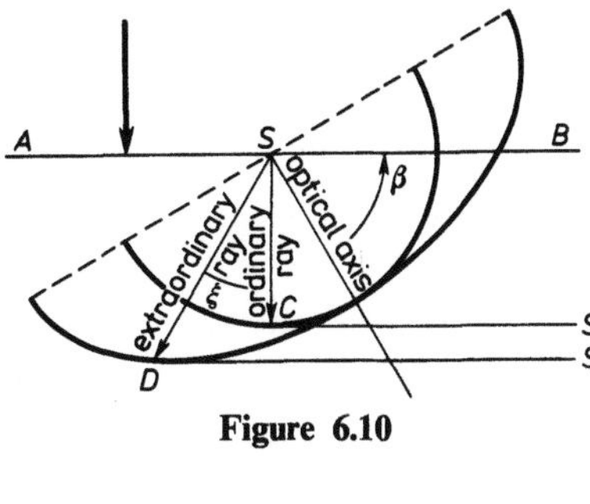

Figure 6.10

related to the ellipsoidal surface of ray velocity. Even for perpendicular incidence (under condition that the optical axis is not perpendicular to the boundary surface) it is deflected from the normal, remaining in the plane of incidence. In this situation the two rays: ordinary and extraordinary make the angle ξ. If the optical axis does not lie in the plane of incidence, then after refraction the ordinary ray of the refracted wave remains in the plane of incidence, while the extraordinary one goes out of this plane.

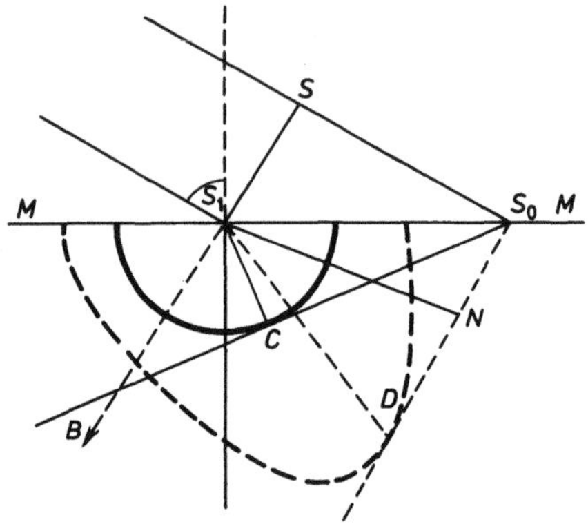

Figure 6.11

Let (Fig. 6.11) MM be the surface of the crystal upon which falls a plane wave S_1S, and S_1B is the direction of the optical axis. The phase envelope of the elementary waves, determining the direction of the ordinary ray of the refracted wave, touches the sphere of ray velocity of centre S_1 at point C lying in the plane of incidence. Thus the ordinary ray S_1C lies in the plane of incidence. The point of tangency D of the envelope with the elementary ellipsoidal surface of ray velocity goes out of this plane; the extraordinary ray S_1D does not lie in the plane of incidence. The normal S_1N to the envelope of the elementary ellipsoidal surface of phase velocity for the refracted wave lies in the plane of incidence, and the phase connected with it propagates in the direction S_1N. (The elements lying outside the plane of the figure are marked with the dashed line.)

6.9. The Change of Polarization Plane by the Passage through a Uniaxial Crystal Plate

Let us choose a system of coordinates in which the z-axis is perpendicular to the surface of the plate and coincides with the direction of propagation of the wave. The optical axis of the plate is assumed to be lying in the yz-plane and directed at the angle δ in relation to the z-axis (Fig. 6.12).

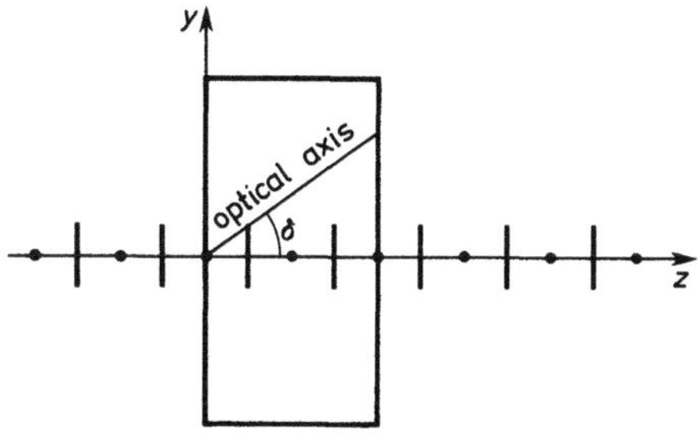

Figure 6.12

We will discuss the case where upon the "air–plate" interface there falls an electromagnetic wave of the induction vector

$$\mathbf{D} = \mathbf{i}\mathbf{D}_{01}\cos\omega\left(t - \frac{z}{c}\right) + \mathbf{j}D_{02}\cos\omega\left(t - \frac{z}{c}\right), \tag{6.65}$$

where $\mathbf{i}$ and $\mathbf{j}$ are the versors of the x- and y-axis of a chosen system of coordinates. It is a linearly polarized wave. Its induction vector makes with the x-axis the angle ψ_0, such that $\tan\psi_0 = D_{02}/D_{01}$. After entering a crystal medium this wave is naturally decomposed in the ordinary wave with vector $\mathbf{D}_1$ perpendicular to the principal plane of the crystal and the extraordinary wave with vector $\mathbf{D}_2$ lying in this plane. In accordance with the results from Sec. 6.4, the first of these waves will propagate with the velocity

$$v'_f = v_o = \frac{1}{\sqrt{\mu\varepsilon^{(o)}\varepsilon_0}}, \tag{6.66}$$

and the spherical sheet of the surface of phase velocity will correspond to it. The other wave will propagate with velocity

$$v_f'' = \sqrt{v_o^2 \cos^2\delta + v_e^2 \sin^2\delta}, \tag{6.67}$$

where $v_e = 1/\sqrt{\mu\varepsilon^{(e)}}$, and the surface of phase velocity assigned to it will be an ellipsoid. For the perpendicular incidence on the plate the both waves will propagate after refraction along the same perpendicular direction $O_i = O_t' = O_t'' = 0$ but with different velocities. That is why they gain different phase variations on traversing the plate of thickness l. If $\delta = \pi/2$, the optical axis remains parallel to the plane of the plate, and then the first, i.e., the ordinary wave, will gain in phase

$$\Delta_1 = \frac{\omega}{c} n_o l, \tag{6.68}$$

where $n_o = \sqrt{\varepsilon^{(o)}}$; the other, i.e. the extraordinary wave, will gain in phase

$$\Delta_2 = \frac{\omega}{c} n_e l, \tag{6.69}$$

where $n_e = \sqrt{\varepsilon^{(e)}}$.

Owing to this, the induction vector of the field after emerging from the plate will have the components:

$$\begin{aligned} D_1 &= D_{01} \cos(\omega t + kz + \Delta_1), \\ D_2 &= D_{02} \cos(\omega t + kz + \Delta_2), \end{aligned} \tag{6.70}$$

and the end of this vector will draw in space a curve whose projection on the xy-plane is

$$\left(\frac{D_1}{D_{01}}\right)^2 + \left(\frac{D_2}{D_{02}}\right)^2 - 2\frac{D_1 D_2}{D_{01} D_{02}} \cos(\Delta_2 - \Delta_1) = \sin^2(\Delta_2 - \Delta_1). \tag{6.71}$$

If the plate thickness is chosen in such a way that

$$\Delta_2 - \Delta_1 = \frac{\omega}{c}(n_e - n_o) l = \pi N \quad (N = 0, 1, 2, 3 \ldots), \tag{6.72}$$

the primary linear polarization will not change. If

$$\Delta_2 - \Delta_1 = \frac{\pi}{2}(2N + 1), \tag{6.73}$$

then an originally linearly polarized wave becomes an elliptically polarized one. Equality (6.73) holds if

$$\frac{\omega}{c}(n_e - n_o)l = \frac{\pi}{2}(2N + 1). \tag{6.74}$$

Equality (6.74) may be transformed into the form

$$l = \frac{2N + 1}{n_e - n_o}\frac{\lambda_0}{4}, \tag{6.75}$$

where $\lambda_0 = 2\pi c/\omega$ is the wavelength in vacuum. If the thickness of a crystallic plate is given by (6.75), the plate is called a *quater-wave plate.*

With the use of a quater-wave plate a primary wave with linear polarization may be transformed into a wave with elliptical polarization and the other way round. This feature of a quater-wave plate is used to distinguish elliptically polarized light from light with partial polarization. Using a polarization instrument only (the Nicol prism – see Sec. 6.11) is not sufficient for telling these two states of light polarization apart.

6.10. Compensators

As the elliptical polarization is a result of the superposition of two mutually perpendicular vibrations with the phase difference $\Delta \neq \pi/2$, in order to transform a wave of this kind into a linearly polarized one, it is necessary to introduce a certain additional phase difference Δ' which together with Δ gives the phase difference equal to π. To achieve this we apply, instead of a quarter-wave plate, an instrument called a *compensator*. It permits to obtain an arbitrary phase difference Δ^*. One of such instruments is the so-called *Babinet compensator.*

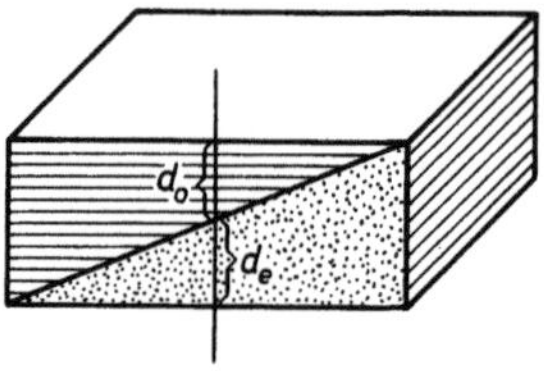

Figure 6.13

It consists of two wedges of quartz cut out in such a way that the optical axis in one of them is perpendicular to the optical axis in the other (Fig. 6.13). Thus the ordinary ray in the upper wedge becomes the extraordinary ray in the lower wedge, and the other way round. Accordingly, at a point where the wave traverses the equal thicknesses of the wedges there is no phase difference between the two rays. If the path in the first wedge is longer than that in the second one, there arises the phase difference between the rays. In this way dependently on the point of incidence we can acquire any phase difference.

The disadvantage of the Babinet compensator lies in the necessity of using very narrow light beams. The last restriction can be removed by the so-called *Bravais compensator.*

It consists of two quartz wedges with the optical axes mutually parallel and a plane-parallel plate of quartz with the optical axis perpendicular to the optical axes in the wedges (Fig. 6.14). The upper wedge can be shifted by means of a micrometer screw. The total thickness of the wedges along the whole length of their contact may be equal or not equal to the thickness of the lower plate.

In the first case the compensator does not introduce any phase difference, in the other we may obtain a required phase difference.

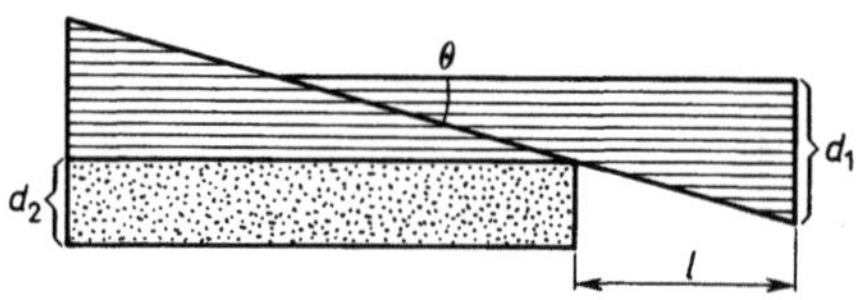

Figure 6.14

The difference in optical paths introduced by the wedges of height d_1, the angle of inclination θ, and the relative horizontal shift l is

$$\Delta' = (n_e - n_o)(d_1 - l\tan\theta). \tag{6.76}$$

The difference in optical paths introduced by a plane-parallel plate of thickness d_2 is

$$\Delta'' = (n_o - n_e)d_2. \tag{6.77}$$

The total difference in optical paths is

$$\Delta^* = (n_e - n_o)(d_1 - d_2 - l\tan\theta). \tag{6.78}$$

For $d_1 = d_2$ we have

$$\Delta^* = (n_o - n_e)l\tan\theta. \tag{6.79}$$

The corresponding phase difference for both the rays emerging from the compensator is

$$\delta = 2\pi\frac{\Delta' + \Delta''}{\lambda}. \tag{6.80}$$

6.11. The Nicol Prism

The Nicol prism is an instrument for obtaining linearly polarized light. It is cut out from crystal of Iceland spar $CaCO_3$, which is an uniaxial crystal. These crystals have a rhombohedron shape (Fig. 6.15), i.e. a solid bounded by six rhombuses with obtuse angles $\alpha = 101°52'$. The direction of optical axis coincides with the

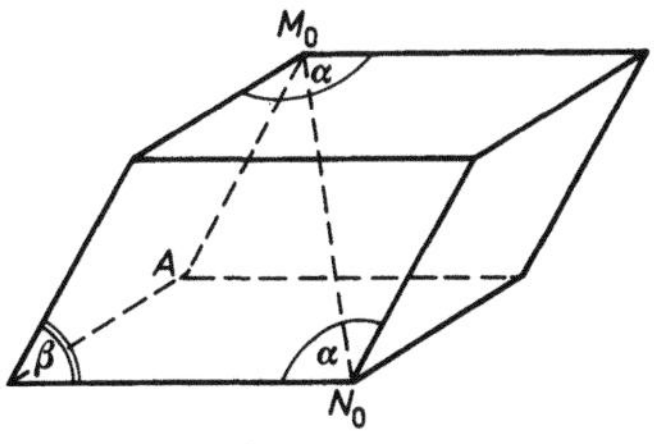

Figure 6.15

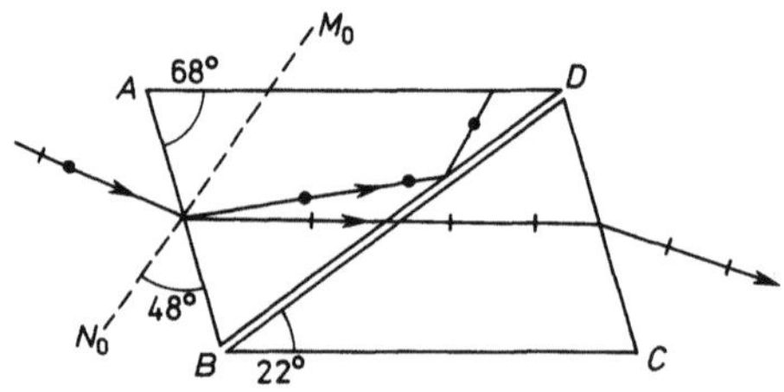

Figure 6.16

direction of the diagonal M_0N_0, which joins corners M_0 and N_0 at which the angles between rhombuses on all three walls are equal α.

The longitudinal section of the crystal is shown in Fig. 6.16. Fore and back edges AB and CD are inclined to edges AD and BC at the angle 68° (instead of natural angle 70°52′). This can be obtained by surface grinding. The optical axis M_0N_0 formes the angle 48° with edge AB. The diagonal plane BD is inclined to AD and BC at the angle 22°. The prism is cut along this plane and then cemented together with Canada balsam.

The refractive index of the Canada balsam is $n = 1.55$, two main refractive indices of the Iceland spar are: $n_o = 1.66$ and $n_e = 1.49$. For the ordinary wave of vector $\mathbf{D}'$ vibrating perpendicularly to the principal plane of the crystal the Canada balsam has lower optical density, for the extraordinary wave of vector $\mathbf{D}''$ vibrating in the principal plane of the crystal – higher optical density. The ordinary wave can enter the Canada balsam only if its angle of incidence is smaller than the angle of the total internal reflection defined by the formula

$$\sin\beta_{cr} = \frac{n}{n_o} = \frac{1.55}{1.66}, \tag{6.81}$$

which gives the angle $\beta_{cr} = 69°10'$.

If a wave of the vector of propagation parallel to AD is incident on the front face AB of the prism, it splits into two waves: ordinary and extraordinary. The ordinary wave falling on the surface of the Canada balsam at the angle 76°30′, greater than the critical angle, suffers internal reflection and is absorbed by the crystal holder. In order to achieve this situation, the primary angle 70°52′ at the vertex A is reduced by grinding to 68°.

The extraordinary wave traverses easily the layer of the balsam and after refraction at the back face it emerges from the prism parallely to the incident ray.

The Nicol prism transforms natural light or partially polarized light into linearly polarized light whose plane of vibration coincides with the principal plane, in which lie the vector of propagation and the optical axis M_0N_0. Rotating the prism around the vector of propagation **s** by a given angle, by the same angle rotates the plane of vibration of the vector **D** in the wave passing through the prism.

6.12. Birefringence Forced by an Electric Field

In a steady strong electric field an isotropic medium becomes an anisotropic one. An explanation for this kind of anisotropy is given by considering the nonlinear relation $\mathbf{D} = \mathbf{D}(\mathbf{E})$. Expanding the functional relations

$$D_i = D_i(E_1, E_2, E_3) \qquad (i = 1, 2, 3)$$

into a series (with respect to the components of the field **E**), and taking the terms up to the third order, we obtain

$$D_i = \varepsilon^{(0)}\delta_{ik}E_k + \alpha_{ikl}E_kE_l + \beta_{iklm}E_kE_lE_m. \tag{6.82}$$

The components of the dielectric tensor may be now defined by

$$\varepsilon_{ik} = \frac{\partial D_i}{\partial E_k},$$

which implies

$$\varepsilon_{ik} = \varepsilon^{(0)}\delta_{ik} + \alpha_{ikl}E_l + \beta_{iklm}E_lE_m. \tag{6.83}$$

For ordinary classical optical fields the terms further than the first in relation (6.82) are so small that they can be neglected, and accordingly for an isotropic medium (for classical sources of radiation) we obtain $\varepsilon_{ik} = \varepsilon^{(0)}\delta_{ik}$, where $\varepsilon^{(0)}$ is the isotropic permittivity of the medium in the absence of the field.

The higher order terms in expansion (6.83) play an essential role when, in addition to the optical wave with the electric vector $\mathbf{E}'$, there exists in a medium a strong polarizing external field $\mathbf{E}_0$ $(|\mathbf{E}_0| \gg E')$. Then in an isotropic medium the tensor of electric permittivity takes the form

$$\varepsilon_{ik} = \varepsilon^{(0)}\delta_{ik} + \beta E_{0i}E_{0k}. \tag{6.84}$$

The lack of the linear term in relation (6.84) is due to the fact that the change of the sense of vector $\mathbf{E}_0$ should not affect the physical situation. That is the property of a free isotropic medium and it is possible only if $\alpha_{ikl} = 0$.

Thus the tensor of electric permittivity of an isotropic medium placed in a strong external electric field depends only on the value of the field, and is expressed by formula (6.84), β being a scalar quantity dependent on the frequency of an optical wave. One of the principal axis of this tensor coincides with the direction of the external electric field, the principal value corresponding to it is

$$\varepsilon_{\parallel} = \varepsilon^{(0)} + \beta E_0^2. \tag{6.85}$$

The two remaining principal values are equal to each other and equal to

$$\varepsilon_{\perp} = \varepsilon^{(0)}, \tag{6.86}$$

and the principal axes corresponding to them lie in a plane perpendicular to the field, their directions being arbitrary. The isotropic medium placed in a strong external electric field behaves, regarding optical properties, like an uniaxial crystal. This phenomenon is known as *Kerr's effect* (*electrooptical effect*).

The medium without the centre of symmetry admits the changes of physical properties following the change of the sense of the vector $\mathbf{E}$, and that is why the linear term in relation (6.83) may be kept. Such a change of optical symmetry in the electric field happens in crystals. Here the dependence of ε_{ik} on $\mathbf{E}$ may be linear (*Pockels effect*). This effect is described by the dielectric tensor of the form

$$\varepsilon_{ik} = \varepsilon_{ik}^{(0)} + \alpha_{ikl}E_l, \tag{6.87}$$

where the set of coefficients α_{ikl} makes up the tensor of the third order symmetrical with respect to the subscripts i and k, i.e. $\alpha_{ikl} = \alpha_{kil}$. Due to this effect a uniaxial crystal becomes biaxial, while an optically isotropic regular crystal becomes uniaxial.

6.13. Elastooptic Phenomenon

The optical symmetry of a medium may undergo a change under the elastic deformation. Under the influence of elastic strains an isotropic medium can become an anisotropic one. Stresses and strains in an elastic solid body are described by second order tensors: the stress tensor σ_{kl} and the strain tensor u_{kl}, their components being related through linear relations. These two tensors are always symmetric, but the principal axes of the ellipsoids corresponding to them do not coincide generally with those of the electric tensor. When we apply stress to the body, the dielectric tensor is modified. In the first approximation we may assume that the components of the dielectric tensor are linearly connected with the components of the strain tensor. And thus

$$\varepsilon_{ik} = \varepsilon^{(0)}\delta_{ik} + a_1 u_{ik} + a_2 \delta_{ik} u_{ii}, \tag{6.88}$$

where a_1, a_2 are the elastooptic constants, $\varepsilon^{(0)}$ is the electric permittivity of a non-deformed body, δ_{ik} the Kronecker delta,

$$u_{ik} = \frac{1}{2}\left(\frac{\partial u_i}{\partial x_k} + \frac{\partial u_k}{\partial x_i}\right),$$

while u_i ($i = 1, 2, 3$) are the components of the strain vector of the medium at a point in question given in the Cartesian coordinates x_k ($k = 1, 2, 3$). (Here $x_1 = x$, $x_2 = y$, $x_3 = z$.)

The dielectric tensor of a deformed crystal has the form

$$\varepsilon_{ik} = \varepsilon_{ik}^{(0)} + a_{iklm} u_{lm}, \tag{6.89}$$

where $\varepsilon_{ik}^{(0)}$ are the components of the tensor of electric permittivity without strains, a_{iklm} the fourth-order tensor symmeric with regard to the pairs of subscripts (i, k) and (l, m).

The number of non-equal zero independent components of this tensor depends on the symmetry of the crystal.

If we choose the axes of the coordinate system so that they coincide with the principal dielectric axes of the tensor $\varepsilon_{ik}^{(0)}$ for an unstressed medium, the ellipsoid of indices may be written by means of the equation

$$\frac{x^2}{\varepsilon_{11}^{(0)}} + \frac{y^2}{\varepsilon_{22}^{(0)}} + \frac{z^2}{\varepsilon_{33}^{(0)}} = 1. \tag{6.90}$$

In the presence of the stresses σ_{ik} the ellipsoid is transformed into another one whose equation may be written as

$$b_{11}x^2 + b_{22}y^2 + b_{33}z^2 + b_{23}yz + b_{31}zx + b_{12}yx = 1. \tag{6.91}$$

In accordance with (6.89), for the linear relation between σ_{ik} and u_{ik}, we obtain

$$\Delta b_{ik} = b_{ik} - \frac{1}{\varepsilon_{ii}^{(0)}} \delta_{ik} = \pi_{iklm}\sigma_{lm}. \tag{6.92}$$

Because of the symmetry in a_{iklm} with regard to the pairs of subscripts (i, k) and (l, m) we can apply simpler matrix denotations. In a_{iklm} the first two subscripts can be replaced by one whose value varies from 1 to 6; the same can be done with the two latter subscripts. The scheme according to which we change the subscripts is given below

Tensor notation	11	22	33	23	32	31	13	12	21
Matrix notation	1	2	3	4		5		6	

To determine the matrix elements we introduce additionaly the multipliers 2 and 4 in the following way:

$$\pi_{iklm} = \pi_{nj} \quad \text{if} \quad n, j = 1,\ 2 \text{ or } 3,$$

$$2\pi_{iklm} = \pi_{nj} \quad \text{if} \quad n, j = 4,\ 5 \text{ or } 6,$$

$$4\pi_{iklm} = \pi_{nj} \quad \text{if} \quad n, j = 4,\ 5.$$

The notation introduced above allow us to rewrite (6.92) in the form of a matrix equation

$$\Delta b_i = \pi_{in}\sigma_n, \tag{6.93}$$

where for the components of the stress tensor σ_{lm} we have used the matrix notation according to the exchange rule given above.

For crystals with regular symmetry the axes x, y, z are equivalent, and consequently the following relations hold between the elastooptic coefficients π_{in}:

$$\begin{aligned} &\pi_{11} = \pi_{22} = \pi_{33}, \\ &\pi_{12} = \pi_{21} = \pi_{23} = \pi_{32} = \pi_{31} = \pi_{13}, \\ &\pi_{44} = \pi_{55} = \pi_{66}. \end{aligned} \tag{6.94}$$

For other values of subscripts (i, n) the matrix elements are $\pi_{ik} = 0$.

For an isotropic medium we have one more condition

$$2\pi_{44} = \pi_{11} - \pi_{12}. \tag{6.95}$$

Considering (6.94) and (6.95) in equations (6.93) we arrive, in the case of uniaxial stress along the x-axis, at

$$\begin{aligned} &\Delta b_1 = \pi_{11}\sigma, \\ &\Delta b_2 = \pi_{12}\sigma, \\ &\Delta b_3 = \pi_{12}\sigma, \\ &\Delta b_4 = \Delta b_5 = \Delta b_6 = 0. \end{aligned} \tag{6.96}$$

In order to find a change in three principal refractive indices we note that $n_0^2 = \varepsilon^{(0)}$, where $\varepsilon^{(0)}$ is the electric permittivity of the initially isotropic medium. Assuming that

$$b_i = \frac{1}{n_i^2}, \tag{6.97}$$

where n_i is a modified refractive index, we are given

$$\Delta n_i = -\tfrac{1}{2}(n_i^3)\Delta b_i, \tag{6.98}$$

which, after combining with (6.96) and putting $n_i \approx n_0$, leads to the required relations for the change of the principal refractive indices

$$\begin{aligned} \Delta n_1 &= -\tfrac{1}{2}(n_0)^3 \pi_{11} \sigma, \\ \Delta n_2 &= -\tfrac{1}{2}(n_0)^3 \pi_{12} \sigma, \\ \Delta n_3 &= -\tfrac{1}{2}(n_0)^3 \pi_{12} \sigma. \end{aligned} \tag{6.99}$$

As seen from (6.99) a regular crystal under an axial stress becomes a uniaxial crystal. The same can be said about an isotropic medium.

If $n_\parallel$ and $n_\perp$ stand for the refractive indices along the directions parallel and perpendicular to the applied external stress σ, we obtain for a wave propagating along the y-axis

$$n_\parallel - n_\perp = \Delta n_1 - \Delta n_2 = -\tfrac{1}{2}(n_0)^3(\pi_{11} - \pi_{12})\sigma. \tag{6.100}$$

Equation (6.100) gives the quantitative description of the birefringence forced by external stress. That birefringence may be observed by putting the body subjected to the stress between crossed Nicol prisms. If light falls normally on a plate of thickness h, and the plate is stressed along the direction lying in the plane of the plate, the axes of the dielectric tensor also lie in this plane. The phase difference between the two components of the electric vector of the incident wave (projected on the direction of the applied stress and on the direction perpendicular to this) is

$$\delta = \frac{\pi h}{\lambda} n_0^3(\pi_{11} - \pi_{12})\sigma. \tag{6.101}$$

This phase difference is proportional to σ and may be different in different areas of the plate. The plate observed through an analyser crossed with polarizer gives an interference pattern associated with the local distribution of stress.

6.14. Optical Anisotropy in Moving Liquids

Homogeneous translatory motion of liquid does not lead to anisotropy: what is needed is the gradient of velocity. The general expression for the dielectric tensor may be then written as

$$\varepsilon_{ik} = \varepsilon^{(0)}\delta_{ik} + \lambda_1\left(\frac{\partial v_i}{\partial x_k} + \frac{\partial v_k}{\partial x_i}\right) + \lambda_2\left(\frac{\partial v_i}{\partial x_k} - \frac{\partial v_k}{\partial x_i}\right), \tag{6.102}$$

where v_i is a component of the vector of velocity, δ_{ik} the Kronecker delta, $\varepsilon^{(0)}$ the electric permittivity for motionless liquid.

For an incompressible liquid ($\operatorname{div}\mathbf{v} = 0$) the two last components in (6.102) are equal to zero, and the liquid does not reveal the phenomenon of optical anisotropy caused by a flow.

Usually $\lambda_2 \ll \lambda_1$, and for the description of the anisotropy phenomenon in moving liquids it is sufficient to use the simplified formula

$$\varepsilon_{ik} = \varepsilon^{(0)}\delta_{ik} + \lambda_1\left(\frac{\partial v_i}{\partial x_k} + \frac{\partial v_k}{\partial x_i}\right). \tag{6.103}$$

The phenomenon described by this formula is called *Maxwell's effect* and is related with the orientating effects produced by the gradients of velocity upon the particles of liquids of anisotropic structure.

It is worth noting that when rotating liquid as a whole, i.e. when $\mathbf{v} = \boldsymbol{\Omega} \times \mathbf{r}$ (where Ω is the angular velocity), the symmetric part of tensor (6.102) with respect to the subscripts i and k is zero, and the antysymmetric term presents the effect caused by the Coriolis force.

6.15. Spatial Dispersion

While studying crystal media it is sometimes necessary to consider not only *time dispersion* but also *spatial dispersion* which is a source of a new class of phenomena, usually not discussed within classical optics of crystals. In the case of spatial dispersion it is necessary to use the relations between the Fourier transforms of appropriate vector field

$$D_i(\omega, \mathbf{k}) = \varepsilon_{ij}(\omega, \mathbf{k}) E_j(\omega, \mathbf{k}), \tag{6.104}$$

where

$$E_j(\omega, \mathbf{k}) = \frac{1}{2\pi^4} \int E_j(\mathbf{r}, t) \mathrm{e}^{-\mathrm{i}(\mathbf{k}\cdot\mathbf{r} - \omega t)} \mathrm{d}\mathbf{r} \mathrm{d}t,$$

$$D_i(\omega, \mathbf{k}) = \frac{1}{2\pi^4} \int D_i(\mathbf{r}, t) \mathrm{e}^{-\mathrm{i}(\mathbf{k}\cdot\mathbf{r} - \omega t)} \mathrm{d}\mathbf{r} \mathrm{d}t.$$

The essence of spatial dispersion lies in the dependence of the tensor ε_{ij} on $\mathbf{k}$. This dependence reflects the fact that the field $\mathbf{D}(\mathbf{r}, t)$ is determined by the values of the field $\mathbf{E}$ given not only at the point $\mathbf{r}$ but also in its neighbourhood. Polarization arises namely not only at the point $\mathbf{r}_0$ where the field effects, but also within a certain vicinity of this point, which is conditioned by the displacement of the particles of the medium. The size of this vicinity depends on the displacement of the particles of the medium in a period of time in which they can "remember" the interaction performed on this region of the medium (in gases it is the mean free path).

Relation (6.104) is basic in the linear approximation of the optics of crystals with spatial dispersion. It is a consequence of the constitutive relation which in the frame of linear electrodynamics has the form

$$D_i(\mathbf{r}, t) = \int \mathrm{d}t' \int \mathrm{d}\mathbf{r} \varepsilon_{ij}(t, t'; \mathbf{r}, \mathbf{r}') E_j(\mathbf{r}', t). \tag{6.105}$$

This equation is in agreement with the causality principle. The vector $\mathbf{D}(\mathbf{r}, t)$ at a moment of time t is determined by the field in the past $t' \leq t$ (hysteresis phenomenon). Polarization does not vanish immediately after removing the acting field, but only after a period of time τ_0 determined by relaxation processes in the medium. If the medium does not change in time (processes proceeding in the medium are stationary), ε_{ij} will depend merely on the difference $t - t'$. If, however, the medium is homogeneous, the dependence of ε_{ij} on spatial variables is of the same kind, i.e. it will be the dependence on the difference $\mathbf{r} - \mathbf{r}'$.

In a medium of properties described above, formula (6.105) for a plane monochromatic wave, i.e. for

$$E_j(\mathbf{r}, t) = E_j(\omega, \mathbf{k})\mathrm{e}^{\mathrm{i}(\mathbf{k}\cdot\mathbf{r} - \omega t)}, \tag{6.106}$$

becomes formula (6.104), and we have

$$\varepsilon_{ij}(\omega, \mathbf{k}) = \int_0^\infty \mathrm{d}\tau \int \mathrm{d}\mathbf{R}\, \mathrm{e}^{-\mathrm{i}(\mathbf{k}\cdot\mathbf{R} - \omega t)} \varepsilon_{ij}(\tau, \mathbf{R}). \tag{6.107}$$

It can be seen that relation (6.104) makes sense for the appropriate Fourier components of fields with an arbitrary dependence upon $\mathbf{r}$ and t.

The dependence ε_{ij} upon ω, i.e. frequency dispersion, is a consequence of a non-local dependence upon time. As the natural frequency of a medium ω_0 lies usually within the interval of frequency ω, the ratio ω_0/ω is of unity order and that is why time dispersion is large in general.

The situation is quite different in the case of spatial dispersion. It is connected with the fact that in a dielectric the wavelength λ from the optical range is much larger than the characteristic size of the neighbourhood of point $\mathbf{r}$, which has its share in the volume integral. In a dielectric that neighbourhood is given in the unit of a, that is in the lattice constant, so that the ratio a/λ ($\approx 10^{-2}$) is very small. It results in comparatively small spatial dispersion in relation to frequency dispersion, and this in turn allows us to make essential simplifications in the analysis of the effects under examination.

For all known cases we can expand $\varepsilon_{ij}(\omega, \mathbf{k})$ into a series of powers of the components of the vector $\mathbf{k}$ and confine ourselves to two or three terms of the expression

$$\varepsilon_{ij}(\omega, \mathbf{k}) = \varepsilon_{ij}(\omega) + \mathrm{i}\gamma_{ijm}(\omega)k_m + \alpha_{ijms}(\omega)k_m k_s, \tag{6.108}$$

or handle in the same way with the inverse tensor

$$\varepsilon_{ij}^{-1}(\omega, \mathbf{k}) = \varepsilon_{ij}^{-1}(\omega) + \mathrm{i}\eta_{ijm}(\omega)k_m + \beta_{ijms}(\omega)k_m k_s. \tag{6.109}$$

The use of the general form of the tensors $\varepsilon_{ij}(\omega, \mathbf{k})$ or ε_{ij}^{-1} or their expansions given above is equivalent in wide frames, the choice is done depending on the suitability of the expressions obtained in this way. The case of resonace frequencies, where some components of the tensors $\varepsilon_{ij}(\omega)$, $\gamma_{ljm}(\omega)$ tend to infinity, is an exception here. In this case we are sometimes forced to use expansion (6.109) instead of (6.108), as the appropriate coefficients of (6.109) are small, which ensures its rapid convergence.

6.16. The Generalized Fresnel Equation for a Medium with Spatial Dispersion

The homogeneous Maxwell equations for a monochromatic plane wave lead to the relation

$$\mathbf{D} = \frac{1}{\mu_0 \omega^2}\left[k^2\mathbf{E} - \mathbf{k}(\mathbf{k}\cdot\mathbf{E})\right]. \tag{6.110}$$

Taking into account the constitutive relation (6.105), we obtain a homogeneous system of equations of the form ($l = 1, 2, 3$)

$$\left[\mu_0\omega^2\varepsilon_{lj}(\omega,\mathbf{k}) - k^2\delta_{lj} + k_l k_j\right]E_j(\omega,\mathbf{k}) = 0. \tag{6.111}$$

The system (6.111) has non-zero solutions ($\mathbf{E} \neq \mathbf{0}$) when its determinant vanishes, i.e. when

$$\left|\mu_0\omega^2\varepsilon_{lj}(\omega,\mathbf{k}) - k^2\delta_{lj} + k_l k_j\right| = 0. \tag{6.112}$$

This is the required equation of dispersion determining the relation between ω and $\mathbf{k}$. This equation, after replacing the vector $\mathbf{k}$ by the following form of it:

$$\mathbf{k} = \omega\sqrt{\mu_0\varepsilon_0}\,\tilde{n}(\omega,\mathbf{s})\mathbf{s}, \tag{6.112a}$$

where $\tilde{n}^2 = k^2/(\omega^2\varepsilon_0\mu_0)$, and $\mathbf{s}$ is a versor, takes the form

$$\tilde{n}^4(\varepsilon_{lj}s_l s_j) - \tilde{n}^2(\varepsilon_{mm}\varepsilon_{lj}s_l s_j - \varepsilon_{lm}\varepsilon_{mj}s_l s_j) + |\varepsilon_{lj}| = 0, \tag{6.113}$$

where $|\varepsilon_{lj}|$ stands for the determinant of the tensor ε_{lj}. In the absence of spatial dispersion this equation is identical to the classical Fresnel equation for phase velocity in anisotropic crystals (cf. Sec. 6.12).

The roots of equation (6.113) refer to the case

$$\mathbf{E}_\perp = \mathbf{E} - \mathbf{k}(\mathbf{k}\cdot\mathbf{E})k^{-2} \neq \mathbf{0},$$

i.e. to waves for which the electric vector has a non-zero component perpendicular to the wave vector **k**. These are transverse waves. In the absence of spatial dispersion they correspond to two waves occurring usually in anisotropic crystals. In the presence of spatial dispersion, there may occur another type of waves for which $\mathbf{E}_\perp = \mathbf{0}$; these are called *longitudinal waves.* They can exist only for discrete values of ω. Such possibility follows from the fact that the equality $\mathbf{D} = \mathbf{0}$ may hold also for $\mathbf{E} \neq \mathbf{0}$ (see formula (6.104)) if ω and **k** fulfil the condition

$$|\varepsilon_{ij}(\omega, \mathbf{k})| = 0. \tag{6.114}$$

This equality yields the dispersion relation for longitudinal waves

$$\omega_\parallel = \omega_\parallel(\mathbf{k}). \tag{6.115}$$

For spatial dispersion, equation (6.112) connected with the waves for which $\mathbf{E}_\perp \neq \mathbf{0}$ can have a few solutions in several regions of spectrum, which provides a larger scale of possibilities compared with the classical optics of crystals. However, even in cases where new solutions will not occur, spatial dispersion leads to quite a few new phenomena among which the most important is *natural optical activity (gyrotropy).*

6.17. Symmetry Properties of the Tensor $\varepsilon_{ij}(\omega, \mathbf{k})$ and of the Derivative Tensors

In general the tensor ε_{ij} is complex even for real values of **k**. A real character of the fields **E** and **D** in the medium results in a real character of the kernel $\varepsilon_{ij}(\tau, \mathbf{R})$ in the integral equation

$$D_i(\mathbf{r}, t) = \int_{-\infty}^{t} \mathrm{d}t' \int \mathrm{d}\mathbf{r}' \hat{\varepsilon}_{ij}(t - t'; \mathbf{r} - \mathbf{r}') E_j(t', \mathbf{r}'). \tag{6.116}$$

This implies that

$$\varepsilon_{ij}(\omega, \mathbf{k}) = \varepsilon^*_{ji}(-\omega, -\mathbf{k}). \tag{6.117}$$

A medium is not optically active (non-gyrotropic) if for any ω and **k** the following relation holds:

$$\varepsilon_{ij}(\omega, \mathbf{k}) = \varepsilon_{ji}(\omega, \mathbf{k}) = \varepsilon_{ij}(\omega, -\mathbf{k}). \tag{6.118}$$

If the tensor ε_{ij} is asymmetric when regarding spatial symmetry, a medium reveals natural optical activity. In such a medium there must exist at least one direction in which there are two non-equivalent senses. This can take place in the absence of the centre of symmetry but not only in this case. A medium which shows natural optical activity is called a *gyrotropic medium*. When $\mathbf{B}_{\text{ext}} \neq \mathbf{0}$, the tensor ε_{lj} is asymmetrical. The general properties of Onsager's kinetic coefficients lead to the relationship

$$\varepsilon_{lj}(\omega, \mathbf{k}, \mathbf{B}_{\text{ext}}) = \varepsilon_{jl}(\omega, -\mathbf{k}, -\mathbf{B}_{\text{ext}}). \tag{6.119}$$

Thus we can speak about the *gyrotropy* produced by the external magnetic field (see Sec. 7.2).

The tensors $\varepsilon_{lj}(\omega)$, $\gamma_{ljm}(\omega)$, $\alpha_{ljms}(\omega)$, which appear in the extensions (6.108) and (6.109), or the tensors inverse to them $\varepsilon_{lj}^{-1}(\omega)$, $\eta_{ljm}(\omega)$, $\beta_{ljms}(\omega)$ (for $\mathbf{B}_{\text{ext}} = \mathbf{0}$) have symmetry properties, which can be derived from symmetry properties (6.117) and (6.118). And here they are:

$$\begin{aligned}
&\varepsilon_{lj}(\omega) = \varepsilon_{jl}(\omega), && \varepsilon_{lj}^{-1} = \varepsilon_{jl}^{-1},\\
&\gamma_{ljm}(\omega) = -\gamma_{jlm}(\omega), && \eta_{ljm} = -\eta_{jlm},\\
&\alpha_{ljms}(\omega) = \alpha_{jlms}(\omega), && \beta_{ljms} = \beta_{jlms},\\
&\alpha_{ljms}(\omega) = \alpha_{ljsm}(\omega), && \beta_{ljms} = \beta_{ljsm}.
\end{aligned} \tag{6.120}$$

In addition, in a medium with the centre of symmetry for $\mathbf{B}_{\text{ext}} = \mathbf{0}$ the tensor elements of the third order are equal to zero

$$\gamma_{ljm} = \eta_{ljm} = 0.$$

For media without the centre of symmetry or for $\mathbf{B}_{\text{ext}} \neq 0$ (6.119) we obtain

$$\begin{aligned}
&\gamma_{11,l} = \gamma_{22,l} = \gamma_{33,l} = 0,\\
&\gamma_{12,l} = -\gamma_{21,l}, \quad \gamma_{31,l} = -\gamma_{13,l},\\
&\gamma_{23,l} = -\gamma_{32,l} \quad (l = 1,\ 2,\ 3)
\end{aligned} \tag{6.121}$$

and the same concerns η_{ljm}.

As can be seen from (6.121), the tensors γ_{ljm} and η_{ljm} have in general nine independent components. They can be presented in the form

$$\gamma_{ljm} = e_{ljs} g'_{sm}, \qquad \eta_{ljm} = e_{ljs} f'_{sm},$$

where e_{ljs} is a unitary *completely antisymmetric pseudotensor* of the third order, while g'_{sm} and f'_{sm} are the pseudotensors of the second order, respectively.

For the sake of convenience we introduce apart from the tensors g'_{ml} and f'_{ml}, the *gyration vectors* **g** and **f** defined by

$$g_m = g'_{ml} k_l, \qquad f_m = f'_{ml} k_l, \tag{6.122}$$

where k_l $(l = 1, 2, 3)$ are the components of the wave vector.

Applying the extensions (6.108) and (6.109), we can rewrite the constitutive relation (6.104) in the form

$$\begin{aligned} D_j &= \varepsilon_{js}(\omega) E_s - \mathrm{i}(\mathbf{g} \times \mathbf{E})_j, \\ E_j &= \varepsilon_{js}^{-1}(\omega) D_s - \mathrm{i}(\mathbf{f} \times \mathbf{D})_j. \end{aligned} \tag{6.123}$$

Since spatial dispersion in gyrotropic media is insignificant, for most cases the above equations are accurate enough. For non-gyrotropic media the expressions (6.108) have to be replaced by

$$\begin{aligned} \varepsilon_{js}(\omega, \mathbf{k}) &= \varepsilon_{js}(\omega) + \left(\frac{\omega}{c}\right)^2 \tilde{n}^2 \alpha_{jslm} s_l s_m, \\ \varepsilon_{js}^{-1}(\omega, \mathbf{k}) &= \varepsilon_{js}^{-1}(\omega) + \left(\frac{\omega}{c}\right)^2 \tilde{n}^2 \beta_{jslm} s_l s_m, \end{aligned} \tag{6.124}$$

which follows from (6.112a).

In a medium with the centre of symmetry, for $\mathbf{B}_{\text{ext}} \neq \mathbf{0}$, the gyration vector is

$$g_l = \kappa_{ls} B_s, \tag{6.123a}$$

where κ_{ls} is the asymmetric tensor.

6.18. Birefringence and the Rotation of the Plane of Polarization

Pure rotation of the plane of polarization upon the passage of a linearly polarized wave through a gyrotropic crystal can be observed only if the direction of light propagation coincides with the optical axis. If this is not the case, the rotation of the plane of polarization takes place apart from the ordinary birefringence phenomenon. This leads to a more complex physical situation.

The phenomenon of rotation of the plane of polarization results in an insignificant change of the surface of phase velocity of the crystal (see Sec. 6.4). For a uniaxial positive crystal which lacks the optical activity this surface represents a sphere and an ellipsoid of revolution (Fig. 6.17), which are mutually tangent at points lying on the optical axis, because along this direction the velocity of the two waves: ordinary and extraordinary is the same. The presence of the optical activity causes the change of that two-shell surface. The surfaces of the shells are not tangent any more, and the distance between them along the optical axis corresponds to the difference in velocities for two circular polarized waves. For other directions we have the following situation – with every normal to the surface of phase velocity are connected two waves, in general elliptically

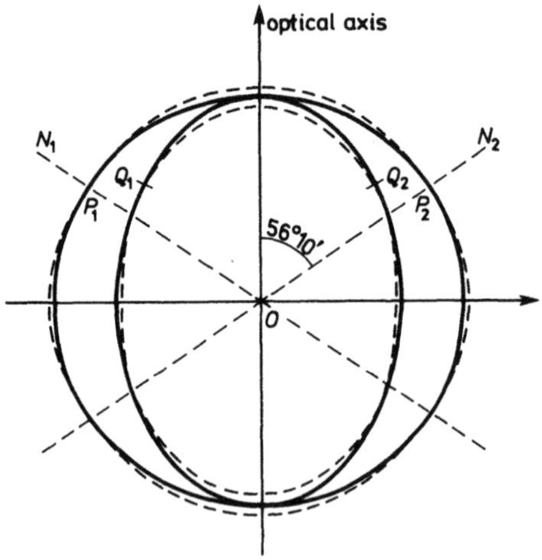

Figure 6.17

polarized. They pass through the crystal without change of their form. The ellipses determining the polarization of these two waves are positioned in such a way that their principal axes make the right angle and coincide with the main directions of vibrations, which would exist in crystal in the absence of optical activity. When a wave propagates along the optical axis, the wave polarization becomes a circular one. In all other directions of wave propagation the ratio κ of the small axis of the ellipse to the long one varies. Thus for example in α-quartz κ depends on the angle between the normal to the surface of phase velocity and the optical axis. For $\beta = 56°10'$, $\kappa = 0$ and it increases together with the increase of the angle, but at the same time the sense of rotation of the vector **D** changes for the opposite one. The situation is illustrated in Fig. 6.18. We can see in this picture how the polarizations of the ordinary and extraordinary wave depend on the direction of propagation. Only in the direction at the angle 56°10′ with respect to the optical axis these waves remain linearly polarized waves.

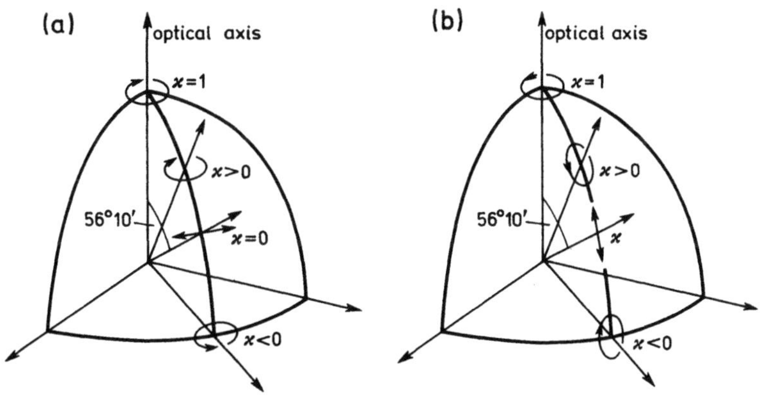

Figure 6.18

Bibliography

1. Born, M. and Wolf, E.: (1965), 'Principles of Optics', Pergamon Press, London.
2. Fedorov, F.J.: (1958), 'Optics of Anisotropic Media' (in Russian), Izd. AN BSSR, Minsk.

3. Fedorov, F.J. and Filippov, V.V.: (1976), 'Reflection and Refraction of Light by Transparent Crystals' (in Russian), Izd. Nauka i Tekhnika, Minsk.
4. Nye, J.F.: (1957), 'Physical Properties of Crystal' Clarendon Press, Oxford.
5. Penkala, T.: (1971), 'Optics of Crystals' (in Polish), PWN, Warszawa.
6. Sommerfeld, A.: (1950), 'Optik, Vorlesungen über theoretische Physik', Bd. IV, Wiesbaden.
7. Wolf, E.M. (ed.): (1971), 'Progress in Optics', Vol. IX, North-Holland, Amsterdam.

7. Gyrotropic Media

7.1. Optical Activity

Optical activity was discovered in 1811 by Avogadro. He established that when linearly polarized light passes through a quartz plate along its optical axis, the plane of polarization turns through the angle proportional to the thickness of the plate. The occurrence of this effect in gases and liquids was confirmed later by Biot. The term "gyrotropy" (*gyros* means circle in Greek) serves to emphasize the effect of rotation of the plane of polarization of light. At present the phenomenon of gyrotropy cannot be resolved itself only into the phenomenon of rotation of the plane of polarization. The ability of rotation of the plane of polarization for linearly polarized light may occur for all directions of light propagation when a medium is isotropic. For crystals it takes place only if light propagates along the optical axis. In many crystals this rotation does not reveal itself at all, which, however, does not prove the lack of gyrotropic properties of the medium.

It is not easy to give the general definition of gyrotropy (optical activity), as there are several kinds of gyrotropy (natural, free and forced), which differ essentially among themselves. The most complete and physically precise definition of the phenomenon or the essence of gyrotropy may be formulated as follows:

Gyrotropy is a property of material media consisting in the fact that a plane homogeneous wave propagating in all or almost all directions reveals, in the absence of dissipation (absorption), elliptical polarization.

In this formulation the phrase "in the absence of absorption" is worth pointing out, since in media exhibiting no gyrotropy but absorbing there may also propagate an elliptically polarized wave (Fedorov, 1976). The phrase "in almost all directions" cannot be omitted either, because there are directions in gyrotropic crystals along which may propagate waves of linear polarization as well as of circular polarization or an arbitrary one.

7.2. Kinds of Gyrotropy

In optically active media (gyrotropic media) electric and magnetic quantities are closely related, and a proper theoretical description of the phenomenon requires, within phenomenological description, the modification of the constitutive relations, which relate field vectors **E** and **H** with induction vectors **D** and **B**.

The electric field of a wave causes a change not only in the electric moment but also in the magnetic polarization of molecules, and this should be taken into account in formulating the constitutive relations.

Optical activity reveals some specific regularities of the electrodynamics of continuous media. Optical activity is related with the molecule structure. This structure however can be revealed in interaction with an electromagnetic wave only if we can consider the spatial extension of molecules. The charged particles which create molecules should react in different ways on the field of a wave of length λ. This is because of the phase difference of the wave at different parts of the molecule and can be described by the ratio λ/a, where a is the transverse size of the molecule. It is closely connected with spatial dispersion discussed in Secs. 6.15, 6.16 and 6.17. It concerns mainly media showing so-called *natural gyrotropy.*

In media with natural gyrotropy the constitutive relations take the form

$$\begin{aligned} \mathbf{D} &= (\varepsilon + \alpha_1 \,\mathrm{curl})\mathbf{E}, \\ \mathbf{B} &= (\mu + \tilde{\alpha}_1 \,\mathrm{curl})\mathbf{H}, \end{aligned} \tag{7.1}$$

where α_1 is the asymmetric real tensor, $\tilde{\alpha}_1$ is the tensor inverse in relation to the tensor α_1, and ε and μ are symmetric real tensors for media without dissipation (absorption).

In the case of a plane wave, equations (7.1) take the form

$$\begin{aligned} \mathbf{D} &= \varepsilon\mathbf{E} - \mathrm{i}\alpha_1(\mathbf{m} \times \mathbf{E}) = \varepsilon_n\mathbf{E}, \\ \mathbf{B} &= \mu\mathbf{H} - \mathrm{i}\tilde{\alpha}_1(\mathbf{m} \times \mathbf{H}) = \mu_n\mathbf{H}, \end{aligned} \tag{7.2}$$

where $\mathbf{m} = n\mathbf{s}$ is the refraction vector, which describes the propagation of a plane wave ($n = c/v$ – appropriate refractive index $\mathbf{s} = \mathbf{k}/k$).

The tensors ε_n and μ_n differ from those occurring in the anisotropic case, because they are complex asymmetric tensors dependent on parameters describing a propagating wave. They consist of 9 independent parameters which are the elements of tensor α.

In gyrotropic media described by the constitutive relations (7.1) or (7.2), a change in the sense of vector $\mathbf{m}$ results in a change in the direction of rotation of vector $\mathbf{E}$ of the propagating wave. The result is that if a linearly polarized plane wave passes through a gyrotropic layer, reflects from an ordinary mirror, and finally passes through the layer once again, then it does not suffer any rotation of the plane of polarization.

Media of *free gyrotropy* are described by the constitutive relations of the form

$$\begin{aligned} \mathbf{D} &= \varepsilon\mathbf{E} + \mathrm{i}(\mathbf{a} \times \mathbf{E}) = \varepsilon_f\mathbf{E}, \\ \mathbf{B} &= \mu\mathbf{H} + \mathrm{i}(\mathbf{b} \times \mathbf{H}) = \mu_f\mathbf{H}, \end{aligned} \tag{7.3}$$

where vectors $\mathbf{a}$ and $\mathbf{b}$ are mutually independent vectors of gyration. Tensors ε_f and μ_f are independent of vector $\mathbf{m}$, i.e., of the velocity and direction of the propagating wave, and remain constant quantities by a fixed wave frequency. These tensors depend on 6 parameters which are the components of the vectors of gyration $\mathbf{a}$ and $\mathbf{b}$. They are *Hermitian tensors.* For transparent media tensors ε and μ in formulas (7.3) remain real symmetric tensors.

Free gyrotropy owns its name to the fact that tensors ε_f and μ_f are independent of the direction of wave propagation and of the wave velocity, and they are fully determined by the properties of the medium. Free gyrotropy is interpreted as a result of existing in a medium its own inner magnetic field, which has molecular (gases, solutions) or intermolecular (crystal bodies) nature, and due to which certain direction connected with the vectors of gyration **a** and **b** can be distinguished in the medium.

The difference between natural (7.1) and free gyrotropy (7.3) reveals itself in propagating light waves in media of these two gyrotropic properties. Although in both cases (7.1) and (7.3) rotation of the polarization plane of linearly polarized light is possible, these two kinds of gyrotropy differ essentially. A linearly polarized wave passing through a layer of a medium of free gyrotropy in the direction parallel to the vector of gyration **a** or **b**, after reflecting from a mirror and passing once again through the layer, undergoes doubled rotation of the plane of polarization. For a layer with natural gyrotropy we had no rotation. All this indicates for the necessity of distinguishing these two kinds of gyrotropy.

A kind of gyrotropy related to free gyrotropy is *forced gyrotropy*. The difference between them lies only in the ground of these occurrence: free gyrotropy is a natural property of the medium while forced gyrotropy is caused by external interactions. The first is caused by "own" internal field existing in the medium, the second should be treated as induced by the external magnetic field.

The tensor of electric permittivity of the medium placed in the external magnetic field becomes dependent on this field

$$\varepsilon = \varepsilon(\mathbf{H}_{\text{ext}}). \tag{7.4}$$

The vector of magnetic field is an axial vector and may be represented by antisymmetric tensor H^X of the form

$$H^X = \begin{bmatrix} 0 & -H_z & H_y \\ H_z & 0 & -H_x \\ -H_y & H_x & 0 \end{bmatrix}. \tag{7.5}$$

It means that tensor ε is a function of the antisymmetric tensor H^X. As known from linear algebra, an arbitrary function of a matrix can be presented in a polynomial form. Since

$$\begin{aligned}(H^X)^2 &= \mathbf{H}\cdot\mathbf{H} - \mathbf{HH},\\ (H^X)^3 &= -(\mathbf{H}\cdot\mathbf{H})H^X,\end{aligned} \tag{7.6}$$

(where $\mathbf{H}\cdot\mathbf{H}$ is a scalar product, and $\mathbf{HH}$ – a dyad), we have

$$\varepsilon(H^X) = \varepsilon + \eta(\mathbf{HH}) + \mathrm{i}\kappa\mathbf{H}^X. \tag{7.7}$$

A strictly imaginary character of the coefficient in front of the tensor H^X in (7.7) provides that the law of conservation of energy (see Sec. 1.8) is satisfied in gyrotropic media. The dyad $\eta(\mathbf{HH})$ in (7.7) is responsible for the *Cotton–Mouton effect*, which consists in inducing the anisotropy of a medium by an external magnetic field.

Expression (7.7) gives the form of the tensor of electric permittivity for a medium with forced gyrotropy. The term $\mathrm{i}\kappa H^X$ is the gyration tensor, which in matrix notation takes the form:

$$H^X = \begin{bmatrix} 0 & -\mathrm{i}\kappa H_z & \mathrm{i}\kappa H_y \\ \mathrm{i}\kappa H_z & 0 & -\mathrm{i}\kappa H_x \\ -\mathrm{i}\kappa H_y & \mathrm{i}\kappa H_x & 0 \end{bmatrix}. \tag{7.8}$$

Tensor of electric permittivity (7.7) for forced gyrotropy is, in respect of its structure, very similar to tensor (7.3): the term $\varepsilon + \eta\mathbf{HH}$ corresponds to tensor ε, vector $\mathrm{i}\kappa\mathbf{H}$ – to the gyration vector $\mathrm{i}\mathbf{a}$.

7.3. Electrodynamics Equations for Gyrotropic Media

The theory of electromagnetic field in gyrotropic media requires that some aspects of traditional electrodynamics should be studied thoroughly and in a more precise way. Electrodynamics equations for gyrotropic media include Maxwell's equations

$$\begin{aligned}\operatorname{curl}\mathbf{E} &= -\dot{\mathbf{B}},\\ \operatorname{curl}\mathbf{H} &= \dot{\mathbf{D}},\end{aligned} \tag{7.9}$$

as well as the constitutive relations (7.1) or (7.3).

The rightness of the statement above is confirmed by a number of arguments of primary character. First of all, it is the only system of equations for which the law of conservation of energy is exactly fulfilled inside an gyrotropic medium as well as on its edge for a plane monochromatic wave incident on the interface of the medium. In gyrotropic media described by the constitutive relations (7.1) or (7.3) dissipation does not occur. Therefore on the interface of such two media the normal component of the total Poynting vector should vanish. It is a necessary condition for the balance of the energy fluxes of the waves while reflecting and refracting.

A general expression for the density of field energy in gyrotropic media should be taken in the form

$$w = \tfrac{1}{2}(\mathbf{D}\varepsilon^{-1}\mathbf{D} + \mathbf{B}\mu^{-1}\mathbf{B}), \tag{7.10}$$

where ε^{-1} and μ^{-1} are respectively the tensors inverse to the tensors of electric and magnetic permittivity of a medium in the absence of gyrotropy. Expression (7.10) for the density of field energy is a more general expression than

$$w = \tfrac{1}{2}(\mathbf{E}\cdot\mathbf{D} + \mathbf{H}\cdot\mathbf{B}), \tag{7.11}$$

which up to now was applied in electrodynamics for media without gyrotropy. This results from the difference between the constitutive relations (1.8) and (7.1) or (7.3). The form of (7.10) for the density of field energy remains valid both for homogeneous and non-homogeneous media, with or without gyrotropy.

The form of Poynting's vector

$$\mathbf{S} = \mathbf{E} \times \mathbf{H} \tag{7.12}$$

does not change neither in gyrotropic nor in non-gyrotropic media. The same applies to the law of conservation of energy for gyrotropic media. It has the form

$$w + \operatorname{div}\mathbf{S} = 0, \tag{7.13}$$

and it is also valid in the absence of dissipation both for homogeneous and non-homogeneous media, when tensors ε, μ and α depend on spatial coordinates.

Taking into account the above changes, the conditions imposed on the field vectors on the interface of two gyrotropic media I and II remain in their ordinary form

$$\begin{aligned} &(\mathbf{E}_\mathrm{I} - \mathbf{E}_\mathrm{II}) \times \mathbf{q} = \mathbf{0}, \qquad (\mathbf{H}_\mathrm{I} - \mathbf{H}_\mathrm{II}) \times \mathbf{q} = \mathbf{0}, \\ &(\mathbf{D}_\mathrm{I} - \mathbf{D}_\mathrm{II}) \cdot \mathbf{q} = 0, \qquad (\mathbf{B}_\mathrm{I} - \mathbf{B}_\mathrm{II}) \cdot \mathbf{q} = 0, \end{aligned} \tag{7.14}$$

where $\mathbf{q}$ is a unit vector perpendicular to the boundary surface.

To sum up, we can state that such formulation of electrodynamics equation for gyrotropic media provides validity of maximum of symmetry properties and dependencies which are in force in classic electrodynamics of non-absorbing media.

7.4. Plane Electromagnetic Waves in a Medium with Free Gyrotropy

For media with free gyrotropy we can write the tensors of the electric and magnetic permittivity ε_f and μ_f in the form

$$\begin{aligned} \varepsilon_f &= \varepsilon_1 + \mathrm{i} a^X, \\ \mu_f &= \mu_1 + \mathrm{i} b^X, \end{aligned} \tag{7.15}$$

where the vectors of gyration $\mathbf{a}$ and $\mathbf{b}$ are shown as antisymmetric tensors:

$$a^X = \begin{bmatrix} 0 & -a_z & a_y \\ a_z & 0 & -a_x \\ -a_y & a_x & 0 \end{bmatrix}, \tag{7.16a}$$

$$b^X = \begin{bmatrix} 0 & -b_z & b_y \\ b_z & 0 & -b_x \\ -b_y & b_x & 0 \end{bmatrix}. \tag{7.16b}$$

The tensors ε_f and μ_f are constant for a fixed wave frequency. Moreover, they are Hermitian tensors. Since the constitutive relations

$$\begin{aligned} \mathbf{D} &= \varepsilon_f \mathbf{E}, \\ \mathbf{B} &= \mu_f \mathbf{H}, \end{aligned} \tag{7.17}$$

are independent in the case of free gyrotropy (the law of energy conservation does not imply any relations between the tensors ε_f and μ_f), we can introduce a simplifying assumption. Within phenomenological theory we can assume that $\mathbf{b} = \mathbf{0}$ and $\mu_1 = J\mu_0$, where J is a unit matrix. Under these assumptions we can write Maxwell's equations for a plane wave in this medium as

$$\begin{aligned} -(\mathbf{m} \times \mathbf{H}) &= \varepsilon_f \mathbf{E} c, \\ \mathbf{m} \times \mathbf{E} &= \mu_0 \mathbf{H} c, \end{aligned} \tag{7.18}$$

where $\mathbf{m} = \frac{c}{\omega}\mathbf{k}$ is the refraction vector, and $\mathbf{k}$ the wave vector. These equations lead to the equality

$$(\mathbf{mm} - \mathbf{m}\cdot\mathbf{m} - \mu_0 \varepsilon_f c^2)\mathbf{E} = \mathbf{0}. \tag{7.19}$$

Assuming that the tensor ε_1 is of the form $\varepsilon_1 = J\varepsilon_1$, and having in view the relation $\mathbf{m} = n\mathbf{s}$ (where n is the refractive index, $\mathbf{s}$ – a unit vector), we can rewrite equality (7.19) in the form

$$(n^2 - n^2\mathbf{ss} - c^2\mu_0\varepsilon_1 - \mathrm{i}\mu_0 c^2 a^X)\mathbf{E} = \mathbf{0}. \tag{7.20}$$

Suppose that the direction of propagation of a plane wave is parallel to the vector of gyration $\mathbf{a} = \eta\mathbf{s}$. Then, for $\mathbf{s}\cdot\mathbf{E} = 0$, we get

$$(\mu_0 c^2 \varepsilon_1 - n^2 - \mathrm{i}\mu_0 \eta c^2 s^X)\mathbf{E} = \mathbf{0}, \tag{7.21}$$

and consequently

$$\mathbf{E} = -\frac{\mathrm{i}\eta(\mathbf{s} \times \mathbf{E})\mu_0 c^2}{c^2\mu_0\varepsilon_1 - n^2} \tag{7.21a}$$

and

$$|\mu_0 c^2 \varepsilon_1 - n^2 - \mathrm{i}c^2 \eta s^X \mu_0| = 0. \tag{7.21b}$$

Vanishing of the determinant (7.21b) leads to the equation

$$(\mu_0 \varepsilon_1 c^2 - n^2)[(\mu_0 c^2 \varepsilon_1 - n^2)^2 - c^4 \eta^2 \mu_0^2] = 0. \tag{7.21c}$$

The first factor in (7.21c) is not zero: $\mu_0 c^2 \varepsilon_1 - n^2 \neq 0$, otherwise (7.21) would yield $\mathbf{s} \times \mathbf{E} = \mathbf{0}$, which together with the condition $\mathbf{s}\cdot\mathbf{E} = 0$ would give $\mathbf{E} = \mathbf{0}$. Thus the following equality may be hold:

$$n^2 = (\mu_0 \varepsilon_1 \pm \eta \mu_0) c^2. \tag{7.22}$$

Taking (7.22) into account in (7.21a) implies the dependencies

$$\mathbf{E} = \pm \mathrm{i}(\mathbf{s} \times \mathbf{E}). \tag{7.23}$$

As can be seen from the above, if the direction of propagation of a plane wave **s** remains parallel to the vector of gyration **a**, then the wave is a left-handed (+) or right-handed (−) circularly polarized wave.

It is worthwhile to fix a condition imposed by the crystal symmetry on the vector of gyration **a**. To this end we use the constitutive relation of the form

$$\mathbf{D} = \varepsilon \mathbf{E} + \mathrm{i}(\mathbf{a} \times \mathbf{E}). \tag{7.24}$$

Let $\check{S}$ be symmetry operator determined by the crystal symmetry. This operator applied to the field vectors gives

$$\begin{aligned} \mathbf{D}' &= \check{S}\mathbf{D}, \\ \mathbf{E}' &= \check{S}\mathbf{E}, \end{aligned} \tag{7.25}$$

whereas the constitutive relation takes the form

$$\mathbf{D}' = \varepsilon \mathbf{E}' + \mathrm{i}(\mathbf{a} \times \mathbf{E}') \tag{7.26}$$

or

$$\check{S}\mathbf{D} = \varepsilon \check{S}\mathbf{E} + \mathrm{i}(\mathbf{a} \times \check{S}\mathbf{E}). \tag{7.26a}$$

Since we must obtain equation (7.26a) using the operator $\check{S}$ to both sides of (7.24) it leads to the equalities:

$$\check{S} \varepsilon \check{S}^{-1} = \varepsilon \tag{7.27a}$$

and

$$\mathbf{a} \times \check{S}\mathbf{E} = \check{S}(\mathbf{a} \times \mathbf{E}). \tag{7.27b}$$

From (7.27b), and from the identity

$$\mathbf{a} \times \mathbf{E} = \frac{1}{|\check{S}|} \check{S}^{-1}(\check{S}\mathbf{a} \times \check{S}\mathbf{E}), \tag{7.28}$$

we get

$$\check{S}\mathbf{a} = |\check{S}|\mathbf{a}. \tag{7.29}$$

From relation (7.29) we conclude that the vector of gyration **a** is bound to be an eigenvector of the symmetry operator $\check{S}$ when the eigenvalues $|\check{S}| = \pm 1$. Hence it is obvious that the vector of gyration **a** should be pointed along the axis of symmetry of the crystal or should be perpendicular to its plane of symmetry.

7.5. Forced Gyrotropy – Faraday's Effect

In a medium with forced gyrotropy the components of the vector of gyration **a** are functions of the intensity of the external magnetic field. In a non-magnetic isotropic medium placed in an external magnetic field $\mathbf{H}_{\text{ext}}$ with direction along the z-axis, the tensor of electric permittivity can be written in the form

$$\varepsilon = \begin{bmatrix} \varepsilon_1 & -i\kappa H_{\text{ext}} & 0 \\ i\kappa H_{\text{ext}} & \varepsilon_1 & 0 \\ 0 & 0 & \varepsilon_1 \end{bmatrix}, \tag{7.30}$$

where ε_1 is the electric permittivity of the medium in the absence of the field.

In such a medium, for an electromagnetic wave of type of a plane wave propagating along the z-axis, we obtain from Maxwell's equations the relationship

$$-\mathbf{k} \times (\mathbf{k} \times \mathbf{E}) = k_0^2 \varepsilon_r \mathbf{E}, \tag{7.31}$$

which, on applying an explicit form of the electric permittivity tensor (7.30), leads for a transverse wave to the following system of equations:

$$\begin{aligned} (k^2 - k_0^2 \varepsilon_{1r}) E_x + i g k_0^2 E_y &= 0, \\ -i g k_0^2 E_x + (k^2 - k_0^2 \varepsilon_{1r}) E_y &= 0, \end{aligned} \tag{7.32}$$

where $g = \kappa H_{\text{ext}}/\varepsilon_0$, and ε_{1r} is the relative electric permittivity.

Zero value of the determinant of this system of equations (which is a condition for existence of non-zero solutions) leads to two independent values of the refractive index

$$n_{\mp} = \sqrt{\varepsilon_{1r} \mp g} \approx \sqrt{\varepsilon_{1r}} \mp \frac{g}{2\sqrt{\varepsilon_{1r}}} = n_0 \mp \frac{g}{2n_0}, \tag{7.33}$$

where $n_0 = \sqrt{\varepsilon_{1r}}$ (the extension of (7.33) obtained under the assumption $g/\varepsilon_{1r} \ll 1$). For these acceptable values of the refractive index the system of equations (7.32) yields two independent solutions, in which E_x and E_y are related in the following way:

$$\frac{E_x}{E_y} = \pm \mathrm{i}. \tag{7.34}$$

These two independent values correspond to plane electromagnetic waves which propagate along the z-axis and are circularly polarized right- and left-handed:

$$\begin{aligned} \mathbf{E}_{\mathrm{I}} &= [E_0, -\mathrm{i}E_0, 0]\exp[\mathrm{i}(\omega t - n_- k_0 z)], \\ \mathbf{E}_{\mathrm{II}} &= [E_0, \mathrm{i}E_0, 0]\exp[\mathrm{i}(\omega t - n_+ k_0 z)]. \end{aligned} \tag{7.35}$$

We can easily prove this by taking the real part of these solutions and proceeding according to the analysis of the composition of vibrations perpendicular to each other. Thus in the first case we obtain

$$\begin{aligned} E_x &= E_0 \cos(\omega t - n_- k_0 z), \\ E_y &= E_0 \cos\left(-\frac{\pi}{2} + \omega t - n_- k_0 z\right), \end{aligned} \tag{7.36}$$

which in the plane $z = 0$ may be shown as in Fig. 7.1b.

In the second case we obtain

$$\begin{aligned} E_x &= E_0 \cos(\omega t - n_+ k_0 z), \\ E_y &= E_0 \cos\left(\frac{\pi}{2} + \omega t - n_+ k_0 z\right), \end{aligned} \tag{7.37}$$

which corresponds with the situation (in the plane $z = 0$) illustrated in Fig. 7.1a.

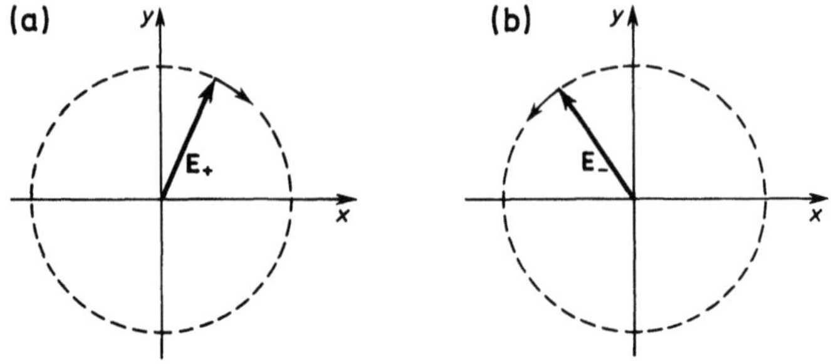

Figure 7.1

The waves under consideration (7.35) differ, except for polarization, also in acceptable values of propagation velocity, which leads to the relative phase lag between them and consequently to the phenomenon of rotation of the plane of polarization of the electromagnetic wave.

The phenomenon of rotation of the plane of polarization can be explained as follows. In all cases a linearly polarized wave entering a medium is decomposed in two waves circularly polarized: left- and right-handed. These two waves with different circular polarization, propagating in an isotropic medium placed in the external magnetic field parallel to the direction of wave propagation, will have different propagation velocities. Due to this, while passing through a medium of length L, each of these waves is subject to a different change of phase

$$\delta_{\mp} = \frac{2\pi L}{\lambda_0} n_{\mp}. \tag{7.38}$$

That difference between phase changes leads to the rotation of the plane of polarization by the angle

$$\psi = \tfrac{1}{2}(\delta_+ - \delta_-) = \frac{\pi L}{\lambda_0}(n_+ - n_-) = \frac{\pi L}{\lambda_0 n_0} g. \tag{7.39}$$

As the value of g is proportional to the external magnetic field H_{ext}, we may rewrite relation (7.39) in the form

$$\psi = CLH_{\text{ext}}, \tag{7.40}$$

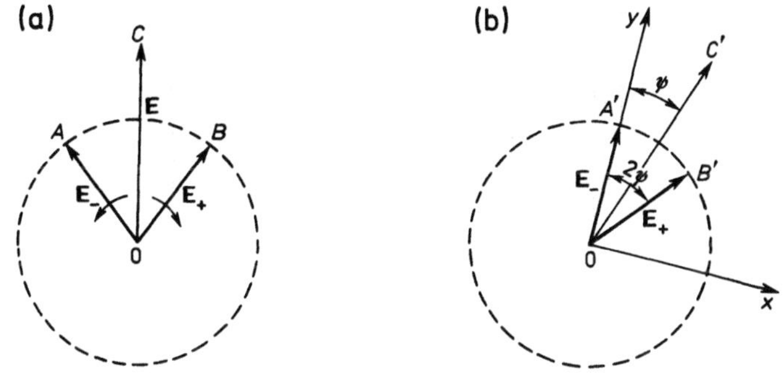

Figure 7.2

where C is a quantity dependent on the wave frequency and temperature of the medium.

The situation discussed above is presented in Fig. 7.2a, which shows the decomposition of the field vector $\mathbf{E}$ at a point where a wave enters the medium, with respect to the observer towards whom the wave is directed. Vectors $\overrightarrow{0A}$ and $\overrightarrow{0B}$ represent the vectors $\mathbf{E}_{\mp}$ for two circularly polarized waves with opposite directions of rotation. Their sum is equal to vector $\overrightarrow{0C}$, which represents the field $\mathbf{E}$ for the primary wave with linear polarization. At the point of emerging (Fig. 7.2b) the vectors $\overrightarrow{0A'}$ and $\overrightarrow{0B'}$ as a result of rotation will change their mutual orientation giving the resultant vector $\overrightarrow{0C'}$. When vector $\overrightarrow{0A'}$ coincides with the perpendicular (y axis), vector $\overrightarrow{0B'}$ makes with the perpendicular the angle 2ψ, while vector $\overrightarrow{0C'}$ the angle ψ.

In the situation illustrated in Fig. 7.2 the right-handed polarized wave has smaller velocity than the left-handed one, so that the field emerging from the medium experiences rotation of the plane of polarization by the angle ψ in the clockwise direction. In general, whether the rotation of the plane of polarization is positive or negative decides the sign of the difference between the phase velocities of right- and left-handed polarized components of the wave.

The phenomenon of rotation of the plane of polarization, in the presence of the external magnetic field parallel to the direction of wave propagation can be observed, apart from an isotropic medium, also in some crystals with cubic symmetry and in many uniaxial and biaxial crystals when light propagates along the optical axis, i.e. along the direction on which birefringence does not occur (see Sec. 6.18).

Bibliography

1. Agranovich, V.M., Ginzburg, V.L.: (1971), 'Crystal Optics with Spatial Dispersion', in.: 'Progress in Optics', Vol. IX, p. 237, Wolf E.M. (ed.), North-Holland, Amsterdam.
2. Chen, H.C.: (1983), 'Theory of Electromagnetic Waves', McGraw-Hill Book Company, New York.
3. Fedorov, F.I.: (1976), 'Theory of Gyrotropy' (in Russian), Nauka i Tekhnika, Minsk.

8. Light Scattering

8.1. General Remarks

The phenomenon of scattering consists in the occurrence of secondary scattered waves with the directions of propagation and frequencies which can differ from the direction of propagation and frequency of the primary wave. This is connected with the change of the polarization state of the medium caused by the primary field, which in consequence leads to the radiation of secondary electromagnetic waves.

According to the character of the changes of the frequency of the scattered waves with respect to the frequency of the incident wave frequency we distinguish two kinds of scattering.

1. *Rayleigh scattering* consists in the occurrence of scattered waves of weak intensity propagating in directions different from the direction of propagation of the primary wave, but without the change of frequency. In Rayleigh scattering there are no changes of the internal states of the particles of the scattering medium. In rarefied gas, where the free path of molecule is significantly longer than the wavelength of the incident wave, scattering takes place independently on each particle. If the free path of the molecule is shorter than the wavelength of the incident wave, scattering may be attributed to two mechanisms: one of them is related to the irregularity in the orientation of the molecules with anisotropic structure, and is caused by the so-called *anisotropy fluctuation*; the other is due to the fluctuations of gas density.

In the case of a monochromatic wave, scattering on the anisotropy fluctuations leads to the occurrence of a blurred line with the maximum for $\omega' = \omega$ and width $\propto h/\tau$, where h is the Planck constant, and τ the time of free passage of the molecule (the orientation of the molecule changes completely after a few collisions).

In scattering on the density fluctuations of the medium, the broadening of the spectral line is smaller. For a wave of length λ those density fluctuations are essential which occur within volumes $\propto \lambda^3$; they change very slowly, which results in narrow spectral lines.

2. Scattering which leads to the change of the frequency of the scattered wave can be explained within the framework of quantum mechanics. However, we can also look at this phenomenon from the phenomenological point of view. The basic type of elementary acts of scattering consists in absorbing a primary quantum of energy $\hbar\omega$ by the scattering system and in simultaneous emission of another quantum $\hbar\omega'$. The frequency of the scattered quantum can be less or greater than ω. In the first case the energy $\hbar(\omega - \omega')$ is absorbed by the system; in the second case the system dissipates the energy $\hbar(\omega' - \omega)$ (which happens, e.g. at the cost of transition into an energetically lower level). In the simplest case of rarefied gas, when scattering takes place on separate molecules, a change of frequency may take place both as a result of the transition of the molecule into a different energetic level and because of the changes of the kinetic energy of the molecules as a whole.

In the case of Rayleigh type scattering (called also *elastic scattering*) there are close phase relations between the incident wave and the scattered wave. Hence this type of scattering is called *coherent*.

In the kinds of scattering which are accompanied by changes of frequency of the scattered wave with respect to the incident wave (i.e. for non-elastic scattering), there occurs a random change of the phase of the scattered wave with respect to the incident wave, and consequently this type of scattering is called *incoherent*.

The scattering of light is associated with an exchange of the momentum and the angular momentum between light and matter, as a result of which light exerts pressure on matter, and under certain conditions matter is subjected to the action of the torque.

The quantitative characteristics describing the scattering properties of the medium are as follows:

1. *Light scattering cross-section* for the molecule (or the coefficient of light scattering divided by the unit of volume of a given medium) which is expressed by the ratio of the power carried by the scattered wave to the power of the wave incident on the molecule.

2. The *extinction coefficient* κ, which describes the attenuation of the primary light beam resulting both from scattering and from the absorption of light by the medium in question. It appears in the law describing the attenuation of the density of the energy flux in the light beam passing through the medium, i.e. in the law

$$I = I_0 e^{-\kappa l}, \tag{8.1}$$

where I_0 is the value of the energy flux entering the medium, and I the value of the energy flux after travelling along the path l .

The scattering of light depends on its polarization and indicates the tensor character of the quantities describing this process.

According to the state of the medium, light scattering proceeds in different ways. Scattering on isolated molecules differs in its course and mechanism from scattering in continuous media.

Among the mechanisms of scattering on isolated molecules we should mention:

(1) light scattering on free electrons,

(2) light scattering on induced electric dipoles,

(3) molecular light scattering on atoms and molecules.

In continuous media scattering may be due to the optical inhomogeneities of the medium, which are generally spread in a chaotic way. They may be caused by the fluctuations of the density of the medium due to the thermic movement of the medium particles, and then we speak of scattering on inhomogeneities caused by fluctuations or of molecular scattering.

Another kind of scattering is due to the inhomogeneities of the medium caused by the existence of foreign microscopic areas in the form of suspended matter of different kinds and sizes. The general formulation of the problem in the theory of light scattering in media containing foreign suspensions as a matter of fact does not essentially differ from the formulations of the electromagnetic theory of light diffraction. The difference between the problem of diffraction and that of scattering is connected with the size of the screen and its distance from the point of observation. At a large distance from the screen of finite dimensions, the field is determined mainly by the incident field. The effect of the screen may be treated as a kind of disturbance and that disturbance gives the *scattered field*.

8.2. Light Scattering on Isolated Particles

A. Light Scattering on Free Electrons

Generally speaking, this type of scattering is not elastic scattering (e.g. the *Compton effect*), but in the range of optical frequencies it may be regarded as such, since the exchange of energy between the photon and the electron is negligible.

If the set of charges is under the influence of an electromagnetic wave, the charges are set in motion. This motion, being an accelerated motion, becomes in turn a source of secondary radiation emitted in all directions: the primary wave is scattered. Let a monochromatic wave of linear polarization fall on the electron. The electric field of the wave can be written in the form

$$\mathbf{E} = \mathbf{E}_0 \cos(\mathbf{k} \cdot \mathbf{r} - \omega t + \alpha). \tag{8.2}$$

We assume that the velocity attained by the electron under the influence of the field of the incident wave is small compared with the light velocity (which is true in most practical cases). Under this assumption the force acting upon the charge is $\mathbf{F}_E = e\mathbf{E}$. (The interaction caused by the magnetic field $\mathbf{F}_B = e(\mathbf{v} \times \mathbf{B})$ can be disregarded, since it is in respect of its value considerably smaller than $\mathbf{F}_E$.) If the charge oscillates about the origin of the coordinate system, it can be considered to be under the influence of the field existing at the origin of the coordinate system, i.e.

$$\mathbf{E} = \mathbf{E}_0 \cos(\omega t - \alpha). \tag{8.3}$$

Under the assumptions formulated above the equation of motion of the electron has the form

$$m\ddot{\mathbf{r}} = -e\mathbf{E}. \tag{8.4}$$

Therefore the dipole moment of the charge oscillating about the origin of the coordinate system $\mathbf{p} = -e\mathbf{r}$ satisfies the equation

$$\ddot{\mathbf{p}} = \frac{e^2}{m}\mathbf{E}. \tag{8.5}$$

To calculate the scattered radiation we use the formula for the intensity of the radiating dipole (see Sec. 1.12). At a large distance R (compared with the wavelength of the wave emitted by the dipole), within the so-called wave zone, the field of the electromagnetic wave emitted by the dipole is given by the formulas

$$\mathbf{H}' = \frac{\omega^2}{cR}(\mathbf{n} \times \mathbf{p}), \quad \mathbf{E}' = (\mathbf{H} \times \mathbf{n})\sqrt{\frac{\mu_0}{\varepsilon_0}}, \tag{8.6}$$

where $\mathbf{n}$ is the versor describing the direction of scattering, and $\mathbf{p}$ the dipole moment of the value at the instant $t - R/c$.

The time averaged intensity of scattered radiation at a solid angle $\mathrm{d}\Omega$ around the direction $\mathbf{n}$ is

$$\mathrm{d}\bar{I} = \tfrac{1}{2}|H|^2 R^2 \mathrm{d}\Omega\sqrt{\frac{\mu_0}{\varepsilon_0}}. \tag{8.7}$$

Dividing this quantity by the time averaged density of the energy flux of the incident wave, which equals

$$|\mathbf{S}| = \tfrac{1}{2}\sqrt{\frac{\varepsilon_0}{\mu_0}}E_0^2, \tag{8.8}$$

we obtain the expression for the light scattering *differential cross-section*

$$\mathrm{d}\sigma = \frac{\mathrm{d}\bar{I}}{|\mathbf{S}|} = \left(\frac{e^2}{mc}\right)^2 \frac{\mu_0}{\varepsilon_0}\sin^2\theta\,\mathrm{d}\Omega, \tag{8.9}$$

where θ is the angle between the vector $\mathbf{E}$ of the primary field and the versor **n**. Hence the effective light scattering cross-section of the electron is

$$\sigma = \frac{8\pi}{3}\left(\frac{e^2}{mc}\right)^2 \frac{\mu_0}{\varepsilon_0}. \tag{8.10}$$

B. Light Scattering on Induced Electric Dipoles

A variable electric field of the light wave induces in the molecules electric dipoles whose value is

$$\mathbf{p} = \alpha \mathbf{E},$$

where α is the polarizability of the molecule. If in a unit of volume there are N molecules subjected to polarization under the influence of the incident wave, the total dipole moment of the unit of volume is

$$\mathbf{P} = N\mathbf{p}.$$

On applying formulas (8.6)–(8.8), we obtain the following expression for the differential scattering cross-section:

$$\mathrm{d}\sigma = \frac{\omega^4}{c^2} N^2 |\alpha|^2 \sin^2\theta \,\mathrm{d}\Omega \frac{\mu_0}{\varepsilon_0}, \tag{8.11}$$

where θ is the angle between **n** and the direction of the field **E**. Hence, on computing the integral over the solid angle, we obtain the *effective light scattering cross-section* of the medium

$$\sigma = \frac{8\pi}{3} |\alpha|^2 \left(\frac{\omega}{c}\right)^4 \varepsilon_0^{-2} N^2. \tag{8.12}$$

It should be remembered (see (1.39)) that

$$\alpha = \frac{e^2(\omega_0^2 - \omega^2 + 2\mathrm{i}\gamma\omega)^{-1}}{m}, \tag{8.13}$$

where ω_0 is the frequency of the free vibrations of the electron, and γ is the damping decrement of the vibrations of the electron.

The scattering cross-section depends on ω, and for ω greatly differing from ω_0 the dependence is given by the factor ω^4. With the increase of this quantity, σ increases very rapidly.

If the incident wave is not polarized, in order to obtain the differential scattering cross-section it is necessary to average (8.11) over all possible directions of **E** in the plane perpendicular to the direction of propagation of the incident wave. Denoting the direction angles of the scattering versor **n** by δ, φ in the coordinate system connected with the vector of propagation **k** and the vector **E** of the incident wave (Fig. 8.1), we obtain the following equality

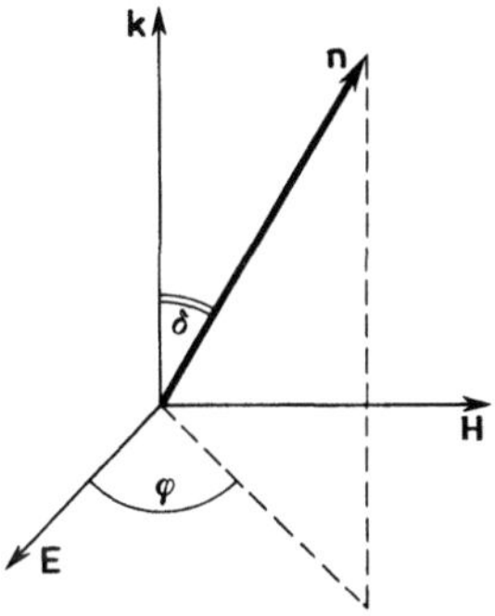

Figure 8.1

for the angle between the direction of scattering and the direction of the induced dipole:

$$\cos\theta = \sin\delta\cos\varphi. \tag{8.14}$$

Then the differential cross-section is

$$d\sigma = \frac{\omega^4}{c^4}|\alpha|^2(1 - \sin^2\delta\cos^2\varphi)d\Omega N^2\varepsilon_0^{-2}. \tag{8.15}$$

Averaging this expression over the angle φ, we finally obtain the formula for the effective cross-section in the case of a non-polarized wave

$$d\sigma = \frac{\omega^4}{2c^4}|\alpha|^2 N^2(1 + \cos^2\delta)d\Omega\varepsilon_0^{-2}, \tag{8.16}$$

where δ is the angle between the direction of scattering and the direction of the primary incident wave.

Moreover, we can find the degree of depolarization of scattered light. To this end let us note that for a fixed direction of $\mathbf{E}$ the vector $\mathbf{E}'$ of the scattered wave lies in the plane determined by the vectors $\mathbf{E}$ and $\mathbf{n}$ (see (8.6)). Thus the vector $\mathbf{E}'$ has a component $\mathbf{E}'_{\parallel}$ lying in the plane of scattering determined by the vectors $\mathbf{n}$ and $\mathbf{k}$, and a component $\mathbf{E}'_{\perp}$ perpendicular to that plane. The field $\mathbf{E}'$ itself and its components $\mathbf{E}'_{\parallel}$ and $\mathbf{E}'_{\perp}$ are given by the expressions

$$\mathbf{E}' = \frac{\omega^2}{\varepsilon_0 c^2 R}\left[\mathbf{p} - \mathbf{n}(\mathbf{n}\cdot\mathbf{p})\right] \tag{8.17}$$

and

$$\begin{aligned} \mathbf{E}'_{\parallel} &= \frac{\omega^2}{R}|\mathbf{p}|\cos\delta\cos\varphi\cdot\mu_0\left[\mathbf{n}\times\left(\mathbf{n}\times\frac{\mathbf{k}}{|\mathbf{k}|}\right)\right]_0 , \\ \mathbf{E}'_{\perp} &= \frac{\omega^2}{R}|\mathbf{p}|\sin\varphi\cdot\mu_0\left(\mathbf{n}\times\frac{\mathbf{k}}{|\mathbf{k}|}\right)_0 , \end{aligned} \tag{8.18}$$

respectively (where indices 0 denote the versors).

In the case of a polarized incident wave, where the direction of scattering $\mathbf{n}$ lies in the plane perpendicular to $\mathbf{E}$ (i.e. where $\varphi = \pi/2$), we have $\mathbf{E}'_{\parallel} = \mathbf{0}$, $\mathbf{E}'_{\perp} \neq \mathbf{0}$ and the intensity connected with the scattered wave is

$$I_{\perp} = \frac{1}{2}\sqrt{\frac{\varepsilon_0}{\mu_0}}|\mathbf{E}_{\perp}|^2 = \frac{1}{2}\left[\frac{\omega^2}{c^2 R}|\mathbf{p}|\right]^2\frac{c}{\varepsilon_0}. \tag{8.19}$$

If $\mathbf{n}$ lies in the plane determined by the vector of propagation and the dipole moment (i.e. if $\varphi = 0$), the intensity is

$$I_{\parallel} = \frac{1}{2}\sqrt{\frac{\varepsilon_0}{\mu_0}}|\mathbf{E}_{\parallel}|^2 = \frac{1}{2}\left[\frac{\omega^2}{c^2 R}|\mathbf{p}|\right]^2\cos^2\delta\cdot\frac{c}{\varepsilon_0}. \tag{8.20}$$

If the incident wave is not polarized, the intensity is

$$I = I_{\perp} + I_{\parallel} = \frac{1}{2}\frac{\omega^4}{c^4 R^2}|\mathbf{p}|^2(1+\cos^2\delta)\frac{c}{\varepsilon_0}. \tag{8.21}$$

From the expression above it can be seen that the intensity of the

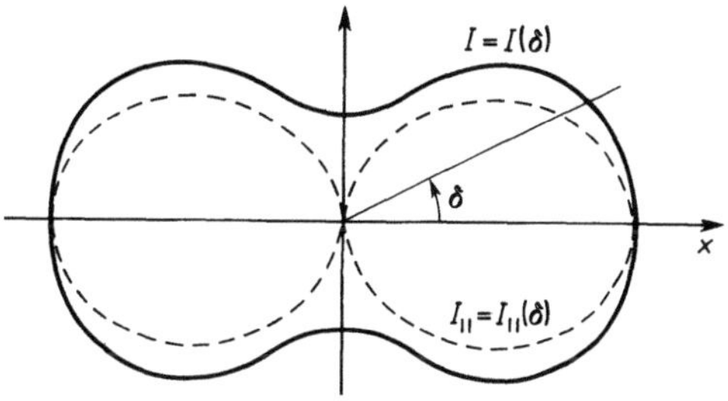

Figure 8.2

light scattered by electric dipoles is inversely proportional to λ^4 (i.e. to the fourth power of the wavelength of the incident wave).

Placing the origin of the reference system of coordinates on the scattering molecule and laying off on the lines passing through that molecule (and forming angles δ with the direction of propagation of the primary beam) the appropriate values of the intensity, we obtain a surface which is the directional characterics of the intensity of the scattered light – it is called the *indicatrix*. We can imagine its shape by rotating the continuous curve shown in Fig. 8.2 about the x-axis, which coincides with the direction of the primary beam.

The continuous curve in Fig. 8.2 represents the dependence of the total intensity of the scattered wave as a function of the angle between the direction of the primary beam and the direction of scattering: $I = I(\delta)$. The dashed line shows the dependence upon the angle defined above of that component of the scattered wave whose electric vector remains parallel to the plane of scattering, i.e. it represents the relation $I_{\parallel} = I_{\parallel}(\delta)$. The dependence of the degree of polarization

$$P = \frac{I_{\perp} - I_{\parallel}}{I_{\perp} + I_{\parallel}} = \frac{\sin^2\delta}{1 + \cos^2\delta} \tag{8.22}$$

on the angle of scattering δ is shown in Fig. 8.3. It can be seen from it that if the angle of observation is $\delta = \pi/2$, we have $I_{\parallel} = 0$ and $P = 1$; this means that the light scattered in directions perpendicular to the direction of propagation of the primary wave is totally polarized. The electric vector and the direction of the scattered

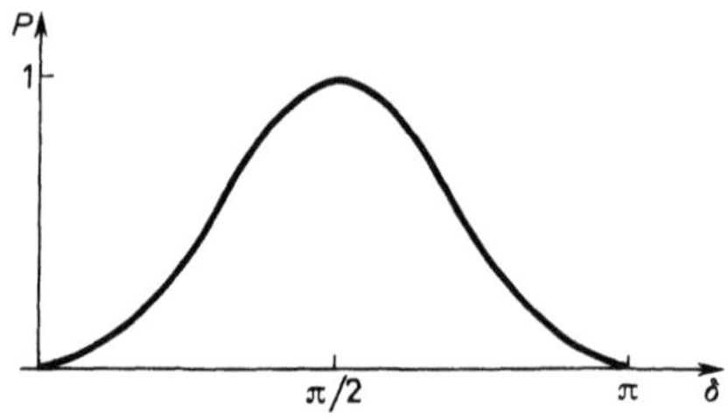

Figure 8.3

wave form, together with the propagation direction of the primary wave, a right-handed system.

The extinction coefficient κ, which describes the attenuation of the primary beam caused by scattering, is calculated by using the relation (see Sec. 1.8) between the intensity of the electromagnetic field and the density of the field energy

$$I = \langle |\mathbf{S}| \rangle = c\rho. \tag{8.23}$$

Differentiating this relation with respect to time and introducing a length element $\mathrm{d}l = c\mathrm{d}t$ lying in the direction of propagation, we obtain

$$\frac{\mathrm{d}I}{\mathrm{d}l} = \frac{\mathrm{d}\rho}{\mathrm{d}t}, \tag{8.24}$$

which defines the rate of the changes of the field energy density. This density change in the field energy of the primary beam is caused by the scattering losses.

Each of the dipoles induced by the primary beam emits the energy $\Delta\xi$ within the time Δt, which gives

$$\frac{\Delta\xi}{\Delta t} = 2\pi \frac{\omega^4}{3c^3} |\mathbf{p}|^2 \varepsilon_0^{-1}. \tag{8.25}$$

For N scattering molecules, where $|\mathbf{p}| = \alpha E$, this change of energy in time is expressed by the formula

$$\frac{\Delta\xi}{\Delta t} = 2\pi \frac{\omega^4 \alpha^2}{3c^3} N E^2 \varepsilon_0^{-1}. \tag{8.26}$$

If a unit of volume contains N_0 molecules, we can write

$$N = VN_0. \tag{8.27}$$

where V is a volume occupied by N scattering molecules.

Using this formula makes it possible to express the time change of the energy density of the primary beam as

$$-\frac{\mathrm{d}\rho}{\mathrm{d}t} = 2\pi \frac{\omega^4}{3c^3} \alpha^2 N_0 E^2 \varepsilon_0^{-1} \tag{8.28}$$

or taking into account (8.23) and the dependence $\rho = \mathbf{E}\cdot\mathbf{D}$, we obtain the following formula for the rate of the intensity changes in the beam along the path traversed by it

$$\frac{\mathrm{d}I}{\mathrm{d}l} = \frac{-4\pi\omega^4}{3c^4\varepsilon_0^2} \alpha^2 N_0 I. \tag{8.29}$$

The sign "minus" is connected with the decrease of the energy of the primary beam, which is transformed into the energy of the scattered wave.

The above equation implies that the intensity of the primary light beam varies exponentially, i.e.,

$$I = I_0 \mathrm{e}^{-\kappa l}, \tag{8.30}$$

where I_0 is the primary intensity of the flux after traversing the path l, l the length of the path covered by the beam, and κ the extinction coefficient equal to

$$\kappa = 4\pi \frac{\omega^4}{3c^4} \frac{\alpha^2}{\varepsilon_0^2} N_0. \tag{8.31}$$

Formulas (8.30) and (8.31) relate the intensity of the primary beam after it has covered the path l with the number of scattering molecules contained in a unit of volume. This allows us, on the basis of transparency studies of gas media, to determine, among other things, the Avogadro number.

The above considerations concern chemically homogeneous media in a liquid or gaseous state. The set of polarized gas molecules can be treated as a large number of totally independent incoherent sources of scattered waves radiation. That is why, in calculating total intensity, we can sum up the intensities given by the squared modules of the amplitudes of the individual scattered waves, disregarding their phase relations.

C. Light Scattering on Anisotropic Molecules

Scattered light polarization observed in pure gases and liquids is connected also with the electron shell of the molecule. The Rayleigh scattering is determined by their polarizability α. For most molecules this quantity, describing the deformation of electron shells, may be different in different directions. In general, it is a second order tensor, which in the system of coordinates (x, y, z) connected with the molecule has the form

$$\begin{bmatrix} \alpha_{11} & \alpha_{12} & \alpha_{13} \\ \alpha_{21} & \alpha_{22} & \alpha_{23} \\ \alpha_{31} & \alpha_{32} & \alpha_{33} \end{bmatrix} . \tag{8.32}$$

The components of the induced dipole moment are related to the appropriate components of the electric vector for the field in question by the relations

$$P_j = \alpha_{jl} E_l \quad (j, l = 1,\ 2,\ 3). \tag{8.33}$$

The polarizability tensor may be univocally presented as the sum of a symmetric tensor and an asymmetric one, the antysymmetric part being responsible for the phenomenon of the polarization plane rotation.

For molecules which are optically non-active the polarizability tensor is always symmetric, i.e., its elements satisfy the equalities

$$\alpha_{jl} = \alpha_{lj}.$$

If starting from the centre of mass of the molecule we lay off along the appropriate directions $\mathbf{s}$ the corresponding vectors $|\alpha| = |\alpha \cdot \mathbf{s}|$, their end-points will form the surface of the ellipsoid of polarizability with axes α_1, α_2, α_3. In the coordinate system whose axes coincide with the axes of the ellipsoid of polarizability the polarizability tensor takes the form

$$\begin{bmatrix} \alpha_1 & 0 & 0 \\ 0 & \alpha_2 & 0 \\ 0 & 0 & \alpha_3 \end{bmatrix}. \tag{8.32a}$$

The form of the ellipsoid of polarizability reflects the structure of the electron shell of the molecule. Namely: for a linear molecule (e.g. H_2, HCl, CO_2) the ellipsoid of polarizability has the elements $\alpha_1 \neq \alpha_2 = \alpha_3$, and for molecules with higher symmetry (say SF_4) the ellipsoid in question reduces to a sphere because $\alpha_1 = \alpha_2 = \alpha_3$.

If the polarizability tensor is reduced to the principal axes, the relations between the components of the induced dipole moment **P** and the field **E** reduce to the form

$$P'_j = \alpha_{jj} E'_j. \tag{8.33a}$$

It is usual to express the incident wave and the intensity of scattered light in a fixed coordinate system X_1, X_2, X_3. The electric dipole moment is calculated in a reference system connected with the molecule x_1, x_2, x_3. Therefore it is necessary to formulate the transformation equations describing the change of the coordinate system. They are of the form

$$X_j = \gamma_{jm} x_m \qquad (j, m = 1, 2, 3), \tag{8.34}$$

where the elements of the transformation matrix $\{\gamma_{jm}\}$satisfying the relations $\gamma_{jm}\gamma_{lm} = \delta_{lj}$ have the sense of direction cosines of the axes of one system with respect to the axes of the other, X_j and x_m being the appropriate components of the position vector in the fixed reference system and in the system connected with the centre of mass of the molecule.

The transformation inverse to (8.34) has the form

$$x_m = \gamma^{\mathrm{T}}_{mj} X_j \tag{8.35}$$

and is characterized by a matrix transposed with respect to the matrix $\{\gamma_{jm}\}$.

If the linearly polarized incident wave is described in the fixed system by the electric vector of the form $\mathbf{E} = [0, 0, E_3]$, then the same wave in the system of the molecule whose axes are the principal axes of the polarizability tensor has, by (8.35), the components $E'_j = \gamma^{\mathrm{T}}_{j3} E_3$.

In accordance with (8.33a) the components of the induced dipole moment are

$$P'_j = \alpha_{jj}\gamma^{\mathrm{T}}_{j3}E_3. \tag{8.36}$$

To find the intensity of the scattered light it is necessary to know the components of the induced dipole moment in the fixed coordinate system. Using (8.34) and (8.36), we obtain for these components the following expressions:

$$\begin{aligned} P_1 &= (\gamma_{11}\alpha_1\gamma^{\mathrm{T}}_{13} + \gamma_{12}\alpha_2\gamma^{\mathrm{T}}_{23} + \gamma_{13}\alpha_3\gamma^{\mathrm{T}}_{33})E_3, \\ P_2 &= (\gamma_{21}\alpha_1\gamma^{\mathrm{T}}_{13} + \gamma_{22}\alpha_2\gamma^{\mathrm{T}}_{23} + \gamma_{23}\alpha_3\gamma^{\mathrm{T}}_{33})E_3, \\ P_3 &= (\gamma_{31}\alpha_1\gamma^{\mathrm{T}}_{13} + \gamma_{32}\alpha_2\gamma^{\mathrm{T}}_{23} + \gamma_{33}\alpha_3\gamma^{\mathrm{T}}_{33})E_3. \end{aligned} \tag{8.37}$$

To simplify the analysis we consider molecules with an axis of symmetry, assuming that the principal values of the polarizability tensor α fulfil the condition

$$\alpha_1 = \alpha_3 \neq \alpha_2, \tag{8.38}$$

under the additional assumption that one of the principal axes of that tensor, x_1, lies in the X_1X_2-plane of the fixed coordinate system. Denoting the angle between the axes x_1 and X_1 by φ, and

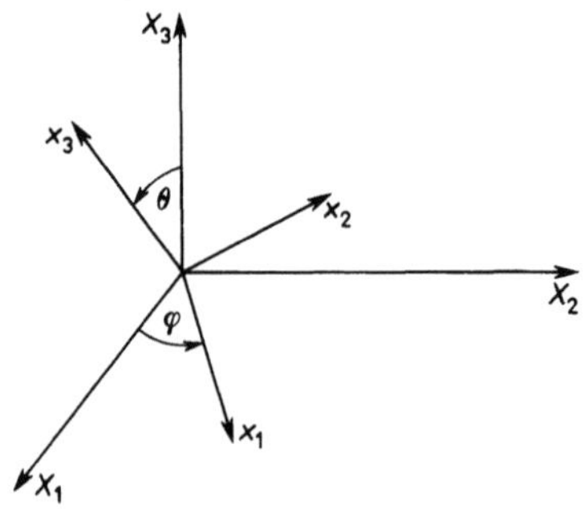

Figure 8.4

the angle between the axes x_3 and X_3 by θ (Fig. 8.4), we may write the transformation matrix as

$$\{\gamma_{jm}\} = \begin{bmatrix} \cos\varphi & -\sin\varphi\cos\theta & \sin\varphi\sin\theta \\ \sin\varphi & \cos\varphi\cos\theta & -\cos\varphi\sin\theta \\ 0 & \sin\theta & \cos\theta \end{bmatrix}.$$

Using this form of the transformation matrix and condition (8.38), we can rewrite relations (8.37) as follows:

$$\begin{aligned} P_1 &= (\alpha_1 - \alpha_2)E_3 \sin\varphi \sin\theta \cos\theta, \\ P_2 &= (\alpha_2 - \alpha_1)E_3 \cos\varphi \sin\theta \cos\ \theta, \\ P_3 &= (\alpha_1 \cos^2\theta + \alpha_2 \sin^2\theta)E_3. \end{aligned} \tag{8.39}$$

Let I_1, I_2, I_3 denote the mean values of the intensities of light scattered by an individual molecule, which are associated with the appropriate components of the induced dipole moment, i.e., let

$$I_j = \frac{\omega^4}{6c^3\varepsilon_0 R^2}\langle|P_j|^2\rangle \quad (j = 1,\ 2,\ 3), \tag{8.40}$$

where the bracket $\langle\ \rangle$ denotes time averaging.

Regarding the direction of propagation of the primary beam of the wave as the X_1-axis and assuming that the electric vector of the wave oscillates in different directions, but is always perpendicular to the axis of the beam, we can state what follows.

The polarizability of isotropic molecules does not depend on the direction, and the induced dipole **P** always coincides with the electric vector of the incident wave. The light scattered in the direction perpendicular to the direction of propagation of the primary beam (the X_2-axis) is fully polarized, because it is associated with the projection of the oscillations of the induced dipole upon the direction perpendicular to the direction of observation (the X_3-axis) and to the direction of propagation of the primary wave (Fig. 8.5).

If a molecule shows anisotropic polarization, the induced dipole **P** does not necessarily coincide with the direction of the electric dipole of the primary wave **E**. Observed along the direction (X_2)

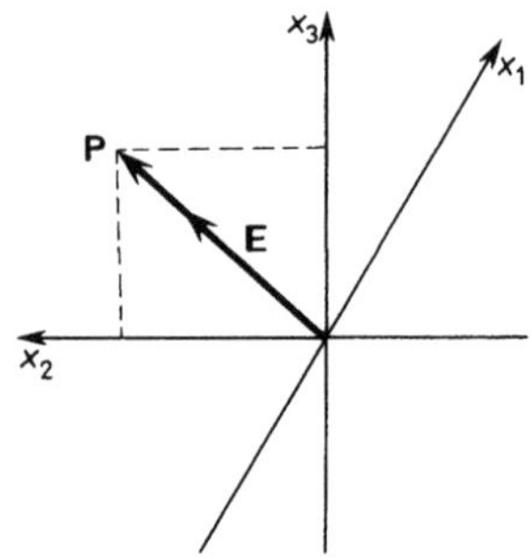

Figure 8.5

perpendicular to the direction of propagation of the primary wave, the scattered wave is connected with the projection of the induced dipole upon the X_1X_3-plane, the dipole having in the general case the components P_1 and $P_3 \neq 0$. For this reason the scattered wave is not polarized (Fig. 8.6).

If we are interested in the intensity of light scattered in gas, i.e. light scattered by a large number of molecules, then in addition to time averaging it is necessary to carry out the averaging over all orientations of the molecules. We assume that the axes of the molecules are distributed in space quite at random and their orientation does not change.

The intensity of light scattered by gas is proportional to the

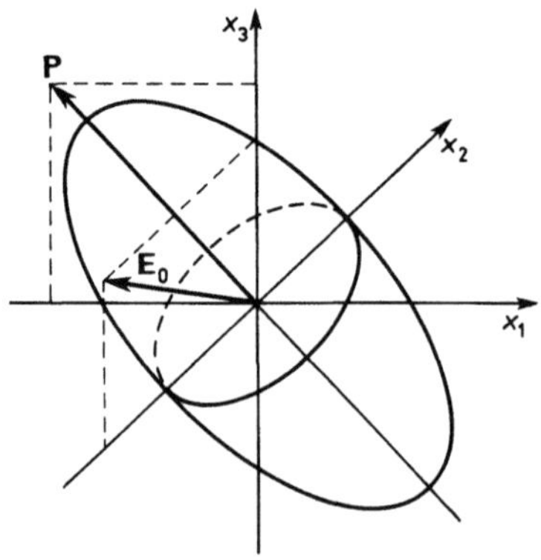

Figure 8.6

number of molecules. It involves the lack of coherence of the waves scattered by the individual molecules.

Carrying out spatial averaging of expressions (8.40) and considering that

$$\overline{(\sin^2\varphi \sin^2\theta \cos^2\theta)^2} = \frac{1}{15},$$

$$\overline{(\alpha_1\cos^2\theta + \alpha_2\sin^2\theta)^2} = \tfrac{1}{9}(\alpha_1 + 2\alpha_2)^2 + \tfrac{4}{45}(\alpha_1 - \alpha_2)^2,$$

we obtain for $\alpha_1 = \alpha_3 \neq \alpha_2$

$$I_1 = I_2 = \frac{1}{30}\frac{\omega^4}{3c^3\varepsilon_0 R^2}(\alpha_1 - \alpha_2)^2 E_0^2,$$

$$I_3 = \frac{\omega^4}{6c^3\varepsilon_0 R^2}\left[\tfrac{1}{9}(\alpha_1 + 2\alpha_2)^2 + \tfrac{4}{45}(\alpha_1 - \alpha_2)^2\right]E_0^2. \tag{8.41}$$

For $\alpha_1 \neq \alpha_3$, and taking into account the additional averaging over all orientations axis x_1 with respect to X_1X_2-plane, expressions (8.41) must be replaced by

$$I_1 = I_2 = \frac{1}{60}\frac{\omega^4}{3c^3\varepsilon_0 R^2}[(\alpha_1 - \alpha_2)^2 + (\alpha_2 - \alpha_3)^2 + (\alpha_3 - \alpha_1)^2]E_0^2,$$

$$I_3 = E_0^2\frac{\omega^4}{6c^3\varepsilon_0 R^2}\left\{\frac{(\alpha_1 + \alpha_2 + \alpha_3)^2}{9} + \right.$$

$$\left. + \tfrac{2}{45}[(\alpha_1 - \alpha_2)^2 + (\alpha_2 - \alpha_3)^2 + (\alpha_3 - \alpha_1)^2]\right\}. \tag{8.42}$$

The quantity

$$\alpha = \tfrac{1}{3}(\alpha_1 + \alpha_2 + \alpha_3) \tag{8.43}$$

in the expressions above is an invariant of the transformation. This term characterizes that part of the scattered light which shows spherical symmetry – it is the isotropic part of scattering.

The second invariant of the transformation

$$\beta^2 = \tfrac{1}{2}[(\alpha_1 - \alpha_2)^2 + (\alpha_2 - \alpha_3)^2 + (\alpha_3 - \alpha_1)^2] \tag{8.44}$$

characterizes the anisotropic part of the scattering. It is large if the differences between the elements α_1, α_2 and α_3 are large.

For isotropic molecules we have $\beta^2 = 0$.

The degree of depolarization of scattered light is given by the formula

$$\Delta = \frac{I_1}{I_3}, \tag{8.45}$$

where intensity I_3 is connected with the projection of the dipole moment on the direction $\mathbf{n} \times \mathbf{k}$ (perpendicular to the direction of observation $\mathbf{n}$ and to the direction of propagation $\mathbf{k}$ of the primary wave), while intensity I_1 is connected with the projection on the direction $\mathbf{n} \times (\mathbf{n} \times \mathbf{k})$.

For the primary wave propagating in the direction of the X_1-axis and linearly polarized in such a way that only one component of the electric field E_3 does not equal 0, the degree of depolarization in a medium consisting of anisotropic molecules is

$$\Delta = \frac{3\beta^2}{4(45\alpha^2 + 4\beta^2)}. \tag{8.46}$$

The degree of depolarization grows with the magnitude of the anisotropy of the molecules, i.e. with the value of β.

For purely isotropic molecules where $\beta^2 = 0$, the degree of depolarization is $\Delta = 0$, which means total polarization of scattered light.

8.3. Scattering on Non-homogeneities Caused by Fluctuations

In pure media in a liquid or gaseous state microscopic inhomogeneities can occur spontaneously. This is due to the fluctuations caused by the chaotic thermic motion of the atoms and molecules composing the medium. In very small volumes the fluctuations may cause deviations of the values of the medium parameters from their mean values. The intensity of fluctuation, and consequently the intensity of the scattered light, increases with the temperature of the medium.

Since the fluctuations are arranged chaotically, they can be treated as a large number of independent, completely incoherent sources of scattered waves, which has an essential bearing on the character of the resultant scattered radiation.

Light scattering caused by the thermic motion of molecules is called *molecular scattering*. In an isotropic transparent medium the thermic motion of molecules disturbs the homogeneity of the medium, producing a deviation from a uniform distribution of molecules in space. The momentary number of molecules in a volume unit can be expressed as

$$N = N_{01} + \Delta N,$$

where N_{01} is the number of molecules in the state of equilibrium, and ΔN is the fluctuation of this number with respect to N_{01}.

Since the density of the medium is proportional to the number of molecules, we can write

$$\rho = \rho_0 + \Delta\rho.$$

We introduce the notion of the relative value of fluctuation

$$\sigma = \frac{\Delta\rho}{\rho}.$$

According to Boltzmann's theorem, the probabilities W and W_0 of two states of the medium when it is subject to fluctuations and when it is in a state of thermodynamic equilibrium are related to the corresponding entropy values by the dependencies

$$S = k \ln W, \qquad S_0 = k \ln W_0, \tag{8.47}$$

where k is the Boltzmann constant.

Formula (8.47) implies the relation

$$W = W_0 \exp[(S - S_0)k^{-1}]. \tag{8.48}$$

If the range of changes of the physical parameters characterizing the system is denoted by $\mathrm{d}\Omega$, the probability of a deviation of the system from the state of thermodynamic equilibrium is expressed by the formula

$$\mathrm{d}W = W_0 \exp\left[(S - S_0)k^{-1}\right]\mathrm{d}\Omega. \tag{8.49}$$

Using the molar quantities, namely, the volume of the gram-molecule V_0 and the entropy of one mole of a substance S^{mol}, we can write the equality

$$S - S_0 = \frac{v\left(S^{\mathrm{mol}} - S_0^{\mathrm{mol}}\right)}{V_0} = \frac{v\Delta S^{\mathrm{mol}}}{V_0}, \tag{8.50}$$

where v is the volume of fluctuation.

In addition to density fluctuation, which is equivalent to the fluctuation of the mole volume $\Delta V = V - V_0$, fluctuation of molecule velocity can also take place. Therefore mole entropy can be regarded as a function of these two kinds of fluctuation:

$$S^{\mathrm{mol}} = f(\Delta V, \Delta T).$$

Expanding entropy in a series with respect to the powers of ΔT and ΔV, we obtain

$$\begin{aligned} S^{\mathrm{mol}} - S_0^{\mathrm{mol}} = {} & \left(\frac{\partial S^{\mathrm{mol}}}{\partial V}\right)_0 \Delta V + \left(\frac{\partial S^{\mathrm{mol}}}{\partial T}\right)_0 \Delta T + \\ & + \frac{1}{2}\left(\frac{\partial^2 S^{\mathrm{mol}}}{\partial V^2}\right)_0 (\Delta V)^2 + \frac{1}{2}\left(\frac{\partial^2 S^{\mathrm{mol}}}{\partial V \partial T}\right)_0 \Delta V \Delta T + \\ & + \frac{1}{2}\left(\frac{\partial^2 S^{\mathrm{mol}}}{\partial T^2}\right)_0 (\Delta T)^2 + \ldots \end{aligned} \tag{8.51}$$

Since we consider deviations from the equilibrium state, in which entropy assumes the maximal value $S_0^{\mathrm{mol}} = S_{\max}^{\mathrm{mol}}$, the following equalities must hold:

$$\left(\frac{\partial S^{\mathrm{mol}}}{\partial V}\right)_0 = 0, \quad \left(\frac{\partial S^{\mathrm{mol}}}{\partial T}\right)_0 = 0.$$

Assuming additionally that the fluctuations of density and temperature are mutually independent, we can further take

$$\left(\frac{\partial^2 S^{\mathrm{mol}}}{\partial V \partial T}\right)_0 = 0.$$

Taking this into account and neglecting the terms of higher orders, we obtain the following expression for the change of entropy:

$$\Delta S^{\mathrm{mol}} = \tfrac{1}{2}\left(\frac{\partial^2 S^{\mathrm{mol}}}{\partial V^2}\right)_0 (\Delta V)^2 + \tfrac{1}{2}\left(\frac{\partial^2 S^{\mathrm{mol}}}{\partial T^2}\right)_0 (\Delta T)^2. \tag{8.52}$$

Since $\Delta S^{\mathrm{mol}} < 0$ (as $S_0^{\mathrm{mol}} = S_{\mathrm{max}}^{\mathrm{mol}}$), the derivatives

$$\left(\frac{\partial^2 S^{\mathrm{mol}}}{\partial V^2}\right)_0, \quad \left(\frac{\partial^2 S^{\mathrm{mol}}}{\partial T^2}\right)_0$$

must be negative.

Introducing the notation

$$\tfrac{1}{2}\left(\frac{\partial^2 S^{\mathrm{mol}}}{\partial V^2}\right)_0 = -a^2, \quad \tfrac{1}{2}\left(\frac{\partial^2 S^{\mathrm{mol}}}{\partial T^2}\right)_0 = -b^2,$$

$$\frac{\Delta V}{V_0} = \sigma, \quad \frac{\Delta T}{T_0} = \tau, \tag{8.53}$$

and using (8.50), we can write (8.52) in the form

$$\Delta S = -(a^2 V_0^2 \sigma^2 + b^2 T_0^2 \tau^2)\frac{v}{V_0}. \tag{8.54}$$

On the basis of (8.54) and (8.49) the probability of density and temperature fluctuation is

$$\mathrm{d}W = W_0 \exp\left[\frac{-(a^2 V_0^2 \sigma^2 + b^2 T_0^2 \tau^2)v}{kV_0}\right] \mathrm{d}\sigma\,\mathrm{d}\tau. \tag{8.55}$$

This expression is the product of two probabilities: of density fluctuation and of temperature fluctuation. Assuming that the fluctuation of density is independent of the fluctuation of temperature, we obtain the following expression for the probability of the fluctuation of density:

$$\mathrm{d}W = A\exp(-a^2 V_0 \sigma^2 v k^{-1})\mathrm{d}\sigma, \tag{8.56}$$

where the constant A is obtained from the normalization condition. Knowing this probability, we can calculate the mean value of the square of the fluctuation of density

$$\overline{\sigma^2} = \int_0^\infty \sigma^2 \mathrm{d}W = \frac{k}{2vV_0 a^2}. \tag{8.57}$$

Hence, on substituting a^2 defined by formula (8.53), we obtain

$$\overline{\sigma^2} = -\frac{k}{vV_0(\partial^2 S/\partial V^2)_0}. \tag{8.57a}$$

In the case of ideal gases, for which

$$\frac{\partial^2 S}{\partial V^2} = -\frac{R}{V_0^2},$$

where R is the gas constant, formula (8.57a) takes the form

$$\overline{\sigma^2} = \frac{kV_0}{Rv}. \tag{8.58}$$

If we now take into account that $R = kN_0$ (where N_0 is the number of molecules in a mole) and introduce the notation $N_{01} = N_0/V_0$ for the number of molecules in a volume unit, the mean value of the square of fluctuation is

$$\overline{\sigma^2} = \frac{1}{N_{01}v}. \tag{8.59}$$

This means that the relative fluctuations in the ideal gas increase with the increase of rarefaction and decrease of the volume in which fluctuation takes place.

The change of the density of the medium implies the change of its permittivity

$$\Delta\varepsilon = \frac{\partial\varepsilon}{\partial V}\Delta V = \frac{\partial\varepsilon}{\partial V}V_0\left(\frac{\Delta V}{V_0}\right) = V_0\left(\frac{\partial\varepsilon}{\partial V}\right)\sigma.$$

If in an element of volume v a disturbance of permittivity takes place, then illuminating it by an electromagnetic field will produce in it a dipole moment equal to

$$p = \Delta\varepsilon E v \tag{8.60}$$

(this is a consequence of the relation $\varepsilon\mathbf{E} = \varepsilon_0\mathbf{E} + \mathbf{P}$ and the fact that the dipole moment of the volume element v is $\mathbf{p} = \mathbf{P}v$). This dipole moment of the volume element v will be a source of secondary radiation. The electric field of the radiated wave will be expressed by the formula

$$E_0 = \frac{Ev}{\varepsilon_0 r}\left(\frac{2\pi}{\lambda}\right)^2 \Delta\varepsilon \sin\theta, \tag{8.61}$$

where r is the distance from the radiating volume, and θ the angle between the direction of the vector $\mathbf{E}$ of the incident wave and the direction of scattering (see (8.17)).

Since light intensity is connected with the mean value of Poynting's vector $\langle|\mathbf{S}|\rangle$, and the value of that vector for a monochromatic wave may be expressed by the intensity of the electric field (see Sec. 1.14), on the basis of formula (8.61) and the equality

$$I = \frac{1}{2}\sqrt{\frac{\varepsilon}{\mu}}\,E^2,$$

where E is the amplitude of the field, we can write the expression for the intensity of the scattered wave

$$I^s = I\frac{v^2}{\varepsilon_0^2 r^2}\left(\frac{2\pi}{\lambda}\right)^4 \overline{(\Delta\varepsilon)^2}\sin^2\theta, \tag{8.62}$$

where for the value of the fluctuation of permittivity we should take its mean square value related to the mean value of the fluctuation of density $\overline{\sigma^2}$,

$$\overline{\Delta\varepsilon^2} = V_0^2\left(\frac{\partial\varepsilon}{\partial V}\right)_0^2 \overline{\sigma^2}. \tag{8.63}$$

Formula (8.62) implies that:

The intensity of scattered light is inversely proportional to the fourth power of the wavelength.

Thus shorter waves will be scattered more intensively than longer ones. Therefore in molecular scattering light will be richer in ultraviolet, which, incidentally, explains the blue colour of the sky.

8.4. Polarization of Light in Scattering on Fluctuations

In Sec. 8.3 we discussed the case of an incident wave polarized linearly. We can extend our considerations to the case of nonpolarized light, which is most frequently emitted by natural sources. It can always be presented as a superposition of two waves polarized in perpendicular planes. Namely, if the direction of propagation of the incident wave coincides with the x-axis, its electric field can be decomposed into the component $\mathbf{E}_y$ along the y-axis and the component $\mathbf{E}_z$ along the z-axis. For each component we can repeat the analysis carried out in Sec. 8.3 and show that scattered light consists of two components. One of them has the electric vector in the plane $(\mathbf{E}_y, \mathbf{r})$ and the other in the plane $(\mathbf{E}_z, \mathbf{r})$ (Fig. 8.7). The total intensity of scattered light may be presented as the sum

$$I^s = I^s_y + I^s_z. \tag{8.64}$$

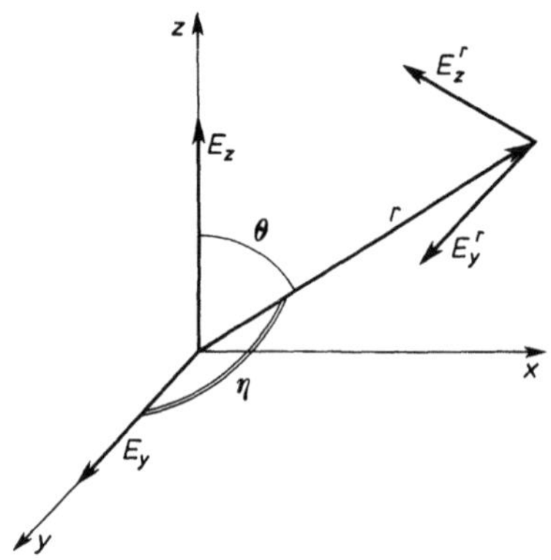

Figure 8.7

Applying (8.62) to the components $\mathbf{E}_z$ and $\mathbf{E}_y$ of the incident wave and considering the different angles formed by these vectors with the direction of scattering $\mathbf{r}$, we obtain formula (8.64) in the form

$$I^s_{\theta,\eta} = IA(\sin^2\theta + \sin^2\eta), \tag{8.65}$$

where $A = (v^2/r^2)(2\pi/\lambda)^4(\Delta\varepsilon)^2\varepsilon_0^{-2}$,

$$\sin\theta = \sin(\mathbf{E}_z, \mathbf{r}) = \sin(\mathbf{z}, \mathbf{r}) \qquad \sin\eta = \sin(\mathbf{E}_y, \mathbf{r}) = \sin(\mathbf{y}, \mathbf{r}).$$

The identity

$$\cos^2(\mathbf{x}, \mathbf{r}) + \cos^2(\mathbf{y}, \mathbf{r}) + \cos^2(\mathbf{z}, \mathbf{r}) = 1$$

gives

$$\sin^2\theta + \sin^2\eta = 1 + \cos^2(\mathbf{k}, \mathbf{r}).$$

Thus the intensity of scattered light in the direction $\mathbf{r}$ is finally expressed by the formula

$$I^s_{\theta,\eta} = IA[1 + \cos^2(\mathbf{k}, \mathbf{r})]. \tag{8.66}$$

In the general case, under different orientations of the vector $\mathbf{r}$, the intensities for different states of polarization of scattered light are not equal to each other. Scattered light is partially polarized. If observation is carried out perpendicularly to the direction of propagation of the primary wave, for example along the y-axis, the component of the intensity of the scattered wave I^s_y associated with that component of the electric vector which lies in the plane $(\mathbf{E}_y, \mathbf{r})$ is equal to zero. The total intensity of the light scattered in the direction in question will be related only with that component of the electric vector of scattered light which lies in the plane $(\mathbf{E}_z, \mathbf{r})$, and it will be $I^s_z = IA$.

Thus, in the direction perpendicular to the direction of the primary wave propagation, light in the *Rayleigh scattering* is totally polarized. The distribution of the intensity of light scattered in space can be illustrated by means of a surface obtained by laying off the appropriate values of I^s along the direction of propagation of the scattered light wave. As shown in Fig. 8.8, the surface obtained in this way is symmetrical with respect to the x-axis, along which the primary wave propagates.

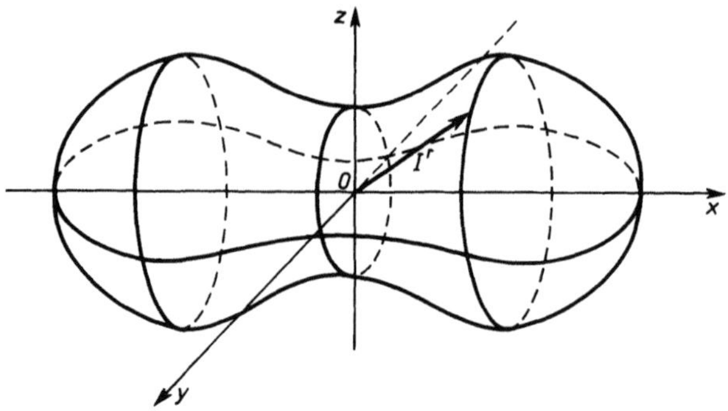

Figure 8.8

Scattering on fluctuations is crucial at the critical state of the medium matter, because fluctuations are then maximal. It is experimentally confirmed that a medium which is in the critical state scatters light so intensively that it becomes opaque itself.

Unfortunately, the mathematical description of this situation is very complex, because fluctuations in different volume elements can no longer be regarded as independent, as was assumed in Sec. 8.3. In the critical state the mean dimensions of fluctuations strongly increase. When those dimensions exceed the wavelength, the medium should be treated as a suspension of other microscopic particles. In a medium of this kind light from all ranges of the spectrum is scattered in a more uniform way.

8.5. Scattering on Suspensions

A real medium is not homogeneous. There may be differences in local intensity, temperature, local non-homogeneities of anisotropy, etc. Consequently the refractive index of the medium depends on the coordinates and on time.

In addition to non-homogeneities of the macroscopic type, which change in time and space at a relatively slow rate, there may also occur microscopic non-homogeneities the medium. Here belong the particles of suspended matter, for which the refractive index and the absorption coefficient differ from each other. Colloidal solutions, particles of dust, etc. can serve as an example.

In the simplest case non-homogeneities are produced by areas of dimensions equal to or greater than the wavelength of light, which form suspensions in gases, liquids or solid bodies. They are emulsion solutions, aerosols, etc. If the concentration of scattering objects is large enough, the physical media mentioned above may become opaque owing to great losses in scattering.

A non-homogeneous medium in which large foreign macroscopic matter is suspended is called a *turbid medium.* Light propagating in such a medium is deflected from the original direction of propagation, which leads to the weakening of the original beam.

The other kind of light scattering media are *colloidal solutions* in which the dimensions of foreign particles are so small that the physical systems are practically transparent. However, during observation which is lateral with respect to the propagation of the light beam, colloidal particles reveal themselves as scattering centres.

In the problem in question, we should consider, in addition to the wavelength, the following quantities: d – the distance between the atoms inside the suspension, a – the size of the suspended particles, l – the mean distance between the centres of the suspended matter in the medium, R – the size of the scattering medium. These quantities provide us with four dimensionless parameters:

$$x_1 = \frac{d}{\lambda}, \quad x_2 = \frac{a}{\lambda}, \quad x_3 = \frac{l}{\lambda}, \quad x_4 = \frac{R}{\lambda}.$$

It is obvious that

$$x_1 < x_2 < x_3 < x_4$$

and additionally

$$x_1 \ll 1, \quad x_4 \gg 1.$$

When

$$x_3 \gg 1,$$

the distances between the suspended particles are considerably larger than the wavelength. In this case the suspended particles may be regarded as incoherent sources of radiation.

The chaotic distribution of the suspended particles leads to violent changes of the phase of the fields scattered by the suspensions, which destroys the effects accompanying interference. This in turn leads to summation of light intensities of the scattered waves.

In addition to the linear characteristics, of great importance in the problem here discussed are also the physical properties of the scattered particles, among which an essential role is played by the complex refractive index

$$m = n - i\kappa, \tag{8.67}$$

where n is the refractive index, and κ the absorption coefficient.

In scattering on particles of suspended matter, it is necessary to distinguish several cases, relating either to the dimensions of the particles and their ratio to the wavelength or to their properties (particles of dielectrics, of metals).

If the dimensions of the suspended particles which determine the non-homogeneity of the medium are considerably smaller than the light wavelengths, the scattering is the Rayleigh scattering. This occurs for

$$x_3 \ll 1.$$

In this case the field of the scattered wave can be regarded as the field of a vibrating dipole induced in a spherical body of suspended matter by the electric field of the primary wave $\mathbf{E}_0$. The value of this dipole is

$$\mathbf{P} = \varepsilon_0 a^3 \left(\frac{m^2 - 1}{m^2 + 1}\right) \mathbf{E}_0, \tag{8.68}$$

where a is the radius of the spherical body of suspended matter, and m is the complex refractive index (see equation (8.60)).

Because of the condition $x_2 \ll 1$, the field of the wave scattered on the suspended matter in question can be regarded as the field of a vibrating dipole at a considerable distance from it.

Vector $\mathbf{E}$ of this field (see (8.17)) is expressed by the formula

$$\mathbf{E} = \frac{(\ddot{\mathbf{P}} \cdot \mathbf{R}_0)\mathbf{R}_0 - \ddot{\mathbf{P}}}{\varepsilon_0 r c_a^2}\, \omega^2, \tag{8.69}$$

where $\mathbf{R}_0$ is a versor directed from the centre of the dipole towards an observer, $\mathbf{r}$ is the distance from the point of observation, and c_a is the light velocity in the medium. The time derivatives of the dipole moment in formula (8.69) are taken for the time $t' = t - r/c_a$, which takes into account the so-called *time lag effect.*

On the strength of the above remark the electric field of the wave scattered on a spherical body of suspended matter is

$$\mathbf{E} = -\frac{k_a^2 a^3}{r}\left(\frac{m^2 - 1}{m^2 + 2}\right) e^{-ik_a r}\left[(\mathbf{E}_0 \cdot \mathbf{R}_0)\mathbf{R}_0 - \mathbf{E}_0\right], \qquad (8.70)$$

where k_a is the length of the wave vector in the medium. If the primary wave is linearly polarized, vector $\mathbf{E}_0$ vibrates along the x-axis, and the wave comes from the positive direction of the z-axis, then the indicatrix of scattering is determined by the expression

$$I(\delta, \varphi) = I_0 \frac{32\pi^4 a^6}{\lambda_a^4 r^2 \varepsilon_0^2 c}\left|\frac{m^2 - 1}{m^2 + 2}\right|^2 (1 - \sin^2\delta \cos^2\varphi), \qquad (8.71)$$

where the angles δ and φ are as in Fig. 8.1. From formula (8.71) it follows that

(1) in the plane perpendicular to the wave vector of the primary wave $(\delta = \pi/2)$ we obtain

$$I \propto \sin^2\varphi, \qquad (8.72)$$

(2) in the plane determined by the direction of the incident wave and its magnetic vector $(\varphi = \pi/2)$ the indicatrix is

$$I = \text{const},$$

(3) in the plane $\varphi = 0$ or $\varphi = \pi$ the indicatrix is described by

$$I \propto \cos^2\delta.$$

The indicatrix of scattering of the natural light is obtained from (8.71) by averaging it over the angle φ. As a result we obtain

$$I(\delta) = I_0 \frac{9\pi^2 v^2}{\lambda_a^4 r^2}\left|\frac{m^2 - 1}{m^2 + 2}\right|^2 \frac{(1 + \cos^2\delta)}{\varepsilon_0 c}. \qquad (8.73)$$

This is called the *Rayleigh formula.* Its essential features are:

a. $I \propto 1/\lambda^4$.

b. $I \propto v^2$ (the intensity of scattered light is proportional to the square of the volume of the scattering particles).

c. Scattering forward and backward is identical.

By larger dimensions of non-homogeneities we have the dependence

$$I \propto \lambda^{-p},$$

where $p < 4$, and p decreases with the increase of the dimensions of the suspended particles. Moreover, the directional characteristics of the intensity of light scattered on large particles is elongated in the direction of propagation of the incident light (Fig. 8.9).

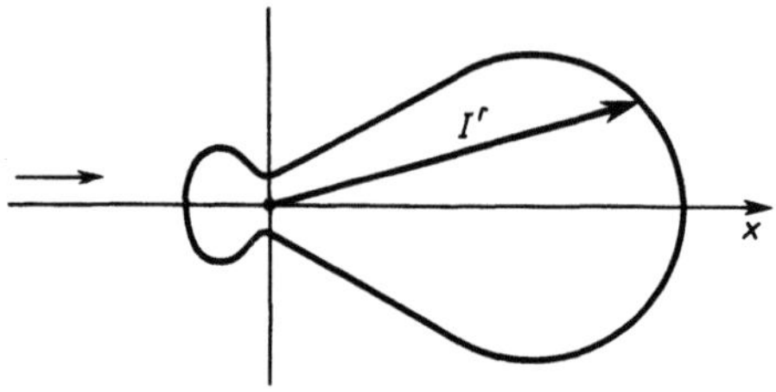

Figure 8.9

This finds an explanation within the framework of diffraction theory: scattered light is the result of a superposition of elementary waves emitted by particles of the scattering medium. The phase difference between these waves depends on the angle of scattering, and is smallest in the direction of the primary wave and greatest in the reverse direction. Its value lies within the range $0-\pi$. The interference of those secondary waves leads to the attenuation of the resultant wave in the reverse direction and to its amplification in the direction of propagation of the primary beam.

If the radii of the suspended particles a are such that

$$x_2 = \frac{a}{\lambda} \gg 10, \tag{8.74}$$

the intensity of the scattered light no longer depends on the wavelength. In the scattered light there is no dominance of any colour (contrary to what occurred in the Rayleigh scattering). The transmitted light and the scattered light have the same spectral composition. Particles of such dimensions appear in clouds, in the fog, in milk – this explains their white colour.

As the original primary light beam passes through a turbid medium, it undergoes attenuation as a result of both scattering and absorption. The change of intensity is described by the dependence

$$I = I_0 \exp(-2\pi N_1 a^2 K z), \tag{8.75}$$

where I_0 is the intensity of light entering the medium, I the intensity of light after passing the layer of thickness z, N_1 the number of suspended particles in a volume unit of the medium, a the radius of a particle, K a coefficient dependent on the density of the medium and on the length of the scattered wave.

In scattering on opaque particles (of metals) the problem of scattering on large areas of non-homogeneity may be regarded as a purely diffractive problem.

The theory of light scattering in media containing foreign metallic inclusions in the form of suspension is formulated in much the same way as the electromagnetic theory of diffraction. Given the original electromagnetic field of the incident wave and known spatial distributions of the electromagnetic properties of the medium, we seek an electromagnetic field at an arbitrary point inside and outside the given area of non-homogeneity.

Knowledge of the primary field scattered by individual particles of the suspended matter enables us to find the total external field as a superposition of the field of the primary wave and the fields of the secondary waves arising from all non-homogeneities.

This approach lies at the basis of the model of single scattering, in the framework of which, for low concentrations of non-homogeneities, the resultant field is a superposition of the primary field and the fields of secondary waves generated by a single scattering of the incident wave on the existing suspensions.

For high concentration of non-homogeneities the field incident on the suspended matter cannot be treated as the field of the primary wave. It is the resultant field of the primary wave and the waves scattered on all the other non-homogeneities. The scattered waves are difficult to determine, because it is not always possible to split the problem into the following:

(a) determining the field of a wave scattered (generated by the primary incident wave) by the suspended matter;

(b) determining the field of a wave scattered by the suspended matter where there is an incident field which is a superposition of the field of the primary wave and the fields of the waves scattered by all suspended particles; the latter fields being calculated at the first stage.

Thus, if the number of non-homogeneities increases, the scattered waves generated by the primary wave on individual suspended particle come across other non-homogeneities on their way and, before reaching the observer, undergo secondary scattering. This is a model of repeated scattering. Whether it is possible to divide the problem into the stages formulated above depends on the dimensions of the areas of matter suspended in the medium and the distances betwen areas, and their ratio to the length of the incident wave.

It is worth noting that if the distances between the suspended particles are comparable with the wavelength, interference is of great importance. Instead of the intensities we have to sum up the vectors of the electromagnetic fields.

8.6. Scattering on Transparent Particles

For transparent particles (dielectrics) with dimensions much larger than the wavelength, scattering can be explained with the use of the laws of refraction of light applied in geometric optics. To determine the indicatrix of scattering on a large transparent sphere we use below the methods of geometric optics. As the basis for our considerations we take Fresnel's formulas for the reflection coefficient $\mathscr{R}$ for two different polarizations of the incident wave:

(a) with the electric vector lying in the plane of incidence,

(b) with the electric vector perpendicular to the plane of incidence, and we denote these two cases by t and p respectively.

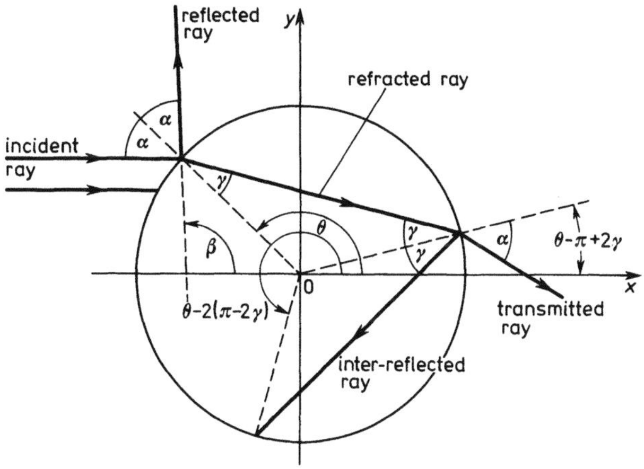

Figure 8.10

If α stands for the angle of incidence and γ for the angle of refraction, then

$$\mathscr{R}_p = \frac{\sin^2(\alpha - \gamma)}{\sin^2(\alpha + \gamma)},$$
$$\mathscr{R}_t = \frac{\tan^2(\alpha - \gamma)}{\tan^2(\alpha + \gamma)}. \tag{8.76}$$

Let a parallel beam of natural light fall on a sphere of radius a in the x-direction (Fig. 8.10). An arbitrary plane which passes through the centre of the sphere and contains the x-direction can be regarded as the plane of incidence. All rays emerging from the sphere will be called *derivative rays.* They show the following properties:

1. All derivative rays and their primary ray lie in the same plane. This permits confining the investigation of the angular distribution of intensity of the scattered light to one chosen plane.

2. All the derivative rays of a primary beam of type p are rays of type p, and the derivative rays of a primary beam of type t are rays of type t.

3. All rays incident on the sphere at the angle α emerge from that sphere at the same angle.

4. If a ray falls on the surface of the sphere at a point with angle θ and $\theta > \pi$, its derivative rays emerge from points with angles θ, $\theta + \mu, \ldots$ If $\theta < \pi$, the derivative rays come from points with angles: $\theta - \mu$, $\theta - 2\mu, \ldots$, where $\mu = \pi - 2\gamma$ (Fig. 8.10).

5. All derivative beams which originate from a given incident beam have the same "thickness" as the incident beam which generates them. Hence

$$I^{(0)} = I^{(1)} + I^{(2)} + I^{(3)} + \ldots, \tag{8.77}$$

where $I^{(0)}$ is the intensity of the primary beam, and $I^{(k)}$ the intensity of the k-th derivative beam. We consider the spatial distribution of the intensity of the reflected beams. The angle of scattering β for the reflected beam is connected with the angle of incidence α through the relation

$$\beta = \pi - 2\alpha. \tag{8.78}$$

We obtain different results according to whether the intensity of the scattered light is examined in the immediate vicinity of the sphere (i.e. for $R - a \ll a$) or far from the sphere ($R \gg a$). R is the distance from the point of observation to the centre of the sphere. In the first case an element of the sphere may be treated as a plane and the intensities of the light reflected by various elements of the sphere will be equal to the products of the intensities of the incident light $I^{(0)}$, $I_t^{(0)}$, $I_p^{(0)}$ and the reflection coefficients $\mathcal{R}$, $\mathcal{R}_t$, $\mathcal{R}_p$, i.e.,

$$I^{(1)} = \mathcal{R} I^{(0)}. \tag{8.79}$$

The light intensity on the dark side equals zero.

To determine the intensity distribution of the scattered light at long distances from the sphere ($R \gg a$) we carry out the following considerations (we take into account only single reflections of the incident light from the surface of the transparent sphere).

We consider on the sphere a strip of width $d\alpha$ (Fig. 8.11). The area of the strip is

$$dS_0 = 2\pi a^2 \sin\alpha \, d\alpha. \tag{8.80}$$

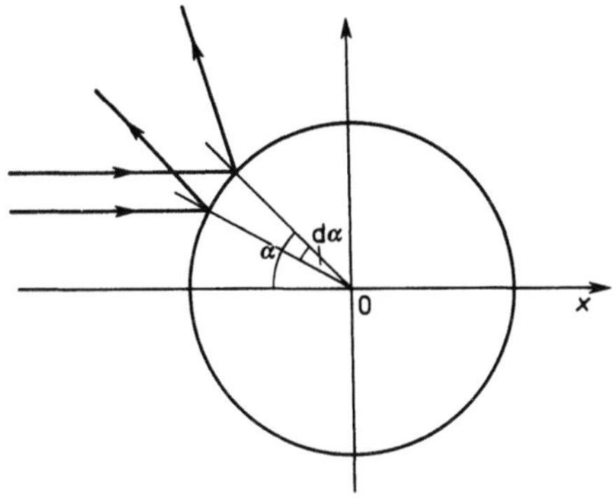

Figure 8.11

The light beam incident on the strip is

$$dF^{(0)} = I^{(0)} \cos\alpha \, dS_0. \tag{8.81}$$

The beam reflected from the strip is

$$dF^{(1)} = \mathscr{R} dF^{(0)} = \mathscr{R} I^{(0)} 2\pi a^2 \cos\alpha \sin\alpha \, d\alpha. \tag{8.82}$$

All the beam will be scattered into the solid angle $d\Omega$ between the angles of scattering β and $\beta + d\beta$. The size of the solid angle is $2\pi \sin\beta \, d\beta$. From the relation $\beta = \pi - 2\alpha$ it follows that $d\beta = -2d\alpha$, which gives

$$d\Omega = 8\pi \cos\alpha \sin\alpha \, d\alpha.$$

This solid angle cuts out on a sphere of radius R a surface element $dS = R^2 d\Omega$ upon which falls the light reflected by a corresponding surface element of the scattering sphere. Hence the intensity of the scattered light at the point of observation is

$$I^{(1)} = \frac{dF^{(1)}}{dS} = \frac{a^2}{4R^2} I^{(0)} \mathscr{R}. \tag{8.83}$$

The intensity of the light reflected from the sphere in a given direction is proportional to the inverse square of the sphere radius and to the reflection coefficient and does not depend on $\cos\alpha$.

The quantity $4R^2/a^2$ characterizes the geometrical change of the primary incident beam. The smaller the sphere the less light power is directed into the same solid angle and the smaller is scattering for a unit solid angle $\Gamma = I^{(1)}R^2$.

For the indicatrices of scattering for beams t and p we obtain the expressions

$$\Gamma_t = \frac{a^2}{4} I_t^{(0)} \mathscr{R}_t, \quad \Gamma_p = \frac{a^2}{4} I_p^{(0)} \mathscr{R}_p. \tag{8.84}$$

The total indicatrix and the degree of polarization in the case of nonpolarized light are respectively

$$\Gamma = \frac{a^2}{8} I^{(0)} (\mathscr{R}_t + \mathscr{R}_p), \tag{8.85}$$

$$P^{(1)} = \left| \frac{\mathscr{R}_p - \mathscr{R}_t}{\mathscr{R}_p + \mathscr{R}_t} \right|. \tag{8.86}$$

The explicit expressions for $\mathscr{R}_p$ and $\mathscr{R}_t$ as functions of the angle of incidence α are of the form

$$\begin{aligned} \mathscr{R}_p &= \left(\frac{\cos\alpha - \sqrt{n^2 - \sin^2\alpha}}{\cos\alpha + \sqrt{n^2 - \sin^2\alpha}} \right)^2, \\ \mathscr{R}_t &= \left(\frac{n^2\cos\alpha - \sqrt{n^2 - \sin^2\alpha}}{n^2\cos\alpha + \sqrt{n^2 - \sin^2\alpha}} \right)^2, \end{aligned} \tag{8.87}$$

and are obtained from equations (8.76).

In Figure 8.12 the indicatrix of scattering is shown. The external

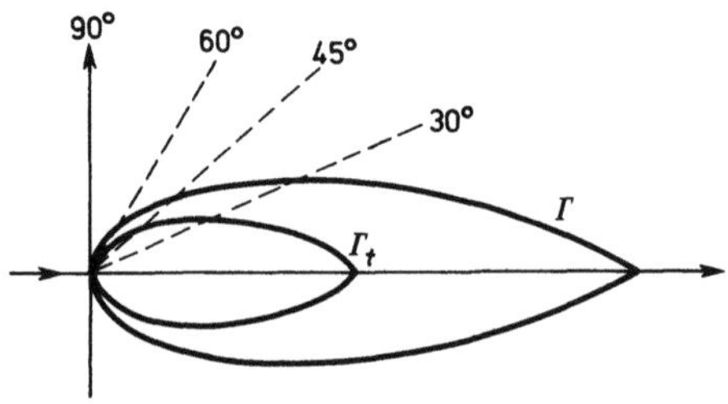

Figure 8.12

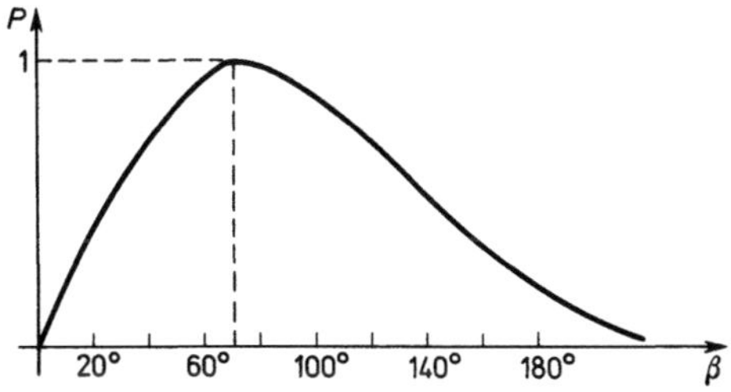

Figure 8.13

curve presents the full indicatrix of scattering Γ. The internal curve represents the indicatrix of the t-beam Γ_t. Figure 8.13 shows the dependence of the degree of polarization upon β.

Bibliography

1. Fabelinskii, I.L.: (1965), 'Molecular Scattering of Light' (in Russian), Izd. Nauka, Moskva.
2. Ivanova, A.P.: (1969), 'Optics of Scattering Media' (in Russian), Nauka i Tekhnika, Minsk.
3. Korolev, F.: (1966), 'Theoretical Optics' (in Russian), Izd. Vysshaya Shkola, Moskva.
4. Landau, L. and Lifshits, E.: (1960), 'Electrodynamics of Continuous Media' (in Polish), PWN, Warszawa.
5. Shifrin, K.S.: (1951), 'Scattering of Light in Turbid Medium' (in Russian), Izd. Tekh.-Teor. Literatury, Moskva.
6. Shishlovskii, A.A.: (1961), 'Applied Physical Optics' (in Russian), Fizmatgiz, Moskva.
7. Stepanov, B.I. and Ivanow, A.P.: (1971), 'Theoretical and Applied Problems of Light Scattering' (in Russian), Izd. Nauka i Tekhnika, Minsk.

9. Physical Principles of Holography

9.1. Introductory Remarks

Below we give only the information which is indispensable for understanding the physical bases of the process of holography. Comprehensive data on the subject can be found in the literature listed at the end of this chapter.

In ordinary photography the image registered on a photographic plate is a suitably scaled intensity distribution in the plane of the object; in this process we lose the information about the distribution of the phase of the wave coming from the object. All detectors are sensitive only to the intensity of light. In order to record information about the phase of the wave it is necessary to make use of the phenomenon of interference. To this end we record the total interference field resulting from the overlapping of two waves: the wave scattered at the object and the reference wave coherent with the first one. Total intensity will depend both on the amplitude and on the phase of each of the waves. From the interference image registered in this way in the plane of a photographic plate – this image is called a *hologram* – we can reconstruct the original wave front by illuminating the hologram with the coherent wave.

Thus holography is a two-stage method of obtaining images, which consists in recording and then reconstructing the primary object wave front.

9.2. Recording and Reconstruction of a Wave Front

A hologram is obtained by registering on a photographic plate the total intensity of light. That intensity is the result of the overlapping of two fields: the object beam U_0 and the reference beam U.

Thus, if

$$U_0 = a(x, y)e^{-i\varphi(x,y)} \tag{9.1}$$

is the wave front which we want to register and

$$U = b(x, y)e^{-i\psi(x,y)} \tag{9.2}$$

is the reference wave, the total intensity of the radiation incident on the plate is

$$\begin{aligned} I(x, y) = |a(x, y)|^2 + |b(x, y)|^2 + \\ + \{ab^*\exp\{-i[\varphi(x, y) - \psi(x, y)]\} + \\ + a^*b\exp\{+i[\varphi(x, y) - \psi(x, y)]\}\}. \end{aligned} \tag{9.3}$$

The first two terms of (9.3) depend only on the intensities of the two overlapping waves, the third – on the difference in their phases. In this way we record information both about the amplitude and about the phase of the wave (9.1).

In the process of recording of essential importance is the quantity ξ called the *exposure*, which is equal to the intensity of light incident on the plate multiplied by the time of exposure t:

$$\xi = I(x_1, y_1)t. \tag{9.4}$$

After chemical treatment the exposed photographic plate becomes an object characterized by a certain *function of amplitude transmittance* defined as

$$T(x_1, y_1) = \frac{\text{amplitude of field just behind the plate at point } (x_1, y_1)}{\text{amplitude of field incident on the plate at point } (x_1, y_1)}, \tag{9.5}$$

where (x_1, y_1) are the coordinates of a point which lies in the plane $z = z_1 = \text{const}$ coinciding with the surface of the photographic plate. Function (9.5) is in general a complex function of the exposure

$$T(x, y) = f\{\xi(x, y)\}. \tag{9.6}$$

Expanding (9.6) in a Taylor series in the neighbourhood of a certain fixed value of exposure ξ_0, which in general is a function of point (x_1, y_1), we obtain with an accuracy up to a linear term

$$T(x_1, y_1) = T_0 + \beta(\xi - \xi_0), \tag{9.7}$$

where $\beta = \mathrm{d}T/\mathrm{d}\xi$ for $\xi = \xi_0$.

The constant β is generally a complex quantity. On the base of (9.3) and (9.4) we can assume that

$$T_0(x_1, y_1) = \beta t\left[|a(x_1, y_1)|^2 + |b(x_1, y_1)|^2\right]. \tag{9.8}$$

Our choice is related to the fact that ξ_0 has to be positive. Moreover, for typical references beams $|a(x, y)|^2 = \text{const}$. In this way the exposure value ξ fluctuates about the value ξ_0, varying slowly from point to point on the hologram. On the basis of (9.3), (9.7) and (9.8) we obtain the following expression for the transmittance function of the photographic plate:

$$T = T_0 + \beta t(U_0 U^* + U_0^* U). \tag{9.9}$$

Expression (9.9) is the equation of the hologram.

In the process of reconstructing the hologram it is illuminated by the reconstructing beam

$$U_r = C(x, y)\mathrm{e}^{-\mathrm{i}\eta(x,y)}. \tag{9.10}$$

We obtain the following expression describing the distribution of the field amplitude just behind the hologram:

$$U_1(x_1, y_1) = T_0 U_r + \beta t(U_0 U^* + U_0^* U) U_r. \tag{9.11}$$

As can be seen from the above, a wave passing through a hologram consists of three components. The first of them is a wave propagating in the same direction as the primary reconstructing wave but with a slightly different amplitude, while the two others are of essential importance in reconstructing the recorded front of the object wave.

And so, if the reconstructing beam and the reference beam are identical, $U_r = U$ and $|U|^2 = \text{const}$, then the second term of the sum

(9.11) reconstructs with high fidelity the distribution of the complex amplitude of the primary wave front of the object beam in the plane (x_1, y_1), i.e. just before recording, and gives the field $U_2 \propto U_0$.

If the reconstructing beam is taken in the form $U_r = U^*$, the third component of (9.11) gives a wave front conjugated with the primary one: $U_3 \propto U_0^*$.

To obtain the primary front of the object wave it is necessary to separate spatially in some way the above three components of (9.11). This is achieved by using holograms in which the direction of propagation of the reference wave does not coincide with the direction on which the object and the photographic plate are situated but forms a certain angle with it (Fig. 9.1).

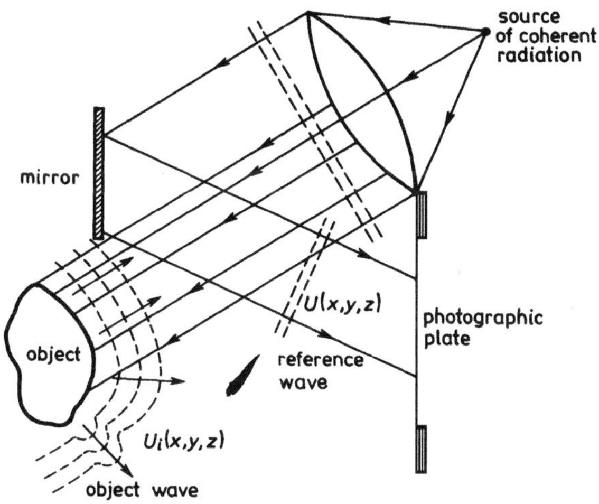

Figure 9.1

9.3. Holography of Point Objects

The linearity of the function of amplitude transmittance of a hologram with respect to exposure (9.7) leads to the linearity of the optical system represented by the holographic system. That is why for a complete description of that system it is enough to know its point response. In other words, it is enough to know how the system behaves in the case of a point object in order to describe its behaviour in the case of an object with any other structure. Two methods of determining the point response of the holographic system are known.

The first of them is a geometric method. It permits us to use the notion of a light ray but at the same time it does not give any information about the resolution of the system. The second method – diffractional – gives in a linear approximation full information about the properties of the system. Both methods can be applied to a system with the same geometry.

A. Geometric Method

Let us consider the geometry shown in Fig. 9.2. A spherical object beam with a complex amplitude U_i propagating from a source point

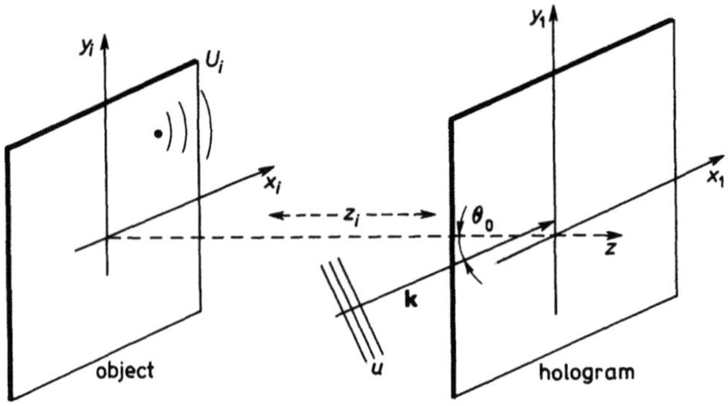

Figure 9.2

with fixed coordinates (x_i, y_i) interferes with a plane reference wave with a complex amplitude U, which falls at a small angle θ_0 and has a wave vector lying in the plane $y = 0$. The amplitudes of those waves have in paraxial approximation $\left(z_i \gg \sqrt{(x_1 - x_i)^2 + (y_1 - y_i)^2}\right)$ in the plane of the hologram (x_1, y_1) the following form:

$$U_i(x_1, y_1) = \exp\left\{\frac{ik}{2z_i}\left[(x_1 - x_i)^2 + (y_1 - y_i)^2\right]\right\}, \qquad (9.12)$$

$$U(x_1, y_1) = \exp(ikx_1\theta_0), \tag{9.13}$$

where $k = 2\pi/\lambda$ and z_i is the distance between the object and the plane of the photographic plate.

The second term of (9.11) has in this case the form

$$T_2(x_1, y_1) = \beta t U_i U^* = A_i \exp\left\{ik\left(\frac{x_1^2 + y_1^2}{2z_i}\right)\right\} \times$$
$$\times \exp\left\{-ikx_1\left(\frac{x_i}{z_i} + \theta_0\right) + iky_1\frac{y_i}{z_i}\right\}, \tag{9.14}$$

where the factor A_i represents all terms independent of (x_1, y_1).

Expression (9.14) can also be presented in the form

$$T_2(x_1, y_1)$$
$$= B_i \exp\left\{+\frac{ik}{2F_H}[(x_1 - x_{1i})^2 + (y_1 - y_{1i})^2]\right\}, \tag{9.15}$$

where $x_{1i} = x_i + z_i\theta_0$, $y_{1i} = y_i$, $F_H = z_i$.

A similar analysis of the third term of (9.11) leads to an expression conjugated with (9.15). Combining the two expressions and introducing the symbols

$$T_2 = T^{(+)}, \qquad T_3 = T^{(-)},$$

we obtain

$$T^{(\pm)} = B_i^{(\pm)} \exp\left\{-\frac{ik}{2F_H}[(x_1 - x_{1i})^2 + (y_1 - y_{1i})^2]\right\}, \tag{9.16}$$

where

$$B_i^{(+)} = B_i, \qquad B_i^{(-)} = B_i^*,$$

$$F_H = \begin{cases} -z_i & \text{for } T^{(+)} \\ z_i & \text{for } T^{(-)}. \end{cases}$$

The function of amplitude transmittance (9.16) defines the so-called *Fresnel lens* or *holographic lens*, equivalent to two ordinary lenses (see Sec. 4.24): a divergent one with focus length $-z_i$ and a convergent one with focus length z_i. The position of the lens centre is defined by the coordinates (x_{1i}, y_{1i}) and depends on the position of the point object. This implies that the hologram of a bulk object is a continuum of Fresnel lenses.

If a plane wave falls on the Fresnel lens made by the hologram, we obtain two points of ray focusing – two focuses situated in the focal plane of the lens. They are two images of a point object, an upright one and a conjugated one, obtained in the holographic process of reconstruction. Generalizing this to the case of an object with an arbitrary structure, we can state that each point of an object is transformed into two point images, and consequently a bulk object produces two images in a hologram reconstruction, as shown in Fig. 9.3. The angle θ_0 should be large enough to prevent the image beams from overlapping each other – this is the so-called *Leith–Upatnicks realization.*

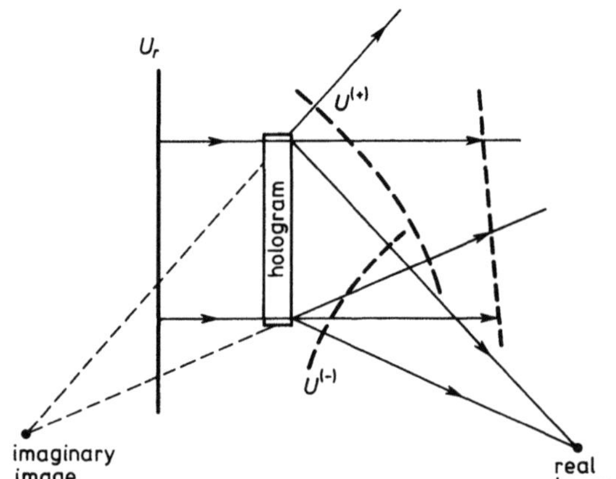

Figure 9.3

If in the reconstruction process a hologram with the function of amplitude transmittance (9.16) is illuminated by a plane wave with a complex amplitude

$$U_r = \exp(\mathrm{i}kx_1\theta_r),$$

we obtain on the basis of (9.11) the wave

$$U^{(+)}(x_1, y_1) = C_i \exp\left\{\frac{ik}{2z_i}\left[(x_1 - x_i')^2 + (y_1 - y_i')^2\right]\right\}, \quad (9.17)$$

where $x_i' = x_i + (\theta_0 - \theta_r)z_i$, $y_i' = y_i$.

The complex amplitude (9.17) describes a spherical wave propagating from the point with coordinates (x_i', y_i') situated at a distance z_i from the hologram. The image of the imaginary point object is displaced along the x-axis by the value $(\theta_0 - \theta_r)z_i$ with respect to the position of the object. For $\theta_0 = \theta_r$ the displacement vanishes.

B. *Diffractional Method*

We assumed above that expression (9.17) describes a spherical wave not only in the plane of the hologram (x_1, y_1) but also beyond that plane. This occurs, however, only for a hologram of unlimited dimensions. In the case of limited dimensions the geometrical method fails because of the appereance of diffractional effects. In order to account for them we use the Fresnel diffractional formula (see Sec. 4.13).

If the hologram occupies a certain finite region A of the plane, the complex amplitude in the plane $z = \text{const}$ behind the plane of the hologram is given, according to formula (4.60), by

$$U(x, y; x_i, y_i) = \frac{\exp(ikz)}{i\lambda z}\iint U^{(+)}(x_1, y_1; x_i, y_i) \times$$

$$\times \exp\left\{\frac{ik}{2z}\left[(x - x_1)^2 + (y - y_1)^2\right]\right\} dx_1 dy_1, \quad (9.18)$$

where $U^{(+)}(x, y; x_i, y_i)$ is the spherical wave given by formula (9.17) and propagating from the point of coordinates (x_i, y_i).

Substituting (9.17) into (9.18) and considering the explicit form of the coefficient C_i, we get

$$U(x, y; x_i, y_i) = \frac{\exp(ikz)}{i\lambda z} \exp\left[ik\frac{x^2+y^2}{2z}\right] \exp\left[ik\frac{x_i^2+y_i^2}{2z_i}\right] \times$$

$$\times \iint_{-\infty}^{+\infty} P(x_1, y_1) \exp\left[\frac{ik(x_1^2+y_1^2)}{2}\left(\frac{1}{z_i}+\frac{1}{z}\right)\right] \times$$

$$\times \exp\left[-ikx_1\left(\frac{x_i}{z_i}+\frac{x}{z}\right)\right] \exp\left[-iky_1\left(\frac{y_i}{z_i}+\frac{y}{z}\right)\right] dx_1\, dy_1, \tag{9.19}$$

where $P(x_1, y_1)$ is the so-called *pupil function*, which is equal to one within the hologram and to zero outside it.

The compatibility of the diffractional method with the geometric method is obtained by assuming in formula (9.19) that $P(x_1, y_1) \equiv 1$. Then, if we have the equality

$$-z_i = z, \tag{9.20}$$

integral (9.19) represents the Dirac delta function. Thus we obtain

$$U(x, y; x_i, y_i) \propto \delta\left(\frac{x_i}{z_i}+\frac{x}{z}\right) \delta\left(\frac{y_i}{z_i}+\frac{y}{z}\right). \tag{9.21}$$

Considering the equation of the image plane (9.20), which coincides in this case with the focal plane of the lens, we obtain the point response of the system, which is not equal to zero only at the point with coordinates $x = x_i$, $y = y_i$, which agrees with (9.17). Hence we can see that in the case of a hologram of unlimited dimensions the two methods yield an identical result.

In the case of a limited hologram the point response in the focal plane $z = -z_i$ has the form

$$U(x, y; x_i, y_i) = U(x - x_i, y - y_i)$$
$$= \iint_{-\infty}^{+\infty} P(x_1, y_1) \exp\left[-\frac{ikx_1}{z}(x - x_i)\right] \times$$
$$\times \exp\left[-\frac{iky_1}{z}(y - y_i)\right] dx_1 dy_1. \tag{9.22}$$

On the basis of (9.22) we can see that the shape of the point response is independent of the position of the point object. Consequently it is enough to determine it for a chosen point of the object, i.e. for coordinates $x_i = (\theta_r - \theta_i)z_i$, $y_i = 0$. In this case we obtain

$$U(x, y) = \iint_{-\infty}^{+\infty} P(x_1, y_1) \exp[-i2\pi(f_x x + f_y y)] dx_1 dy_1. \tag{9.23}$$

Thus, as a result of diffractional effects, a point object is mapped not into a point but into a spot. The dimensions of that spot determine the resolving power of the diffractional system; by decreasing the dimensions of the hologram we lose the information about the details of the object.

Quantitative data on the subject are given by an analysis of a simple example.

If a hologram has the shape of a rectangular plate with dimensions L_x and L_y, its pupil function has the form

$$P(x_1, y_1) = \text{rect}\left(\frac{x_1}{L_x}\right) \text{rect}\left(\frac{y_1}{L_y}\right), \tag{9.24}$$

where the function $\text{rect}(x)$ equals 1 for $|x| \leq \frac{1}{2}$ and equals 0 outside this region.

On the grounds of the above considerations we obtain

$$U(x, y) = L_x L_y \,\text{sinc}\left(\frac{xL_x}{\lambda z_i}\right) \text{sinc}\left(\frac{yL_y}{\lambda z_i}\right). \tag{9.25}$$

Thus in the case of a holographic system with plane reference beams and an object at a distance z_i from the hologram, whose dimensions are L_x and L_y, it is impossible to distinguish the details of the object which are smaller than Δx, Δy, where

$$\Delta x = \frac{2\lambda z_i}{L_x}, \quad \Delta y = \frac{2\lambda z_i}{L_y}. \tag{9.26}$$

Δx and Δy are the distances between the first zeros of the sincus function, and they determine the broadening of the point response.

Substituting in expression (9.26) $\lambda = 0.63\ \mu m$, $z_i = 10$ cm and $L_x = L_y = 1$ mm, we obtain approximately $\Delta x = 0.1$ mm.

9.4. Light Properties Required in Holography

In order to produce a hologram, the reference wave falling directly on the photographic plate has to be coherent with the light scattered by the surfaces of the object whose hologram we want to obtain. In the case of a large object this condition is fulfilled only if the source of radiation has a sufficient degree of spatial coherence (i.e. the angular dimensions of the source are sufficiently small) and a sufficiently long coherence path (i.e. the source produces near-monochromatic radiation).

The coherence length of the light used should be greater than the maximal difference between the paths covered by the reference wave and the object wave.

The spatial coherence of a classical source is evaluated by the quality of the interference image, which is produced by the overlapping of the light beams coming from different points of the source. The smaller the difference in the paths of the corresponding rays, the higher the quality in question. If the angular dimensions of the source are θ, the difference between the path of a ray from the central part of the source and that of a ray from its side part is

$$\Delta \approx \frac{d\theta}{4}, \tag{9.27}$$

where d is the diameter of the source.

If $\Delta \approx \lambda/2$ or $\Delta = \lambda/2$, the interference image will be blurred. The sharpness of the image will be preserved for $\theta \ll 2\lambda/d$. This enables us to estimate the distance from the source at which we may expect this condition to be fulfilled. It turns out that this distance should be

$$L \gg \frac{d^2}{2\lambda}. \tag{9.28}$$

To obtain radiation showing the required time coherence of the wave, we may filter the light produced by a classical source of radiation (a mercury discharge lamp, an incandescent lamp) through a narrow-band monochromator. Spatial coherence of the light wave can be obtained by increasing the distance from the source or by focusing its radiation at a hole with a sufficiently small diameter. All this, however, weakens the usable power of the source.

The problem of radiation coherence resolves itself into the problem of spectral luminosity of the source, i.e., of the power emitted by a surface unit of the source into a unit solid angle, from a unit interval of the wavelengths. All these features: high spectral luminosity of the source, small angular divergence of the beam and monochromatic radiation are found in a laser. For this reason the development of holography became possible only with the advent of laser light.

To end these remarks it is worth while to mention a feature which distinguishes in essential way a hologram from a photograph. In an ordinary photograph to each point of the object corresponds a defined point of the image. It is different in holography – the object whose hologram we want to obtain scatters light, which falls upon the whole surface of the plate. Thus any point of the object is connected with the whole surface of the hologram. Looking at a photograph at different angles, we cannot see what is happening behind the object placed in the foreground. A hologram reconstructs not a two-dimensional image of the object but the wave scattered by it, i.e., it reconstructs the primary object wave; this enables us, by changing the point of observation in the region of that wave field, to see the object at different angles, giving the impression of its spatial character.

Bibliography

1. Cathey, W.T.: (1974), 'Optical Information Processing and Holography', J. Wiley & Sons Inc., New York, London, Sydney, Toronto.
2. Parrent, G.B. and Thompson, B.J.: (1969), 'Physical Optics Notebook', Prentice-Hall, N.Y.
3. Pluta, M. (ed.): (1980), 'Optical Holography' (in Polish), PWN, Warszawa.
4. Stroke, G.W.: (1966), 'An Introduction to Coherent Optics and Holography', Academic Press, New York.

10. Nonlinear Phenomena in Optics

10.1. Introductory Remarks

Nonlinear optical phenomena are phenomena whose course depends on the intensity of radiation. It is known (see Sec. 1.2) that the constitutive relations in the form of linear dependence between the polarization and the intensity of the electric field ($\mathbf{P} = \kappa \mathbf{E}$) are the first approximation. That approximation remains valid as long as the ratio of the field $\mathbf{E}$ of the incident wave to the value $\mathbf{E}_0$ of the internal field responsible for binding the electron with the atom or with the medium is a small quantity. For the hydrogen atom we have the field $|\mathbf{E}_0| \approx 10^9$ V/cm, while for continuous media (semiconductors) this value is $|\mathbf{E}_0| \approx 10^7$ V/cm. The intensity of the wave fields obtained from the quantum generators may be for optical frequencies of the order 10^6 V/cm. Fields of such intensities cannot be disregarded in the interaction of the charged molecules of the medium.

In the case of optical fields of such intensities linear constitutive relations must be replaced by more complex equations of the form $\mathbf{P} = \mathbf{P}(\mathbf{E})$, where $\mathbf{P}(\mathbf{E})$ is in general a functional. In every case we can use for the description of optical phenomena the expansion of polarization in a series with respect to the parameter $E/E_0 < 1$:

$$\mathbf{P} = \mathbf{P}^{(\mathrm{L})} + \mathbf{P}^{(\mathrm{NL})}, \tag{10.1}$$

where the linear polarization is

$$\mathbf{P}^{(\mathrm{L})} = \int_0^\infty \kappa(t')\mathbf{E}(t - t')\,\mathrm{d}t', \tag{10.2}$$

and the nonlinear polarization is

$$\mathbf{P}^{(\mathrm{NL})} = \mathbf{P}^{(2)} + \mathbf{P}^{(3)} + \dots, \tag{10.3}$$

where

$$\mathbf{P}^{(2)} = \int_0^\infty \mathrm{d}t' \int_0^\infty \mathrm{d}t'' \kappa(t', t''){:}\ \mathbf{E}(t - t')\mathbf{E}(t - t' - t'')$$

$$\mathbf{P}^{(3)} = \int_0^\infty \mathrm{d}t' \int_0^\infty \mathrm{d}t'' \int_0^\infty \mathrm{d}t''' \kappa(t', t'', t''') \vdots$$

$$\vdots\, \mathbf{E}(t - t')\mathbf{E}(t - t' - t'')\mathbf{E}(t - t' - t'' - t'''). \tag{10.4}$$

Expressions (10.4) are the natural generalization of dependence (10.2) to the nonlinear case and find justification within quantum theory. Relations (10.2) and (10.4) are valid in media which do not show spatial dispersion, and contain tensors of electric susceptibility: of the second order $\kappa(t')$ for linear polarization, of the third order $\kappa(t', t'')$ for square polarization, i.e. dependent on the second power of the field **E**, and the fourth order $\kappa(t', t'', t''')$ for cubic polarization, i.e. dependent on the third power of the field **E**. These tensors are determined by the physical properties of the material media.

Between the individual terms of sum (10.3) the following relations take place:

$$\left|\mathbf{P}^{(n)}\right| \gg \left|\mathbf{P}^{(n+1)}\right|, \tag{10.5}$$

owing to which polarizations of higher order reveal themselves only if the lower order of nonlinear polarization is equal to zero. This depends in an essential way on the symmetry properties of the medium. (Square polarization is zero in a medium with a centre of symmetry.)

One of the simplest cases is a medium without dispersion (i.e., a medium whose polarization does not show any inertia and is determined by the value of the field at the same instant). In this case the time dependence of the tensors of electric susceptibility takes the form

$$\kappa(t') = \kappa^{(0)}\delta(t'),$$

$$\kappa(t',t'') = \kappa^{(2)}\delta(t')\delta(t''), \tag{10.6}$$

$$\kappa(t',t'',t''') = \kappa^{(3)}\delta(t')\delta(t'')\delta(t''').$$

This means that there is a simple functional relation between **P** and **E**

$$\mathbf{P} = \kappa^{(0)}\mathbf{E} + \kappa^{(2)}\colon\mathbf{EE} + \kappa^{(3)}\vdots\,\mathbf{EEE}. \tag{10.7}$$

If several monochromatic waves propagate in the medium, the electric field vector of the resultant electromagnetic field can be written in the form

$$\mathbf{E} = \tfrac{1}{2}\sum\left[\mathbf{E}_j(\mathbf{r})\exp(\mathrm{i}\omega_j t) + \text{c.c.}\right]. \tag{10.8}$$

The linear polarization of the medium is excited independently by each of the monochromatic waves, giving

$$\mathbf{P}^{(\mathrm{L})} = \sum_j \kappa(\omega_j)^{(0)}\left[\mathbf{E}_j(\mathbf{r})\exp(\mathrm{i}\omega_j t) + \text{c.c.}\right], \tag{10.9}$$

where

$$\kappa(\omega_j)^{(0)} = \int_0^\infty \kappa(t')\exp(-\mathrm{i}\omega_j t')\,\mathrm{d}t'. \tag{10.9a}$$

Nonlinear polarization is more complex. For square polarization it will contain components with different frequencies

$$\mathbf{P}^{(2)} = \sum_m\sum_n \left\{\kappa^{(2)}(\omega_m + \omega_n)\colon \mathbf{E}_m(\mathbf{r})\mathbf{E}_n(\mathbf{r})\exp\left[\mathrm{i}(\omega_m + \omega_n)\right] + \right.$$
$$\left. + \kappa^{(2)}(\omega_m - \omega_n)\colon\mathbf{E}_m(\mathbf{r})\mathbf{E}_n(\mathbf{r})\exp\left[\mathrm{i}(\omega_m - \omega_n)t\right]\right\}, \tag{10.10}$$

where

$$\kappa^{(2)}(\omega_m + \omega_n)$$
$$= \iint_0^\infty \kappa(t',t'')\exp\{\mathrm{i}[-(\omega_m + \omega_n)t' - \omega_n t'']\}\,\mathrm{d}t'\mathrm{d}t'',$$

$$\kappa^{(2)}(\omega_m - \omega_n)$$
$$= \iint_0^\infty \kappa(t',t'')\exp\{\mathrm{i}[-(\omega_m - \omega_n)t' - \omega_n t'']\}\,\mathrm{d}t'\mathrm{d}t''. \tag{10.10a}$$

In the case of cubic polarization the number of polarization components of different frequencies is still greater, because now they are generated no by two but by three monochromatic waves. If the frequencies of the interacting waves are ω_m, ω_n and ω_k, they can induce components of cubic polarization with frequencies

$$\pm\omega_m \pm \omega_n \pm \omega_k > 0. \tag{10.11}$$

Different polarization components may in turn generate secondary waves of frequency equal to the frequency of a given component of polarization. In this way in a nonlinear medium there may be changes in the spectrum of frequency of the primary electromagnetic field incident on the medium.

Nonlinear effects which accompany propagating of electromagnetic waves in a material medium may be connected not only with the interaction of light with optical electrons but also with its interaction with another kind of motion in that medium. An electromagnetic wave can induce electrostriction effects, and these in turn may lead to the occurrence of local tensions proportional to the intensity of the wave field, which results in the dependence of the refractive index on intensity. Thus, for example, an electromagnetic field of the form

$$\mathbf{E} = \mathbf{E}_1 \exp[\mathrm{i}(\omega_1 t - \mathbf{k}_1 \cdot \mathbf{r})] + \mathbf{E}_2 \exp[\mathrm{i}(\omega_2 t - \mathbf{k}_2 \cdot \mathbf{r})] \tag{10.12}$$

produces variable pressure, which generates an acoustic wave with the frequency

$$\Omega = \omega_1 - \omega_2. \tag{10.13}$$

This acoustic wave modulates in turn the electric permittivity of the medium and causes reverse interaction on the electromagnetic field. This results in different behaviour of the light wave according to its intensity.

The process discussed above, in which the field of the light wave excites a secondary motion causing the scattering of the wave, is called *forced scattering.*

The scattering of light on acoustic waves takes place also for weak light intensities, the only difference being that acoustic waves are generated by external factors. It is an example of the so-called *parametric optical effects*, i.e., effects occurring in media whose parameters are changed in a certain definite way by means of external forces. We should include among them, e.g. *Kerr's effect* (see Sec. 6.12), *Faraday's effect* (see Sec. 7.5), etc.

Since the parametric effects are determined by the same physical properties of the medium as the optical effects dependent on the intensity of the light wave, and also because they show a reverse action of the medium upon the field of the light wave, we can treat the problem of the interaction of the field with the medium and of the medium with variable parameters with the field in a uniform way, and regard "parametric optics" as one of the sections of nonlinear optics. All the phenomena of nonlinear optics understood in this way are described by appropriate tensors of the third and higher orders, which relate the interaction of electromagnetic waves with each other and with waves of another kind.

Of the numerous optical effects which depend on the intensity of the light wave, we discuss here only the following:

1. generation of waves with higher harmonics,
2. self-focussing of strong light beams,
3. phase conjugation,
4. optical bistability,
5. nonlinear surface effects.

10.2. Polarization of a Dielectric Induced by a Wave of High Intensity

In discussing linear polarization of a medium (Sec. 1.5) we regarded electrons as harmonic oscillators. The resonance frequencies of the oscillators were chosen in such a way that they should correspond to the observed atomic spectral lines. In reality the motion of valence electrons takes place in the Coulomb field of the nucleus, and if there are appreciable deviations from the position of the stable orbit, it is necessary to take into account the anharmony of the oscillators. The greater is the value of the electric field of the light wave incident on the atom, the greater are those deviations.

The elastic force which acts on the valence electron in the case of wave fields which are strong but weaker than the Coulomb field E_0 can be described by the expression

$$\mathbf{F} = -k\mathbf{r} - qr\mathbf{r}, \tag{10.14}$$

where k, q are constants, and $r = |\mathbf{r}|$ is the distance from the centre of action of the force. In the presence of an external electric field $\mathbf{E}(t)$ the equation of motion of the electron has the form

$$m\ddot{\mathbf{r}} + m\gamma\dot{\mathbf{r}} + k\mathbf{r} + qr\mathbf{r} = e\mathbf{E}(t). \tag{10.15}$$

Defining the polarization vector by the equality

$$\mathbf{P} = Ne\mathbf{r}, \tag{10.16}$$

where N is the number of electrons in a volume unit, we obtain the equation

$$\ddot{\mathbf{P}} + \gamma\dot{\mathbf{P}} + \omega_0^2\mathbf{P} + \frac{q}{e^2N^2m}|\mathbf{P}|\mathbf{P} = \frac{e^2N}{m}\mathbf{E}(t), \tag{10.17}$$

which is a nonlinear equation.

If we assume that the field $\mathbf{E}(t)$ is harmonic and that its frequency ω is such that $|\omega - \omega_0| \gg \gamma$ (i.e. far from the resonance frequency), we can disregard in (10.17) the term which describes attenuation since it is essential only for ω close to ω_0.

Assuming that $\mathbf{E}(t)$ is weaker than the Coulomb field $\mathbf{E}_0$, we can regard the nonlinear term in (10.17) as small and solve the equation by using the method of successive approximations. Equation (10.17) will then assume the form

$$\ddot{\mathbf{P}}_1 + \omega_0^2\mathbf{P}_1 = \frac{e^2N}{m}\mathbf{E}(t), \tag{10.17a}$$

and its solution (under the assumption $\mathbf{E}(t) = E_0\exp(i\omega t)$) is

$$\mathbf{P}_1(t) = \frac{e^2N}{m(\omega_0^2 - \omega^2)}\mathbf{E}(t). \tag{10.17b}$$

The equation for the second correction of $\mathbf{P}$ (t) is obtained by substituting in the nonlinear term in (10.17) the value $\mathbf{P}_1$ instead of $\mathbf{P}$. This gives

$$\ddot{\mathbf{P}}_2 + \omega_0^2\mathbf{P}_2 = -\frac{qe^2|\mathbf{E}|}{m^3(\omega_0^2 - \omega^2)^2}\mathbf{E}(t). \tag{10.17c}$$

For $\mathbf{E} = \mathbf{E}_0\exp(i\omega t)$ the above equation has a solution of the form

$$\mathbf{P}_2(t) = -\frac{qe^2}{m^3(\omega_0^2 - 4\omega^2)(\omega_0^2 - \omega^2)^2}|\mathbf{E}_0|\mathbf{E}_0\exp(i2\omega t). \tag{10.17d}$$

This part of the vector of polarization depends nonlinearly on the field amplitude and varies in time with a frequency twice as high as the primary wave.

The form of the polarization vector after two successive approximations is

$$\mathbf{P}(t) = \frac{e^2N}{m(\omega_0^2 - \omega^2)}\mathbf{E}_0\exp(i\omega t) -$$

$$-\frac{qe^2}{m^3(\omega_0^2 - 4\omega^2)(\omega_0^2 - \omega^2)^2}|\mathbf{E}_0|\mathbf{E}_0\exp(i2\omega t). \tag{10.18}$$

Thus in a strong wave field the character of polarization of the medium undergoes a change: the vector of polarization is a nonlinear function of the field, and, moreover, in a strong field with frequency ω it is no longer a harmonic function of field frequency but can be presented as the sum of two terms: the first having the frequency of the incident field and the second with double the frequency.

This vector becomes the source of secondary radiation. Thus, as a result of nonlinear polarization, there arise in the medium two electromagnetic waves of frequencies ω and 2ω.

10.3. Maxwell's Equations in a Nonlinear Medium

In the region of optical frequencies we can regard, for most media, the magnetic permeability of the vacuum as the magnetic permeability of the medium. The physical properties of the media are then described by the dependence of the vector of polarization **P** of those media upon the intensity of the electric field. In the nonlinear case this dependence is determined by expressions (10.3) and (10.4), and also (10.2).

The above remarks enable us to write Maxwell's equations in a nonlinear nonmagnetic medium in the form

$$\begin{aligned} \operatorname{curl}\mathbf{E} &= -\dot{\mathbf{B}}, \\ \operatorname{curl}\mathbf{H} &= \dot{\mathbf{D}}, \\ \operatorname{div}\mathbf{D} &= 0, \\ \operatorname{div}\mathbf{B} &= 0 \end{aligned} \tag{10.19}$$

(under the assumption that conduction currents and spatial charge do not occur in it). These relations, together with the constitutive relations

$$\begin{aligned} \mathbf{B} &= \mu_0 \mathbf{H}, \\ \mathbf{D} &= \varepsilon_0 \mathbf{E} + \mathbf{P}^{(L)} + \mathbf{P}^{(NL)} \end{aligned} \tag{10.19a}$$

(where μ_0 and ε_0 are appropriate permittivities of the vacuum) form the basis for the description of nonlinear phenomena in nonmagnetic media.

In the general case it is not possible to solve the system of equations (10.19) and (10.19a). However, for numerous cases interesting in practice we succeed in introducing reasonable simplifications which facilitate the analysis of the system.

One of the simplest cases is that of a homogeneous medium without dispersion and dissipation. In such a medium polarization **P** and field **E** agree in time, and the relation $\mathbf{P} = \mathbf{P}(\mathbf{E})$ becomes a simple functional formula, which may be presented as an expansion with respect to the powers of the field components

$$P_j = \varepsilon_0 (\chi_{jm} E_m + \chi_{jml} E_m E_l + \chi_{jmln} E_m E_l E_n + \ldots), \tag{10.20}$$

where χ_{jm}, χ_{jmn}, χ_{jmln} are the components of the tensors of electric susceptibility of an appropriate order. In this expansion high order terms usually have much lower values than the first term, so that in a transparent medium (without dissipation) the relation

$$\chi_{jm}E_m \gg \chi_{jmn}E_mE_n \gg \chi_{jmnl}E_mE_nE_l \ldots \tag{10.21}$$

is satisfied comparatively well. This enables us to neglect the high order terms in expansion (10.20) and confine ourselves to two, or possibly three, terms of the expansion.

The tensors of the electric susceptibility of the medium χ_{jm}, χ_{jml}, χ_{jmln} depend in an essential way on the symmetry of the medium, and if the medium has a centre of symmetry, all the components of the tensor χ_{jml} of the medium are equal to zero. Square polarization does not occur in it, and the first nonlinear term in expansion (10.20) will be the components of cubic polarization $\varepsilon_0\chi_{jmln}E_mE_lE_n$.

10.4. Nonlinear Interaction of Electromagnetic Fields

Nonlinearity creates a complex situation. However, using relations (10.5) or (10.21) and the assumptions of weak dispersion and dissipation of the medium, we can classify nonlinear wave effects with respect to the order of nonlinearity and the number of monochromatic waves propagating in the medium.

In the case of functional relation $\mathbf{P} = \mathbf{P}(\mathbf{E})$ a convenient characteristics of a nonlinear medium is the differential permittivity

$$\tilde{\varepsilon}(E) = \frac{\partial D}{\partial E}.$$

A. Square Polarization

1. As can be seen from (10.10), each of the monochromatic waves ($\omega_m = \omega_n$) induces square polarization with double frequency

$$\mathbf{P}^{(2)}(\omega_m + \omega_m) = \varepsilon_0\chi(\omega_m + \omega_m):\mathbf{E}_m\mathbf{E}_m \tag{10.22a}$$

and with zero frequency (static field)

$$\mathbf{P}^{(2)}(\omega_m - \omega_m) = \varepsilon_0\chi(\omega_m - \omega_m):\mathbf{E}_m\mathbf{E}_m^*. \tag{10.22b}$$

2. Two different monochromatic waves with frequencies $\omega_m \neq \omega_n$ induce polarization with the summary frequency

$$\mathbf{P}^{(2)}(\omega_m + \omega_n) = \varepsilon_0 \chi_0(\omega_m + \omega_n) : \mathbf{E}_m \mathbf{E}_n \tag{10.23a}$$

and the differential frequency

$$\mathbf{P}^{(2)}(\omega_m - \omega_n) = \varepsilon_0 \chi(\omega_m - \omega_n) : \mathbf{E}_m \mathbf{E}_n. \tag{10.23b}$$

As a consequence there appear in the nonlinear medium monochromatic waves with combined frequencies.

3. If one of the interacting fields is a static field $\mathbf{E}_n^0$ $(\omega_n = 0)$, square polarization will have the frequency ω_m and take the form

$$\mathbf{P}^{(2)}(\omega_m + 0) = \varepsilon_0 \chi(\omega_m + 0) : \mathbf{E}_m \mathbf{E}_n^0. \tag{10.24}$$

This is the *Pockels linear electrooptic effect*, which consists in the change of the electric permittivity of the medium (see Sec. 6.12).

B. Cubic Polarization

Cubic polarization reveals itself in media with a centre of symmetry. In this case the number of nonlinear wave processes is larger. Three monochromatic waves with different frequencies can now together decide about polarization.

1. Each of the propagating monochromatic waves induces polarization with triple frequency $(\omega_m = \omega_n = \omega_k)$

$$\mathbf{P}^{(3)}(\omega_m + \omega_m + \omega_m) = \varepsilon_0 \chi(\omega_m + \omega_m + \omega_m) \vdots \mathbf{E}_m \mathbf{E}_m \mathbf{E}_m, \tag{10.25a}$$

and also nonlinear polarization with the frequency of the incident wave

$$\mathbf{P}^{(3)}(\omega_m + \omega_m - \omega_m) = \varepsilon_0 \chi(\omega_m + \omega_m - \omega_m) \vdots \mathbf{E}_m \mathbf{E}_m \mathbf{E}_m^*. \tag{10.25b}$$

The latter case is the so-called *dynamical Kerr effect*, in which the differential permittivity depends on the wave intensity

$$\tilde{\varepsilon}(\omega_m) = \varepsilon^{(0)}(\omega_m) + \varepsilon_0 \chi(\omega_m) \mathbf{E}_m \mathbf{E}_m^*. \tag{10.26}$$

2. In the interaction of three waves of which two have the same frequencies there arises cubic polarization with frequency ω_j

$$\mathbf{P}^{(3)}(\omega_j + \omega_k - \omega_k) = \varepsilon_0 \chi(\omega_j + \omega_k - \omega_k) \vdots \mathbf{E}_j \mathbf{E}_k \mathbf{E}_k^*. \qquad (10.27)$$

In this case there is interaction of the wave of frequency ω_k with the one of frequency ω_j. It is called *cross modulation.* The electric permittivity takes the form

$$\tilde{\varepsilon}(\omega_j) = \varepsilon^{(0)}(\omega_j) + \varepsilon_0 \chi(\omega_j, \omega_k) : \mathbf{E}_k \mathbf{E}_k^*. \qquad (10.27a)$$

If $\omega_k = 0$, we obtain from (10.27a) the so-called *square electrooptic effect.* The dielectric tensor of the medium then depends only on frequency ω_j.

3. In the general case, where there is cubic nonlinearity, three waves of frequencies ω_j, ω_k, ω_m induce polarization with frequencies

$$\omega_s = \pm\omega_j \pm \omega_k \pm \omega_m. \qquad (10.28)$$

10.5. Slow-Variable Amplitude Method in Nonlinear Optics

Taking into account the constitutive relations (10.19a), we can replace Maxwell's equations (10.19) by the equation

$$\operatorname{curl}\operatorname{curl}\mathbf{E} + \varepsilon_0 \mu_0 \frac{\partial^2 (\mathbf{E} + \varepsilon_0^{-1} \mathbf{P}^{(L)})}{\partial t^2} = -\mu_0 \frac{\partial^2 \mathbf{P}^{(NL)}}{\partial t^2}. \qquad (10.29)$$

We solve equation (10.29) under appropriate boundary and initial conditions.

Substituting in (10.29) the expansions $\mathbf{E}$ and $\mathbf{P}$ in monochromatic components, we obtain for each of them an appropriate equation of the Helmholtz type:

$$\operatorname{curl}\operatorname{curl}\mathbf{E}_j - \frac{\omega_j^2}{c^2} \kappa_j^{(L)} \mathbf{E}_j = \mu_0 \omega_j^2 \mathbf{P}_j^{(NL)}, \qquad (10.30)$$

where the subscript j at $\mathbf{E}_j$ and $\mathbf{P}_j^{(NL)}$ denotes that these quantities are connected with the frequency ω_j, while the symbol $\kappa_j^{(L)}$ stands for the relative permittivity tensor in linear approximation.

To solve equation (10.30) in nonlinear optics we apply approximate methods. One of them, most frequently used, is the so-called *slow-variable amplitude method.* If the medium is weak nonlinear, i.e. if

$$\frac{|\mathbf{P}^{(\mathrm{NL})}|}{|\mathbf{E}|} = \eta \ll 1, \tag{10.31}$$

and shows low absorption, the amplitudes of monochromatic waves incident on the medium will change slightly on a path of the order of the wavelength. This means that the amplitudes remain slow-variable functions of spatial coordinates. In view of this the field and the polarization of the medium may be written in the form

$$\begin{aligned} \mathbf{E}_j &= \mathbf{e}_j A_j(\eta \mathbf{k}_j \mathbf{r}) \exp(\mathrm{i}\mathbf{k}_j \mathbf{r}_j), \\ \mathbf{P}_j &= \mathbf{p}_j P_j(\eta \mathbf{k}_j \mathbf{r}) \exp(\mathrm{i}\mathbf{k}_p \mathbf{r}_p), \end{aligned} \tag{10.32}$$

where $\mathbf{e}_j$ and $\mathbf{p}_j$ are versors, A_j and P_j slow-variable amplitudes, and $\mathbf{k}_j$ and $\mathbf{k}_p$ the wave vectors of the wave field and of the polarization of the nonlinear medium.

The wave vector of nonlinear polarization $\mathbf{k}_p$ is related to the wave vectors of waves polarizing the medium through the relation

$$\mathbf{k}_p = \sum_s l_s \mathbf{k}_s \qquad (l_s = 0, \ \pm 1). \tag{10.33}$$

If a nonlinear medium covers the half-space $z > 0$ and the wave vectors of the interacting waves have the direction of the z-axis: $\mathbf{k}_j = \mathbf{e}_z k_j$, and the vectors $\mathbf{e}_j$ determining wave polarization satisfy the equation

$$\mathbf{k}_j \times (\mathbf{e}_j \times \mathbf{k}_j) + \omega_j^2 \varepsilon_0 \mu_0 \kappa_j^{(\mathrm{L})} \mathbf{e}_j = \mathbf{0}, \tag{10.34}$$

then, applying slow-variable amplitudes

$$\left| \frac{\mathrm{d}^2 A_j}{\mathrm{d}z^2} \right| \ll k_j \left| \frac{\mathrm{d}A_j}{\mathrm{d}z} \right| \tag{10.35}$$

and substituting (10.32) into equation (10.30), we arrive at the equations

$$\frac{\mathrm{d}A_j}{\mathrm{d}z} + \alpha_j A_j = \mathrm{i}\frac{\omega_j \gamma_j}{2} |P_j^{(\mathrm{NL})}| \exp[\mathrm{i}(k_p - k_j)z], \tag{10.36}$$

where

$$\alpha_j = \frac{\omega_j \mu_0 \kappa_j'':\mathbf{e}_j\mathbf{e}_j}{2\sqrt{\kappa_j':\mathbf{e}_j\mathbf{e}_j}\sin\beta},$$

$$\gamma_j = \frac{\mu_0(\mathbf{p}_j\mathbf{e}_j)\dfrac{1}{c}}{\sqrt{\kappa_j':\mathbf{e}_j\mathbf{e}_j}\sin\beta}. \tag{10.36a}$$

β is the angle between the direction of the z-axis and the versor $\mathbf{e}_j$. We assume additionally that

$$\kappa_j^{(\mathrm{L})} = \kappa_j' + \mathrm{i}\kappa_j'', \tag{10.36b}$$

i.e., we permit dissipation due to the occurrence of $\kappa_j'' \neq 0$. The exponential factor $\exp[(\mathrm{i}k_p - \mathrm{i}k_j)z]$ in equations (10.36) has an essential effect on the change of amplitude of the wave. Its greatest stable changes will take place if the phase matching condition is fulfilled

$$\mathbf{k}_p - \mathbf{k}_j = \mathbf{0}, \tag{10.37}$$

where $\mathbf{k}_p$ is defined by (10.33). If this condition is not fulfilled, quick oscillations will appear, which do not favour cummulative changes of the amplitude.

In dispersive media the matching condition (10.37) is in general fulfilled by a small number of waves, usually not greater than three of four. Hence, in many problems, the amplitudes of waves with combined frequencies for which condition (10.37) is not fulfilled are so small that they can be neglected, which considerably reduces the number of equations (10.36).

10.6. Phase Matching Condition and the Methods of its Realization

We explain the phase matching condition (10.37) taking as an example a medium with square nonlinear polarization. Let us assume that the medium fills the half-space $z > 0$ and that from the side $z < 0$ a plane monochromatic wave of the form

$$\mathbf{E} = \mathbf{e}_x E_0 \exp[\mathrm{i}(kz - \omega t)] \tag{10.38}$$

falls upon it. According to (10.22) this wave causes two-component nonlinear polarization of the medium, one of the components is constant in time and the other is of the form

$$\mathbf{P}^{(2)} = \varepsilon_0 \chi(\omega + \omega){:}\mathbf{e}_x \mathbf{e}_x |E(z,\omega)|^2 \exp[\mathrm{i}(2k(\omega)z - 2\omega t)]. \tag{10.39}$$

It is assumed that the vector $\mathbf{e}_x$ fulfils equation (10.34), the wave vector of the wave (10.38) in this medium being $\mathbf{k}(\omega) = [0, 0, k(\omega)]$. (The medium is dispersive and the quantity k (10.34) is a function of ω: $k(\omega)$.)

The presence of nonlinear polarization (10.39) will cause the occurrence in the medium of two waves of the form

$$\mathbf{E}_2 = \mathbf{e}_x E_2(z,\omega) \exp[\mathrm{i}(k(\omega)z - \omega t)], \tag{10.40a}$$

$$\mathbf{E}_1 = \mathbf{e}_2 E_1(z, 2\omega) \exp[\mathrm{i}(k(2\omega)z - 2\omega t)]. \tag{10.40b}$$

The second wave, (10.40b), is a secondary wave generated by nonlinear polarization (10.39), and its amplitude for $z = 0$ is $E(0, 2\omega) = 0$.

There is also a component of nonlinear polarization which varies in time with frequency ω and is described by the expression

$$\mathbf{P}^{(2)} = \chi(2\omega - \omega){:}\mathbf{e}_1 \mathbf{e}_x E_1(z, 2\omega) E_2^*(z, \omega) \times$$
$$\times \exp[\mathrm{i}(k(2\omega) - k(\omega)) z_n]. \tag{10.41}$$

This component is produced by the interaction of the waves (10.40a) and (10.40b). It is, however, a much smaller quantity than (10.39) in view of a small value of the amplitude of the secondary wave $E_1(z, 2\omega)$. For small values of variable z in the approximation of the slow-variable amplitude the quantity (10.41) may be neglected. It becomes significant for large values of z under the conditions of phase matching (10.37).

The application of the results from Sec. 10.5 in the case under consideration gives the following system of equations:

$$\frac{\mathrm{d}E_2(z,\omega)}{\mathrm{d}z} + \alpha_2 E_2(z,\omega) = \mathrm{i}\frac{\omega\gamma_2}{2}\mathbf{e}_x\chi(2\omega-\omega){:}\mathbf{e}_x\mathbf{e}_x \times$$

$$\times E_1(z,2\omega)E_2^*(z,\omega)\exp[\mathrm{i}(k(2\omega)-2k(\omega))z], \quad (10.42a)$$

$$\frac{\mathrm{d}E_1(z,2\omega)}{\mathrm{d}z} + \alpha_1 E_1(z,2\omega) = \mathrm{i}\frac{2\omega\gamma_1}{2}\mathbf{e}_1\chi(\omega+\omega){:}\mathbf{e}_x\mathbf{e}_x \times$$

$$\times \mid E_2(z,\omega)|^2\exp[\mathrm{i}(2k(\omega)z-2k(2\omega))z]. \quad (10.42b)$$

The above equations contain on their right-hand sides the exponential factor $\exp[\pm\mathrm{i}(k(2\omega)-2k(\omega))z]$. The presence of these factors causes oscillatory changes of the amplitude values $E_1(z,\omega)$ and $E_2(z,2\omega)$ accompanying the changes of variable z. Monotonous changes of these amplitudes occur if

$$k(2\omega)-2k(\omega)=[k(2\omega)-k(\omega)]-k(\omega)=k_p-k(\omega)=0. \quad (10.43)$$

In media with natural dispersion it is not always possible to satisfy the condition $k(2\omega)=2k(\omega)$. One of the ways to do that is to use media which can be regarded in linear approximation as anisotropic. This is obtained by matching the phase velocities of two interacting waves with different polarizations in a chosen direction of wave propagation. To this end we can use certain uniaxial crystals in which two waves may propagate in the same direction with different velocities, different polarizations and the same frequency. Figure 10.1 shows the appropriate surfaces of the refractive indices n_o (for the ordinary wave) and n_e (for the extraordinary wave) for two frequencies ω and 2ω in the case of a negative crystal ($n_e < n_o$). It can be seen from the figure that in the direction determined by the angle θ_{match} with respect to the optical axis we have the equality $n_o(\omega)=n_e(2\omega)$. Consequently, in a negative uniaxial crystal matching condition (10.43) for the frequencies ω and 2ω can be realized. In order to make use of this property for practical realization we must cut out a crystalline plate in the way that the optical axis should form the angle θ_{match} with the normal to the surface of the crystal. If a plane wave with frequency ω falls on

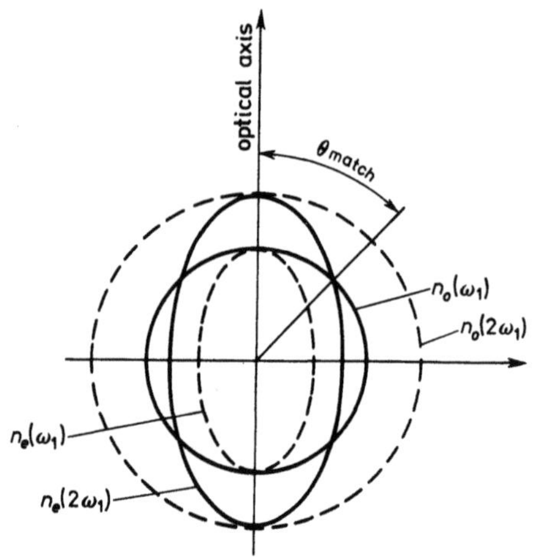

Figure 10.1

such a crystal in the direction normal to the boundary surface and its polarization is the polarization of the ordinary wave, then there appears in the crystal the extraordinary wave with frequency 2ω matched in phase with the primary wave.

10.7. Generation of the Second Harmonic

If we neglect dissipation $(\alpha_j = 0, j = 1, 2)$ and the dispersion of the medium is weak:

$$\frac{\gamma_1}{2}\chi(\omega+\omega) \vdots \mathbf{e}_x\mathbf{e}_x\mathbf{e}_1 = \frac{\gamma_2}{2}\chi(2\omega-\omega) \vdots \mathbf{e}_1\mathbf{e}_x\mathbf{e}_x = \Gamma, \qquad (10.44)$$

then, for the situation discussed in Sec. 10.6 equations (10.36) take the form

$$\frac{\mathrm{d}E_1(z,2\omega)}{\mathrm{d}z} = 2\omega\,\mathrm{i}\Gamma\left[E_2(z,\omega)\right]^2\exp(-\mathrm{i}\Delta kz), \qquad (10.45\text{a})$$

$$\frac{\mathrm{d}E_2(z,\omega)}{\mathrm{d}z} = \omega\,\mathrm{i}\Gamma E_1(z,2\omega)E_2^*(z,\omega)\exp(+\mathrm{i}\Delta kz), \qquad (10.45\text{b})$$

where

$$\Delta k = k(2\omega) - 2k(\omega).$$

Under the boundary conditions in the plane $z = 0$

$$E_1(0, 2\omega) = 0,$$
$$E_2(0, \omega) = E_0,$$

in the case of weak nonlinearity the amplitude of the secondary wave $E_1(z, 2\omega)$ will be small in comparison with the amplitude of the primary wave $E_2(z, \omega)$:

$$|E_1(z, 2\omega)| \ll |E_2(z, \omega)|. \tag{10.46}$$

Therefore we can assume that $E_2(z, \omega) \approx E_0$ and seek for this situation solutions (10.45a). This is called the *constant pump approximation.*

If $\Delta k = 0$, i.e. if the phase matching condition is fulfilled, we have

$$E_1(z, 2\omega) = \mathrm{i}(2E_0^2 \Gamma \omega) z, \tag{10.46a}$$

which means that the amplitude of the secondary wave with frequency 2ω increases proportionally to the increase of z (the straight line in Fig. 10.2). Solution (10.46a), however, has a restricted

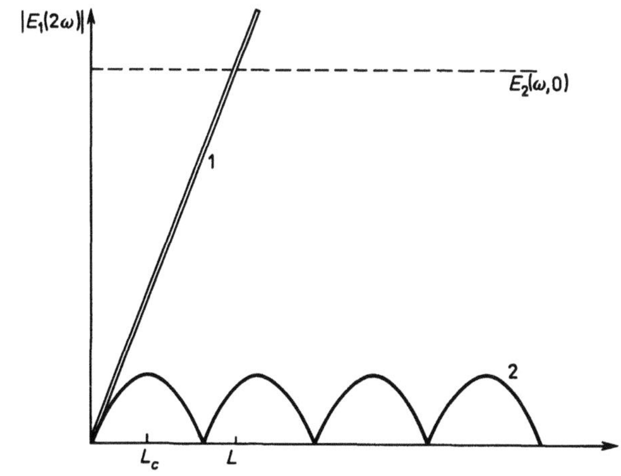

Figure 10.2

region of validity; it is subject to restriction (10.46). This implies that it remains valid for

$$z \ll \frac{1}{2\Gamma E_0 \omega} = L. \tag{10.46b}$$

On the length L we can expect a considerable flow of energy directed from the primary to the secondary wave.

For $\Delta k \neq 0$ the solution of equation (10.45a) is

$$E_1(z, 2\omega) = \frac{4\omega_i \Gamma E_0^2}{\Delta k} \exp\left(\mathrm{i}\frac{\Delta kz}{2}\right) \sin\left(\frac{\Delta kz}{2}\right). \tag{10.47}$$

The length on which the amplitude $|E_1(z, 2\omega)|$ reaches its first maximum is

$$L_c = \frac{\pi}{\Delta k}. \tag{10.47a}$$

It is called the *coherent interaction length.* After covering this distance the amplitude of the secondary wave has the value

$$|E_2(L_c, 2\omega)| = \frac{4E_0^2 \Gamma \omega}{\Delta k}, \tag{10.47b}$$

but condition (10.46) has to be fulfilled as before. This imposes a restriction on the phase mistuning Δk (curve 2 in Fig. 10.2).

If condition (10.46) is not fulfilled, we have to consider the complete system of equations (10.45). It is then convenient to use real functions. Therefore we write the complex amplitudes of the wave in the form

$$E_1(z, 2\omega) = |E_1(z, 2\omega)| \exp[\mathrm{i}\varphi_1(z, 2\omega)], \tag{10.48a}$$

$$E_2(z, \omega) = |E_2(z, \omega)| \exp[\mathrm{i}\varphi_2(z, \omega)]. \tag{10.48b}$$

Substituting these equations in systems (10.45) and separating the real and the imaginary parts, we get

$$\frac{\mathrm{d}|E_1(z, 2\omega)|}{\mathrm{d}z} = -\Gamma |E_2(z, \omega)|^2 \sin\Phi \cdot 2\omega, \tag{10.49a}$$

$$\frac{\mathrm{d}|E_2(z,\omega)|}{\mathrm{d}z} = \Gamma|E_1(z,2\omega)||E_2(z,\omega)|\sin\Phi\cdot\omega, \tag{10.49b}$$

$$\frac{\mathrm{d}\Phi}{\mathrm{d}z} = \Delta k + \omega\Gamma\left(2|E_1(z,2\omega)| - 2\frac{|E_2(z,\omega)|^2}{|E_1(z,2\omega)|}\right)\cos\Phi, \tag{10.49c}$$

where

$$\Phi = 2\varphi_2(z) - \varphi_1(z) + \Delta kz. \tag{10.49d}$$

For solving the system of equations (10.49) it is convenient to use the laws of conservation which follow from these equations. They can be formulated as

$$|E_1(z,2\omega)|^2 + 2|E_2(z,\omega)|^2 = \text{const}, \tag{10.50a}$$

$$-\Delta k|E_1(z,2\omega)|^2 + 2\omega\Gamma|E_2(z,\omega)|^2|E_1(z,2\omega)|\cos\Phi = \text{const}. \tag{10.50b}$$

For $\Delta k = 0$ (i.e. when the phase matching condition is fulfilled) (10.50b) implies $\cos\Phi = 0$. Assuming $\sin\Phi = -1$ we get from system (10.49) the equations

$$\frac{\mathrm{d}|E_2(z,\omega)|}{\mathrm{d}z} = -\Gamma|E_1(z,2\omega)||E_2(z,\omega)|\cdot\omega, \tag{10.51a}$$

$$\frac{\mathrm{d}|E_1(z,2\omega)|}{\mathrm{d}z} = \Gamma|E_2(z,\omega)|^2\cdot 2\omega, \tag{10.51b}$$

$$\Phi = -\frac{\pi}{2}. \tag{10.51c}$$

On applying the law of conservation (10.50a) and the boundary condition (10.45c) equations (10.51) give solutions of the form

$$|E_1(z,2\omega)| = \sqrt{2}E_0\tanh\left[(\Gamma E_0\sqrt{2}\omega)z\right], \tag{10.52a}$$

$$|E_2(z,\omega)| = E_0\cosh^{-1}(\sqrt{2}\omega_0\Gamma Ez). \tag{10.52b}$$

The dependence of the amplitudes upon the distance covered is shown in Fig. 10.3.

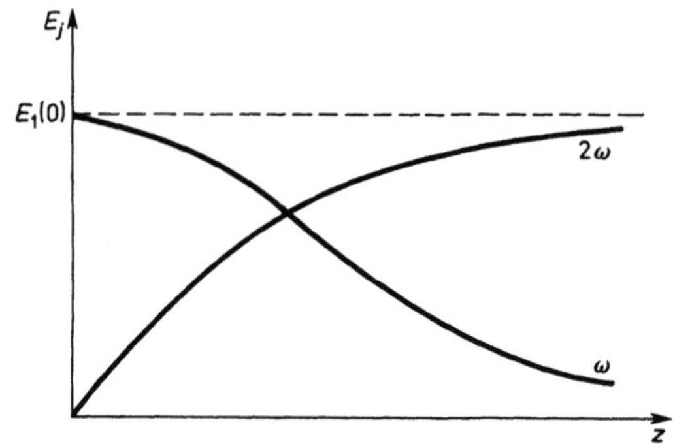

Figure 10.3

10.8. Parametric Amplification and Generation

In a medium with square nonlinear polarization there may interact at the same time three waves with frequencies $\omega_1 < \omega_2 < \omega_3$ which are connected by the relation

$$\omega_3 = \omega_1 + \omega_2. \tag{10.53}$$

These waves will induce in the medium nonlinear polarization with the same frequencies

$$\mathbf{P}_1^{(2)} = \chi(\omega_3 - \omega_2):\mathbf{e}_3\mathbf{e}_2 E_3(z)E_2^*(z)\exp[\mathrm{i}(k_3 - k_2)z]\,\varepsilon_0, \tag{10.54a}$$

$$\mathbf{P}_2^{(2)} = \chi(\omega_3 - \omega_1):\mathbf{e}_3\mathbf{e}_1 E_3(z)E_1^*(z)\exp[\mathrm{i}(k_3 - k_1)z]\,\varepsilon_0, \tag{10.54b}$$

$$\mathbf{P}_3^{(2)} = \chi(\omega_2 + \omega_1):\mathbf{e}_2\mathbf{e}_1 E_2(z)E_1(z)\exp[\mathrm{i}(k_2 + k_1)z]\,\varepsilon_0, \tag{10.54c}$$

where $\mathbf{P}_j^{(2)}$ $(j = 1, 2, 3)$ are the vectors of nonlinear polarization, which vary in time with the frequencies ω_1, ω_2, and ω_3. E_j are the complex amplitudes of the interacting waves, and k_j the wave vectors dependent in the presence of dispersion on the frequency of each wave $(k_j = k(\omega_j))$. The vectors $\mathbf{e}_j$ determine the directions of the electric vectors of the interacting waves.

In the case in question, for $\alpha_j = 0$, equations (10.36) take the form

$$\frac{dE_3}{dz} = i\frac{\omega_1\gamma_1}{2}\chi(\omega_3-\omega_2)\vdots \mathbf{e}_3\mathbf{e}_1\mathbf{e}_1 E_3(z)E_2^*(z)\times \\ \times\exp[i(k_3-k_2-k_1)z], \tag{10.55a}$$

$$\frac{dE_2}{dz} = i\frac{\omega_2\gamma_2}{2}\chi(\omega_3-\omega_1)\vdots \mathbf{e}_2\mathbf{e}_3\mathbf{e}_1 E_3(z)E_1^*(z)\times \\ \times\exp[i(k_3-k_1-k_2)z], \tag{10.55b}$$

$$\frac{dE_3}{dz} = i\frac{\omega_3\gamma_3}{2}\chi(\omega_1+\omega_2)\vdots \mathbf{e}_3\mathbf{e}_1\mathbf{e}_2 E_1(z)E_2(z)\times \\ \times\exp[i(k_1+k_2-k_3)z]. \tag{10.55c}$$

Again, if the dispersion of the medium is weak, we can put

$$\Gamma = \frac{\gamma_j}{2}\chi(\omega_j)\vdots \mathbf{e}_1\mathbf{e}_2\mathbf{e}_3 \quad (j=1,2,3) \tag{10.56a}$$

and introduce the notation

$$\Delta k' = k_3 - k_1 - k_2. \tag{10.56b}$$

From the set of equations (10.55) we obtain three laws of conservation called the *Manley–Rowe relations*

$$\frac{1}{\omega_1}[E_1(z)^2 - E_1(0)^2] = \frac{1}{\omega_2}[E_2(z)^2 - E_2(0)^2] \\ = \frac{1}{\omega_3}[E_3(0)^2 - E_3(z)^2]. \tag{10.57}$$

These relations allow us to formulate the following important conclusion:

If the energy of the wave with maximal frequency ω_3 decreases, it passes at the same time into the two waves with lower frequencies ω_1 and ω_2 (i.e. if the amplitude $E_3(z)$ decreases, the amplitudes $E_1(z)$ and $E_2(z)$ increase).

From (10.57) follow the inequalities

$$-\frac{\omega_1}{\omega_2}\mathrm{E}_2(0)^2 + E_1(0)^2 \leq E_1(z)^2 \leq E_1(0)^2 + \frac{\omega_1}{\omega_3}E_3(0)^2, \tag{10.58a}$$

$$0 \leq E_2(z)^2 \leq E_2(0)^2 + \frac{\omega_2}{\omega_3}E_3(0)^2, \tag{10.58b}$$

$$0 \leq E_3(z)^2 \leq E_3(0)^2 + \frac{\omega_3}{\omega_2}E_2(0)^2. \tag{10.58c}$$

Two cases are considered.

A. Into a medium which is bounded by the plane $z = 0$ (a nonlinear medium fills the half-space $z > 0$) enter three waves with the amplitudes: $E_1(0) \gg E_2(0)$, $E_3(0)$. In this case we can conclude from inequality (10.58) that the wave with the lowest frequency ω_1 and with the largest initial amplitude during propagation in the nonlinear medium will preserve its amplitude practically without change. The amplitude values for the waves with higher frequencies ω_2 and ω_3 remain of the same order as they were at the input, i.e. very small.

Therefore in equations (10.55) we can put $E_1(z) = E_1(0)$, and obtain for the amplitudes of the other two waves the expressions

$$|E_j(z)| = E_j(0)\cos\eta z + B_j\sin\eta z, \tag{10.59}$$

where

$$\eta = \left[\left(\frac{\Delta k'}{2}\right)^2 + \omega_2\Gamma^2 E_1(0)^2\omega_3\right]^{\frac{1}{2}},$$

which show that the strong wave with the lowest frequency does not amplify the waves with higher frequencies. The amplitudes of the latter may merely experience periodic spatial beats whose period depends on $\Delta k'$, i.e. on desynchronization.

B. If we have at the input $E_3(0) \gg E_2(0)$ and $E_1(0)$, then for small z we can assume that $E_3(z) = E_3(0)$ and $E_1(z)$, $E_2(z) \ll E_3(0)$. In view of the above considerations and of the dissipative term (10.36) and additionally with $\alpha_1 = \alpha_2 = \alpha$, equations (10.55) yield solutions of the form

$$E_j(z) = [E_j(0)\cosh\rho z + B_j \sinh\rho z]\exp(-\alpha z), \qquad (10.60)$$

where

$$\rho = \left[\Gamma^2 E_3(0)^2 \omega_1\omega_2 - \left(\frac{\Delta k'}{2}\right)^2\right]^{\frac{1}{2}}.$$

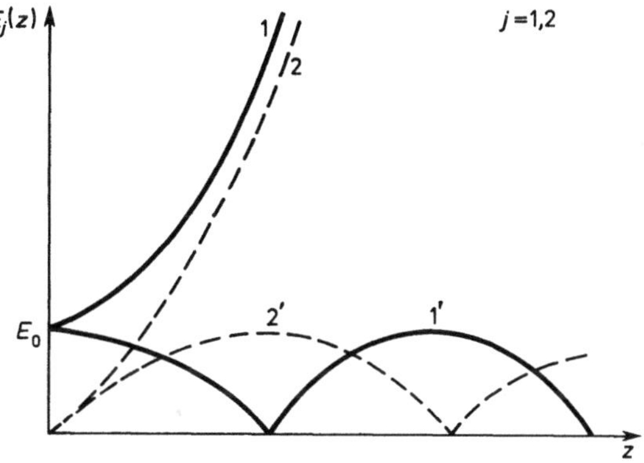

Figure 10.4

It can be shown that an exponential increase of the amplitudes of waves with lower frequencies (Fig. 10.4, curves 1 and 2) takes place only if

$$E_3(0)^2 > \frac{1}{\Gamma^2\omega_1\omega_2}\left[\alpha^2 + \left(\frac{\Delta k'}{2}\right)^2\right]. \qquad (10.60a)$$

Inequality (10.60a) establishes the parametric amplification threshold. It shows that the greater the dissipation α and the phase mistuning $\Delta k'$ the higher the threshold. Below the threshold (10.60a) the amplitudes of the waves with frequencies ω_1 and ω_2 remain oscillating quantities with low external values (curves 1′ and 2′ in Fig. 10.4).

We avail ourselves of the above phenomenon in parametric amplification of weak waves, and also in constructing tunable sources of coherent radiation.

If a strong wave with a fixed frequency ω_3 falls on a nonlinear crystal with square polarization, the wave can begin to interact with two waves of lower frequencies ω_1 and ω_2. This interaction will be most effective if the following two conditions are fulfilled:

$$\omega_3 = \omega_1 + \omega_2 \quad \text{and} \quad k(\omega_3) = k(\omega_2) + k(\omega_1). \tag{10.61}$$

The amplitudes of the waves with frequencies ω_1 and ω_2 may have the nature of quantum fluctuations or of thermic radiation. As a result of parametric amplification the crystal may emit coherent radiation with frequencies for which condition (10.61) is fulfilled. Changing the dispersive properties of the crystal along the direction of wave propagation (by rotating the crystal, changing its temperature, or by compression or a constant external electric field), we can cause conditions (10.61) to be fulfilled by other frequencies ω_1' and ω_2'. Then in place of the previous frequencies the near frequencies will undergo parametric amplifications. The radiation emerging at the output of the crystal will have a different wavelength.

10.9. Self-focusing Phenomenon

Self-focusing is conditioned by the dependence of the electric permittivity of a medium on the intensity of the electric field of the propagating wave. In an isotropic medium we can write this relation as

$$\varepsilon = \varepsilon_1 + \varepsilon_2 \langle E^2 \rangle, \tag{10.62}$$

where $\langle E^2 \rangle$ is the time-averaged value of the square of the electric field intensity. This quantity is equal to half the square of the field amplitude, i.e. $\langle E^2 \rangle = \frac{1}{2}|E|^2$.

The increase of the electric permittivity proportional to $|E|^2$ leads to the self-focusing of the beam if the wave intensity decreases monotonically with the increase of the distance from the axis of the beam. The further away from the axis of the beam we are, the quicker the wave front propagates, since $v = (\mu_0 \varepsilon)^{-1/2}$. In consequence the primary plane front becomes a concave front and the beam becomes convergent.

Of course such a self-focusing effect may be observed only with the use of waves with very high intensities, e.g. in the case of laser beams, which may be treated approximately as the so-called *Gaussian beams*, whose characteristic feature is that the distribution of field intensity in the cross-section of the beam has the form of a two-dimensional Gaussian distribution for the cross-section of the beam. Namely, for a beam propagating in the positive direction of the z-axis, under the assumption that it has a plane wave front in the plane $z = 0$, we can write

$$\psi(x, y, 0) = A \exp\left(-\frac{x^2 + y^2}{\rho_0^2}\right), \tag{10.63}$$

where the parameter ρ_0 is the effective width of the beam in the plane $z = 0$. It is equal to the distance between the points where its value decreases e-fold. If the components of the field vectors of the laser beam can really be described by an expression similar to (10.63), then according to the conditions described above we can expect the self-focusing effect in material media.

In view of this feature, and also of the general application of laser beams, it is worthwhile to study their properties a little more thoroughly. This will lead us to a better understanding of the self-focusing effect.

10.10. Gaussian Beams in Free Space

We seek an expression describing the propagation of a light beam with the Gaussian intensity distribution. We postulate a shape in the plane $z = 0$ in the form (10.63) and we are interested in its functional dependence on the variable z. To this aim we use the Fresnel diffractional integral.

With its help we can reconstruct from our knowledge of the field in the plane $z = 0$ the field for an arbitrary value of variable z. The result is

$$\psi(x', y', z) = \frac{\mathrm{i}A}{\lambda z} \exp(\mathrm{i}kz) \iint\limits_{-\infty}^{+\infty} \exp\left(-\frac{x^2 + y^2}{\rho_0^2}\right) \times$$

$$\times \exp\left\{\mathrm{i}\frac{k}{2z}[(x' - x)^2 + (y' - y)]\right\} \mathrm{d}x\,\mathrm{d}y. \tag{10.64}$$

The double integral obtained above can be presented as the product of the integral over the variable x and the integral over the variable y. The first integral equals

$$\int_{-\infty}^{+\infty} \exp\left(-\frac{x^2}{\rho_0^2}\right)\exp\left[\mathrm{i}\frac{k}{2z}(x'-x)^2\right]\mathrm{d}x$$

$$= \sqrt{\pi}\,\frac{\rho_0\sqrt{2z}}{\sqrt{2z-\mathrm{i}k\rho_0^2}}\exp\left(-\frac{k^2\rho_0^2-2\mathrm{i}kz}{4z^2+k^2\rho_0^2}\,x'^2\right). \quad (10.65)$$

(The second integral is obtained from (10.64) by replacing the variables x' and x by y' and y.)

On the ground of the above the total field for an arbitrary value of the variable z may be written in the form

$$\psi(x,y,z) = \frac{2\mathrm{i}\pi\rho_0^2 A}{\lambda(2z-\mathrm{i}k\rho_0^2)}\exp(\mathrm{i}kz)\exp\left(\frac{2\mathrm{i}kzr^2}{4z^2+k^2\rho_0^4}\right)\times$$

$$\times\exp\left(-\frac{k^2\rho_0^2 r^2}{4z^2+k^2\rho_0^4}\right), \quad (10.66)$$

where $r^2 = x^2 + y^2$.

In expression (10.66) the first factor $\exp(\mathrm{i}kz)$ describes the phase of the plane wave, the second the curvature of the wave front, and the last the intensity distribution in a cross-section of the beam.

We can determine the effective width of the beam for an arbitrary variable z, defining it as the distance (in the transverse direction of the beam) at which the field amplitude decreases e-fold in comparison with its maximal value.

The square of the effective width of the beam is given by the expression

$$[\rho(z)]^2 = \rho_0^2\left[1+\left(\frac{2z}{k\rho_0^2}\right)^2\right]. \quad (10.67)$$

As can be seen from (10.67), the Gaussian wave which propagates along the z-axis increases its effective width.

We can also determine the angle of divergence of the beam by using the relation

$$\theta = \lim_{z \to \infty} \frac{\rho(z)}{z} = \frac{\lambda}{\pi \rho_0}. \tag{10.68}$$

To give a full description of the beam it is necessary to know one more parameter – the radius of the curvature of its wave front. This radius can be determined with the use of the construction shown in Fig. 10.5. The wave phase is constant on the surface determined by the quantity

$$\frac{2kzr^2}{4z^2 + k^2\rho_0^4},$$

which appears in formula (10.66)

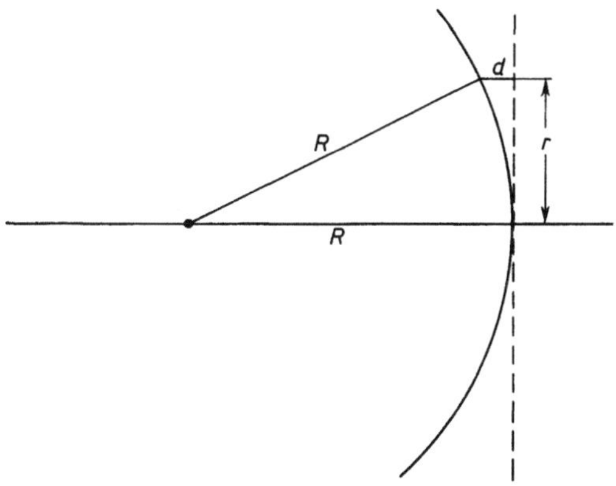

Figure 10.5

The phase shift between that surface at a distance r from the beam axis and the plane tangent to it at a point which lies on the beam axis $(r = 0)$ is

$$kd = \frac{2kzr^2}{4z^2 + (k\rho_0^2)^2}. \tag{10.69}$$

Approximating locally the wave front of a Gaussian beam by a spherical surface of radius R (see Fig. 10.5), we may write the equality

$$R^2 = (R - d)^2 + r^2. \tag{10.70}$$

Hence, having applied the inequality $d \ll R$, we obtain from (10.69) the expression

$$R = \frac{r^2}{2d} = z\left[1 + \left(\frac{\pi \rho_0^2}{\lambda z}\right)^2\right]. \tag{10.71}$$

It is an approximate expression for the radius of curvature of the wave front.

10.11. Approximate Description of the Self-focusing of a Light Beam in a Nonlinear Medium

With the use of Maxwell's equations we can examine the behaviour of a light beam in the medium described by (10.62). To this end it is necessary to know the changes of the electric vector of a monochromatic wave along the path covered by that wave in a nonlinear medium. We assume that the direction of wave propagation coincides with the z-axis, while its electric vector $\mathbf{E}$ oscillates in the direction of the y-axis. Moreover, we assume cylindrical symmetry of the beam with respect to the direction of propagation, i.e., we consider a wave of the form

$$\mathbf{E} = \mathbf{e}_y E(r, z) \exp[\mathrm{i}(\omega t - kz)], \tag{10.72}$$

where $\mathbf{e}_y$ is the unit vector of the y-axis, and $r = \sqrt{x^2 + y^2}$ the distance from the beam axis.

The quantity to be determined is the wave amplitude $E(r, z)$. Maxwell's equations in a medium with $\varepsilon = \varepsilon_1 + \varepsilon_2 |\mathbf{E}|^2$ take the form

$$\begin{aligned} \operatorname{curl} \mathbf{H} &= \frac{\partial}{\partial t}\left(\varepsilon_1 + \varepsilon_2 |\mathbf{E}|^2\right)\mathbf{E}, \\ \operatorname{curl} \mathbf{E} &= -\mu \frac{\partial \mathbf{H}}{\partial t}. \end{aligned} \tag{10.73}$$

Eliminating the vector $\mathbf{H}$ in (10.73), we obtain the equation

$$\operatorname{curl}\operatorname{curl}\mathbf{E} + \mu_0\varepsilon_1 \frac{\partial^2 \mathbf{E}}{\partial t^2} = -\mu\varepsilon_2 \frac{\partial^2}{\partial t^2}|\mathbf{E}|^2\mathbf{E}. \tag{10.74}$$

Making use of a cylindrical system of coordinates whose z-axis coincides with the direction of wave propagation, we obtain from (10.74) a system of three scalar equations

$$\frac{\partial^2 E}{\partial z \partial r} = ik\frac{\partial E}{\partial r}, \tag{10.75a}$$

$$\frac{\partial^2 E}{\partial r^2} + \frac{\partial^2 E}{\partial z^2} - 2ik\frac{\partial E}{\partial z} = -k^2\left(\frac{\varepsilon_2}{2\varepsilon_1}\right)|\mathbf{E}|^2 E, \tag{10.75b}$$

$$\frac{1}{r}\frac{\partial E}{\partial r} = \frac{\partial^2 E}{\partial r^2}. \tag{10.75c}$$

Assuming that self-focusing takes place at distances z greater than the cross-section of the beam, we can take

$$\frac{\partial^2 E}{\partial z^2} \ll \frac{\partial^2 E}{\partial r^2}. \tag{10.76}$$

Owing to this assumption and to (10.75c), equation (10.75b) can be given in the form

$$\frac{1}{r}\frac{\partial E}{\partial r} - 2ik\frac{\partial E}{\partial z} + k^2\left(\frac{\varepsilon_2}{2\varepsilon_1}\right)|\mathbf{E}|^2 E = 0. \tag{10.77}$$

If the field variations in the transverse directions of the beam are small compared with the oscillating character of the dependence on the variable z, we can additionally disregard in (10.77) the term $\frac{1}{r}\frac{\partial E}{\partial r}$, which leads to the equation

$$\frac{\partial E}{\partial z} + ik\frac{\varepsilon_2}{4\varepsilon_1}|\mathbf{E}|^2 E = 0. \tag{10.77a}$$

The solution of this equation is

$$E(z) = E(0)\exp\left[-ik\left(\frac{\varepsilon_2}{4\varepsilon_1}\right)|\mathbf{E}|^2 z\right]. \tag{10.78}$$

The propagation constant k has been replaced by the quantity $(k\varepsilon_2/4\varepsilon_1)|\mathbf{E}|^2$.

To describe the initial process of beam self-focusing we can slightly modify formula (10.78), taking into account the dependence of E on r, i.e. we can write

$$E(r,z) = E(0)\exp\left[-\mathrm{i}\frac{k\varepsilon_2}{4\varepsilon_1}|\mathbf{E}(r,z)|^2 z\right]. \tag{10.79}$$

Hence we obtain the equality

$$\frac{1}{r}\frac{\partial E(r,z)}{\partial r} = E(r,z)\left(-\mathrm{i}k\frac{\varepsilon_2}{4\varepsilon_1}\right)z\frac{1}{r}\frac{\partial}{\partial r}|\mathbf{E}(r,z)|^2. \tag{10.80}$$

Introducing the parameter a through the relation

$$\frac{1}{a^2} = -\frac{1}{4|\mathbf{E}|^2}\frac{1}{r}\frac{\partial}{\partial r}|\mathbf{E}(r,z)|^2$$

(for a Gaussian beam the parameter a is equal to $\rho_0/2$, i.e. to half the effective width of the beam), and assuming that it is a constant quantity, we may rewrite equation (10.77) as follows:

$$\frac{\partial E(r,z)}{\partial z} = \frac{\varepsilon_2}{2a^2\varepsilon_1}|\mathbf{E}(r,z)|^2 E(r,z)z - \frac{\mathrm{i}k\varepsilon_2}{4\varepsilon_1}|\mathbf{E}(r,z)|^2 E. \tag{10.81}$$

The first term on the right-hand side of the equation causes the increasing of intensity accompanying the increasing of z, the second term is responsible for the phase variation of the complex amplitude sought.

Equation (10.81) envisages an infinite increase of intensity around z, i.e., focusing of the beam in an infinitesimally thin thread, but it does not show the existence of any focusing threshold. Beam narrowing occurs at an arbitrary E.

To obtain the expression for the beam power which is necessary for the initiation of self-focusing, we have to take into account the effect of beam broadening. This effect may be characterized with the use of the distance z_d at which the radius of the beam cross-section is doubled. As follows from (10.67), the distance z_d is

$$z_d = \frac{\sqrt{3}\pi\rho_0^2}{\lambda}, \tag{10.82}$$

where λ is the wavelength.

The focal length of the beam z_f may be defined as an interval Δz on which the change of amplitude of the wave is equal to the amplitude itself $\Delta E(r,z) = E(r,z)$. Taking

$$\frac{\partial E(r,z)}{\partial z} = \frac{E(r,z)}{z_f},$$

and equating it with the first term of the right-hand side of equation (10.81), we obtain

$$z_f \approx \frac{a}{|E|}\sqrt{\frac{2\varepsilon_1}{\varepsilon_2}}. \tag{10.83a}$$

In the case of a Gaussian beam the carried power is $P \approx \pi\rho_0^2\varepsilon_1|E|^2 v$, while the parameter a equals $\frac{1}{2}\rho_0$. The focal length for a beam of this type is

$$z_f = \rho_0^2\varepsilon_1\sqrt{\frac{\pi v}{2\varepsilon_2 P}}, \tag{10.83b}$$

where v is the velocity of the wave.

Beam narrowing starts when the self-focusing effect compensates broadening, i.e. when $z_d = z_f$. It follows hence that the process has its starting point when the beam power has exceeded the critical level $P_{cr} = \dfrac{\varepsilon_1^2\lambda^2 v}{6\pi\varepsilon_2}$.

10.12. Interaction of Four Waves and Dynamic Holograms

For simplicity we shall consider only an isotropic medium. A medium of this kind, as one having a centre of symmetry, has zero square nonlinearity. Only some elements of the tensor of cubic susceptibility of that medium are not equal to zero. They satisfy the relations

$$\begin{aligned}
&\chi_{1111} = \chi_{2222} = \chi_{3333} = \chi, \\
&\chi_{2233} = \chi_{3322} = \chi_{3311} = \chi_{1133} = \chi_{1122} = \chi_{2211} = \chi_1, \\
&\chi_{2323} = \chi_{3232} = \chi_{1313} = \chi_{3131} = \chi_{1212} = \chi_{2121} = \chi_2, \\
&\chi_{2332} = \chi_{3223} = \chi_{3113} = \chi_{1331} = \chi_{1221} = \chi_{2112} = \chi_3, \\
&\chi = \chi_1 + \chi_2 + \chi_3.
\end{aligned} \tag{10.84}$$

The nonlinear components of the polarization vector are expressed in this medium by the components of the electric vector in the following way:

$$P_j = \chi(E_x E_x + E_y E_y + E_z E_z)E_j \quad (j = x, y, z). \tag{10.85}$$

If on an isotropic medium filling the area $0 \leq z \leq L$ fall three plane waves with the electric vectors described by the expressions

$$\mathbf{E}_m = \mathbf{e}_m E_{0m} \exp\left[\mathrm{i}\left(\mathbf{k}_m \cdot \mathbf{r} - \omega_m t\right)\right], \tag{10.86}$$

then as a result of nonlinear polarization induced by those waves the following wave is generated in it:

$$\mathbf{E}_4 = \mathbf{e}_4 E_4(z) \exp\left[\mathrm{i}\left(\mathbf{k}_4 \cdot \mathbf{r} - \omega_4 t\right)\right]. \tag{10.87}$$

The frequency of this wave may be one of many possible (10.28). For the present we will consider the case where

$$\omega_4 = \omega_1 + \omega_2 + \omega_3, \tag{10.88}$$

i.e. the case of a wave for which the phase matching condition (10.37) is satisfied. This condition takes here the form

$$\mathbf{k}_4 = \mathbf{k}_1 + \mathbf{k}_2 + \mathbf{k}_3. \tag{10.89}$$

The nonlinear polarizations with frequencies ω_j $(j = 1, 2, 3, 4)$ which occur in the medium are given by

$$\begin{aligned}
\mathbf{P}^{(\mathrm{NL})}_{\omega_1} &= \varepsilon_0 \chi(-\omega_4 + \omega_2 + \omega_3) \vdots \\
&\vdots \left[(\mathbf{E}_4 \mathbf{E}_2^*)\mathbf{E}_3^* + (\mathbf{E}_4 \mathbf{E}_3^*)\mathbf{E}_2^* + (\mathbf{E}_2^* \mathbf{E}_3^*)\mathbf{E}_4\right], \\
\mathbf{P}^{(\mathrm{NL})}_{\omega_2} &= \varepsilon_0 \chi(-\omega_4 + \omega_1 + \omega_3) \vdots \\
&\vdots \left[(\mathbf{E}_4 \mathbf{E}_1^*)\mathbf{E}_3^* + (\mathbf{E}_4 \mathbf{E}_3^*)\mathbf{E}_1^* + (\mathbf{E}_1^* \mathbf{E}_3^*)\mathbf{E}_4\right],
\end{aligned}$$

$$\mathbf{P}^{(\mathrm{NL})}_{\omega_3} = \varepsilon_0 \chi(-\omega_4 + \omega_1 + \omega_2) \vdots \tag{10.90}$$
$$\vdots \left[(\mathbf{E}_4\mathbf{E}_1^*)\mathbf{E}_2^* + (\mathbf{E}_4\mathbf{E}_2^*)\mathbf{E}_1^* + (\mathbf{E}_1^*\mathbf{E}_2^*)\mathbf{E}_4\right],$$
$$\mathbf{P}^{(\mathrm{NL})}_{\omega_4} = \varepsilon_0 \chi(-\omega_1 - \omega_2 - \omega_3) \vdots$$
$$\vdots \left[(\mathbf{E}_1\mathbf{E}_2)\mathbf{E}_3 + (\mathbf{E}_1\mathbf{E}_3)\mathbf{E}_2 + (\mathbf{E}_2\mathbf{E}_3)\mathbf{E}_1\right].$$

Each of the monochromatic waves with frequency ω_j ($j = 1,2,3,4$) satisfies in a nonlinear medium equation (10.30). For plane waves with constant amplitudes on the surface of the constant phase this equation reduces to

$$(\mathbf{S}_j \cdot \nabla) E_j(z) + \alpha_j E_j(z)$$
$$= \mathrm{i} \frac{\omega_j^2}{2|\mathbf{k}_j| c^2} \varepsilon_0^{-1} \mathbf{P}^{(\mathrm{NL})}(\omega_j) \cdot \mathbf{e}_j \exp(-\mathrm{i}\mathbf{k}_j \cdot \mathbf{r}), \tag{10.91}$$

where $\mathbf{S}_j = \dfrac{\mathbf{k}_j}{|\mathbf{k}_j|}$ has the direction of the wave vector, $z = \mathbf{S}_j \cdot \mathbf{r}$ and

$$\alpha_j = \frac{\omega_j^2 \varepsilon''(\omega_j)}{2c^2 |\mathbf{k}_j|}.$$

$\varepsilon''(\omega_j)$ is the imaginary part of the electric permittivity of the medium $\varepsilon(\omega_j) = \varepsilon'(\omega_j) - \mathrm{i}\varepsilon''(\omega_j)$.

From the form of equation (10.91) we conclude that the amplitudes of the plane waves interacting with a nonlinear medium may change only along the direction of propagation.

It is worthwhile to consider three different cases of four wave interaction (Fig. 10.6).

A. We seek a wave with the wave vector $\mathbf{k}_4 = \mathbf{e}_z k_4$ generated by three waves (10.85) incident on a nonlinear medium and having their wave vectors $\mathbf{k}_m$ orientated in such a way that

$$\mathbf{k}_4 - \mathbf{k}_3 - \mathbf{k}_2 - \mathbf{k}_1 = \mathbf{e}_z \Delta k \tag{10.92}$$

and under the assumption that Δk is a very small quantity (insignificant phase mistuning). This leads to the following equation for the amplitude of the generated wave:

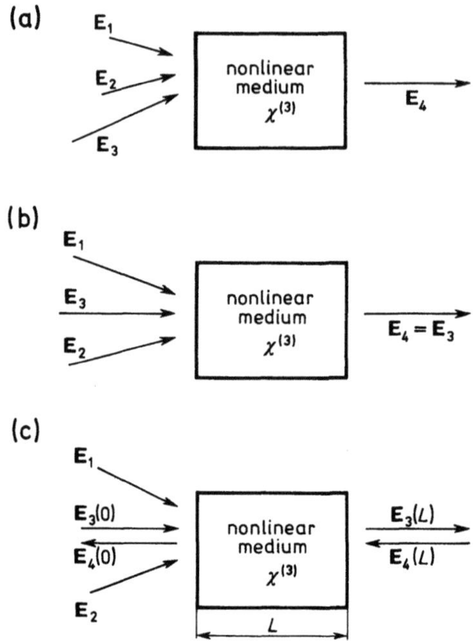

Figure 10.6

$$\frac{dE_4}{dz} + \alpha_4 E_4 = \frac{i\omega_4\gamma_4}{2}\chi(-\omega_1-\omega_2-\omega_3)\vdots[(\mathbf{E}_1\mathbf{E}_2)(\mathbf{E}_3\mathbf{e}_4) + (\mathbf{E}_1\mathbf{E}_3)(\mathbf{E}_2\mathbf{e}_4) + (\mathbf{E}_2\mathbf{E}_3)(\mathbf{E}_1\mathbf{e}_4)]\exp(-i\Delta kz), \quad (10.93)$$

where $\gamma_4 = \dfrac{\omega_4}{2c^2|\mathbf{k}_4|}$.

Assuming that the amplitudes of waves $\mathbf{E}_m$ $(m = 1, 2, 3)$ are large and change insignificantly while crossing the medium, we can regard them in (10.93) as equal to the amplitude values at the input. The boundary condition $E_z(0) = 0$ then leads to the equation

$$E_4(z) = \frac{i\omega_4\gamma_4}{2(\alpha_4 - i\Delta k)}\chi(\omega_4)\vdots[(\mathbf{E}_1\mathbf{E}_2)(\mathbf{E}_3\mathbf{e}_4) + (\mathbf{E}_1\mathbf{E}_3)(\mathbf{E}_2\mathbf{e}_4) + (\mathbf{E}_2\mathbf{E}_3)(\mathbf{E}_1\mathbf{e}_4)][\exp(-i\Delta kz) - \exp(-\alpha_4 z)]. \quad (10.94)$$

As can be seen from (10.94), for $\Delta k \neq 0$ the amplitude $E_4(z)$ undergoes oscillating variations. It is only for $\Delta k = 0$, the damping $\alpha_4 = 0$ being disregarded, that we can observe its linear growth as the path is being covered.

B. If the generated wave is to have the same frequency and the same direction of propagation as one of the incident waves, e.g. the wave $\mathbf{E}_3 = \mathbf{E}_4$ (i.e. if $\omega_4 = \omega_3$ and $\mathbf{k}_4 = \mathbf{k}_3$), then (10.88) implies $\omega_2 = -\omega_1$.

The equation for the amplitude $\mathbf{E}_3$ takes the from

$$\frac{\mathrm{d}E_3(z)}{\mathrm{d}z} + \alpha_3 E_3(z) = \frac{\mathrm{i}\omega_3\gamma_3}{2}\chi(\omega_3)\vdots$$
$$\vdots[(\mathbf{E}_1^*\mathbf{E}_2) + 2(\mathbf{E}_1^*\mathbf{e}_3)(\mathbf{E}_2^*\mathbf{e}_3)]\,E_3(z)\exp(-\mathrm{i}\Delta kz), \quad (10.95)$$

where $\Delta\mathbf{k} = \mathbf{k}_1 + \mathbf{k}_2$.

If we assume that $E_3(z=0) = E_3(0)$ and the amplitudes $E_1(z)$ and $E_2(z)$ are constant in the region of interacting, equation (10.95) has the solution

$$E_3(z) = E_3(0)\exp[-\alpha_3 z - g_3(z)], \quad (10.96)$$

where

$$g_3(z) = \frac{\omega_3\gamma_3}{2}\chi(\omega_3)E_1^*E_2^*\vdots[(\mathbf{e}_1\mathbf{e}_2)(\mathbf{e}_3\mathbf{e}_3) +$$
$$+ 2(\mathbf{e}_1\mathbf{e}_3)(\mathbf{e}_2\mathbf{e}_3)][1 - \exp(-\mathrm{i}\Delta kz)]\frac{1}{\Delta k}. \quad (10.96a)$$

The real part $g_3(z)$ $[\mathrm{Re}\, g_3(z)]$ for $\Delta k > 0$ will describe the amplification of the wave with the amplitude $E_3(z)$.

If $E_1(z) = E_2(z) = E_{10}$ and $\Delta k = 0$, then

$$g_3(z) = \mathrm{i}\omega_3\gamma_3\chi(\omega_3)(E_{10}^*)^2\vdots[(\mathbf{e}_1\mathbf{e}_2)(\mathbf{e}_3\mathbf{e}_3) + 2(\mathbf{e}_1\mathbf{e}_3)(\mathbf{e}_2\mathbf{e}_3)]\,z. \quad (10.96b)$$

C. If all the interacting waves have the same frequency $\omega_j = \omega$ $(j = 1, 2, 3, 4)$, and

$$\mathbf{k}_1 + \mathbf{k}_2 = \mathbf{0} \quad \text{and} \quad \mathbf{k}_3 + \mathbf{k}_4 = \mathbf{0},$$

nonlinear polarization of the third order with frequency ω has six components with different wave vectors:

$$\begin{aligned}\mathbf{P}^{(\mathrm{NL})} = \varepsilon_0\chi\vdots\{&\mathbf{E}_1(\mathbf{k}_1)\mathbf{E}_2(\mathbf{k}_2)\mathbf{E}_3^*(\mathbf{k}_3)\exp[-\mathrm{i}(\mathbf{k}_3\cdot\mathbf{r})] + \\ &+ \mathbf{E}_1(\mathbf{k}_1)\mathbf{E}_2^*(\mathbf{k}_2)\mathbf{E}_3(\mathbf{k}_3)\exp[\mathrm{i}(2\mathbf{k}_1 + \mathbf{k}_3)\cdot\mathbf{r}] + \\ &+ \mathbf{E}_1^*(\mathbf{k}_1)\mathbf{E}_2(\mathbf{k}_2)\mathbf{E}_3(\mathbf{k}_3)\exp[\mathrm{i}(\mathbf{k}_3 - 2\mathbf{k}_1)\cdot\mathbf{r}] + \\ &+ \mathbf{E}_1(\mathbf{k}_1)\mathbf{E}_2(\mathbf{k}_2)\mathbf{E}_4^*(\mathbf{k}_4)\exp[-\mathrm{i}\mathbf{k}_4\cdot\mathbf{r}] + \\ &+ \mathbf{E}_1^*(\mathbf{k}_1)\mathbf{E}_2(\mathbf{k}_2)\mathbf{E}_4(\mathbf{k}_4)\exp[\mathrm{i}(\mathbf{k}_4 - 2\mathbf{k}_1)\cdot\mathbf{r}] + \\ &+ \mathbf{E}_1(\mathbf{k}_1)\mathbf{E}_2^*(\mathbf{k}_2)\mathbf{E}_4(\mathbf{k}_4)\exp[\mathrm{i}(2\mathbf{k}_1 + \mathbf{k}_4)\cdot\mathbf{r}]\}. \quad (10.97)\end{aligned}$$

The equations for the amplitudes $E_3(z)$ and $E_4(z)$ (for $\alpha_3 = \alpha_4 = \alpha$) take the form

$$\begin{aligned}\frac{\mathrm{d}E_3}{\mathrm{d}z} + \alpha E_3 &= \mathrm{i}\kappa E_4^*, \\ \frac{\mathrm{d}E_4}{\mathrm{d}z} + \alpha E_4 &= \mathrm{i}\kappa E_3^*,\end{aligned} \quad (10.98)$$

where

$$\kappa = \frac{\omega_3\gamma_3}{2}\chi(\omega_3)E_1E_2\vdots[(\mathbf{e}_1\mathbf{e}_2)(\mathbf{e}_3\mathbf{e}_4) + (\mathbf{e}_1\mathbf{e}_4)(\mathbf{e}_2\mathbf{e}_3) + + (\mathbf{e}_1\mathbf{e}_3)(\mathbf{e}_2\mathbf{e}_4)]. \quad (10.98a)$$

The remaining components of nonlinear polarization with frequency 3ω have been neglected because of the absence of phase matching. Under the boundary conditions

$$E_3(0) = E_{30} \quad \text{and} \quad E_4(L) = 0,$$

the solutions of equations (10.98) are the expressions

$$\begin{aligned}E_3(z) &= E_{30}[\cos(\kappa z) + \tan(\kappa L)\sin(\kappa z)]\exp(-\alpha z), \\ E_4(z) &= \mathrm{i}E_{30}(\sin(\kappa z) - \tan(\kappa L)\cos(\kappa z)]\exp(-\alpha z).\end{aligned} \quad (10.99)$$

Expressions (10.99) show that for $z = 0$ the amplitude $E_4(z)$ assumes the value

$$|E_4(0)| = E_{30}\tan(\kappa L), \tag{10.100}$$

and for $\kappa L > \pi/4$ it may be greater than the amplitude of the incident wave. The value κ (10.98a) is determined by the amplitudes ($E_1(z)$ and $E_2(z)$) and by the polarizations ($\mathbf{e}_1$ and $\mathbf{e}_2$) of the pumping waves $\mathbf{E}_1$ and $\mathbf{E}_2$. Also the polarizations of the generated waves ($\mathbf{e}_3$ and $\mathbf{e}_4$) have an effect on κ. Increasing L, we can increase κL. For $z = L$ the amplitude of the wave $\mathbf{E}_3$ is

$$E_3(L) = E_{30}\frac{1}{\cos(\kappa L)}, \tag{10.101}$$

which means that it also undergoes amplification, and increases with the growth of L.

In view of the condition $\mathbf{k}_3 = -\mathbf{k}_4$ we can assert that the phase of the wave $\mathbf{E}_4$ has its spatial part conjugate with the spatial part of the phase of the wave $\mathbf{E}_3$. Thus, if the wave $\mathbf{E}_3$ has its phase factor of the form $\exp[\mathrm{i}k_3z - \mathrm{i}\omega t]$, the phase factor of the wave $\mathbf{E}_4$ is $\exp[\mathrm{i}(-k_3z - \omega t)]$. From that possibility of conjugating the spatial part of the wave phase arises a new section of nonlinear optics, called *conjugate phase optics*.

Wave interaction in a nonlinear medium with cubic polarization can be interpreted physically as follows.

Two out of three waves (with the same frequency) entering a nonlinear medium undergo interference and each of them generates either a static or a dynamic phase grating. The third wave is scattered on these gratings and gives a secondary scattered wave. We can distinguish the following possibilities (we consider only static gratings).

1. The gratings produced by the waves with wave vectors $\mathbf{k}_1$ and $\mathbf{k}_3$ scatter the wave with wave vector $\mathbf{k}_2$. Then it is possible to generate the waves with wave vector

$$\mathbf{k}_4 = \mathbf{k}_2 \pm (\mathbf{k}_1 - \mathbf{k}_3). \tag{10.102}$$

2. The gratings produced by the waves with wave vectors $\mathbf{k}_2$ and $\mathbf{k}_3$ scatter the wave with wave vector $\mathbf{k}_1$ and generate waves with wave vectors

$$\mathbf{k}_4 = \mathbf{k}_1 \pm (\mathbf{k}_2 - \mathbf{k}_3). \tag{10.103}$$

3. The gratings produced by the wave with wave vectors $\mathbf{k}_1$ and $\mathbf{k}_2$ scatter the wave with wave vector $\mathbf{k}_3$ and generate waves with wave vectors

$$\mathbf{k}_4 = \mathbf{k}_3 \pm (\mathbf{k}_1 - \mathbf{k}_2). \tag{10.104}$$

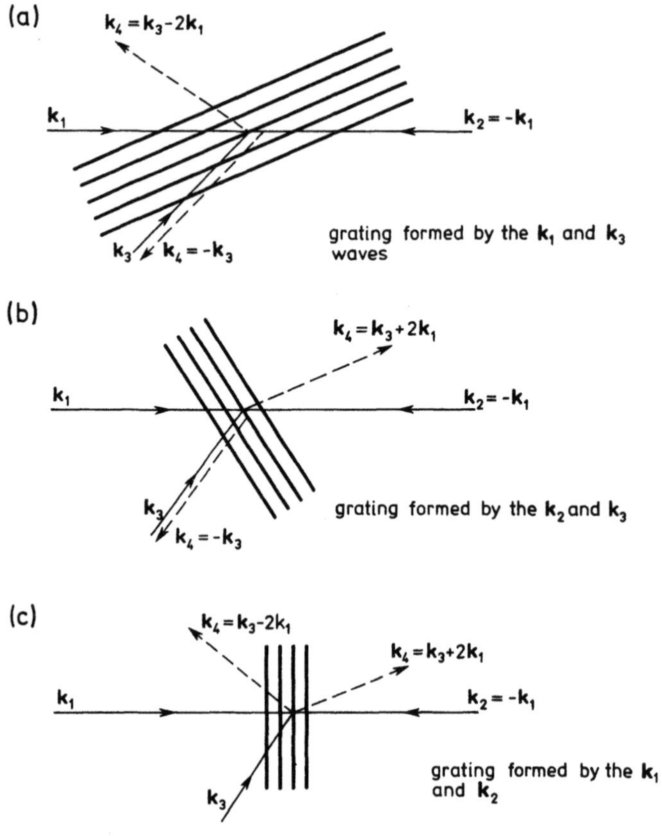

Figure 10.7

All these three situations are illustrated in Fig. 10.7, for $\mathbf{k}_2 = -\mathbf{k}_1$.

In general we can expect generation of three different waves with frequency ω and wave vectors

$$\begin{aligned} \mathbf{k}_4 &= \mathbf{k}_1 + \mathbf{k}_2 - \mathbf{k}_3, \\ \mathbf{k}_4 &= \mathbf{k}_1 - \mathbf{k}_2 + \mathbf{k}_3, \\ \mathbf{k}_4 &= -\mathbf{k}_1 + \mathbf{k}_2 + \mathbf{k}_3. \end{aligned} \tag{10.105}$$

Of the three possibilities (10.105) only those can be realized for which the phase matching condition is fulfilled. When $\mathbf{k}_2 = -\mathbf{k}_1$, then, by (10.105), we can expect the generation of waves with wave vectors

$$\begin{aligned} \mathbf{k}_4 &= -\mathbf{k}_3, \\ \mathbf{k}_4 &= \mathbf{k}_3 - 2\mathbf{k}_1, \\ \mathbf{k}_4 &= \mathbf{k}_3 + 2\mathbf{k}_1. \end{aligned} \tag{10.105a}$$

However, it is only for the wave with wave vector $\mathbf{k}_4 = -\mathbf{k}_3$ that phase matching takes place and the matching condition is fulfilled.

The above interpretation reveals a close connection between wave interaction in a nonlinear medium and holography. Both in holography and in cases of nonlinear wave interaction the emerging wave $\mathbf{k}_4$ arises from the scattering of one of the waves $(\mathbf{k}_2$ or $\mathbf{k}_1)$ on an interferogram formed by the object wave $\mathbf{k}_3$ and one of the pumping waves $\mathbf{k}_2$ or $\mathbf{k}_1$, where $\mathbf{k}_4 = -\mathbf{k}_3$.

A real object can be represented by a whole set of waves $\mathbf{k}_3$ and the image of the object can be reconstructed by appropriate generated waves.

As can be seen from expressions (10.97), we may present the components of nonlinear polarization with frequency ω in the form

$$\begin{aligned} \mathbf{P}^{(\mathrm{NL})}(\mathbf{k}_4 = -\mathbf{k}_3, \omega) = A(\mathbf{E}_1\mathbf{E}_3^*)\cdot\mathbf{E}_2 + B(\mathbf{E}_2\mathbf{E}_3^*)\cdot\mathbf{E}_1 + \\ + C(\mathbf{E}_1 E_2)\cdot\mathbf{E}_3^*, \end{aligned} \tag{10.97a}$$

where A, B, C are the tensors of second order and depend on the angles between the appropriate vectors.

Expression (10.97a) may be interpreted as follows: the functional relations $A(\mathbf{E}_1\mathbf{E}_3^*)$ and $B(\mathbf{E}\mathbf{E}_3^*)$ describe static gratings produced by the interference of the appropriate wave $\mathbf{E}_1$ (or $\mathbf{E}_2$) with the wave

$\mathbf{E}_3$, while the functional relation $C(\mathbf{E}_1\mathbf{E}_2)$ describes a dynamic grating varying with frequency 2ω.

Only one of these gratings can be effectively realized by the proper choice of interacting waves (only one of the tensors A, B, C is not equal to zero).

10.13. Optical Phase Conjugation

The phenomenon of phase conjugation is defined as a process in which the spatial part of the phase of the emerging wave is conjugate in the sense of complex conjugation with the phase of the entering wave. In other words, it is a process of reversing the entering phase. It may occur in generating differential frequency, in parametric amplification or in mixing four waves (in a nonlinear cubic medium).

In the previous section we have shown that a wave emerging from a nonlinear medium in which there occurs a degenerated interaction of waves has its spatial phase conjugate with the phase of the incident wave ($\mathbf{E}_4^*(\mathbf{r})\exp(-i\omega t) = \mathbf{E}_3(\mathbf{r})\exp(-i\omega t)$). Therefore the nonlinear medium in this process may be used as a phase-conjugating mirror.

The action of the phase-conjugating mirror can be explained in the simplest way by means of an example. Thus, if a wave incident on such a mirror is a divergent spherical wave propagating from a point source, then after reflecting connected with phase conjugation it becomes a convergent spherical wave moving along the path covered by the incident wave.

If the wave with a conjugate phase propagates in the opposite direction to that of the incident wave, it can be used for corrections of the disturbances of the front of the incident wave; this is illustrated by the following example.

If wave 1 meets on its way a non-homogeneity in the form of a glass cylinder (Fig. 10.8a), then having passed the obstacle it assumes the shape of wave front 2. This wave, reflected next from a conventional mirror, has the shape of front 3, and after having

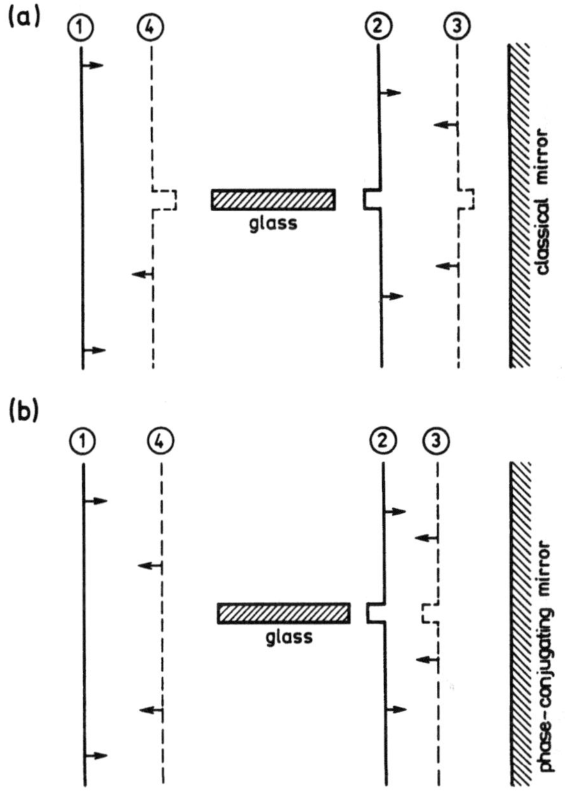

Figure 10.8

passed again through the cylinder, it will have a double delay of wave front 4 in that of its parts which has propagated through the non-homogeneity region. This behaviour of the wave is a consequence of the dispersive properties of the medium.

Removing the distortion of the wave vector with the use of phase conjugation is shown in Fig. 10.8b. The phase-conjugating mirror gives wave front 3 reversed in relation to wave front 2. Wave front 3, having passed the cylinder, becomes identical with the front of primary wave 1. Apart from being applied to remove the distortion of the wave front the phenomenon of a conjugate wave may be used in the construction of lasers. It enables us to construct high power lasers whose beam properties are comparable with those of the beam of a one-mode oscillator.

It is possible to construct a laser with one phase-conjugating mirror which permits obtaining a better quality of the beam with mode stability and additional amplification of its power. Another example is the amplification of a laser beam. The quality of the beam is seriously affected during its passage through the amplifying medium. However, if a phase-conjugating mirror is used to direct the beam backwards through that medium, the light obtained at the output has the properties of the original light beam introduced into the amplifying medium.

The mathematical description of the conjugate phase in the case of an arbitrary monochromatic wave is as follows.

Let the wave

$$\begin{aligned} E_1(\mathbf{r}, t) &= \mathrm{Re}[\psi(\mathbf{r})\exp(\mathrm{i}\omega t - \mathrm{i}kz)] \\ &= \mathrm{Re}[A_1^{\omega}(\mathbf{r})\exp(\mathrm{i}\omega t)] \end{aligned} \tag{10.106}$$

propagate in a lossless dispersive medium in the direction taken as the z-axis. The relations $\psi(\mathbf{r})$ and $A_1^{\omega}(\mathbf{r})$ reflect the spatial modulation of the wave used for transmitting information or acquired by interaction with the dispersive medium or with diffractional obstacles.

If in a certain space region close to, say, z_0 a field $E_2(\mathbf{r}, t)$ is generated, its description being the expression

$$\begin{aligned} E_2(\mathbf{r}, t) &= \mathrm{Re}[\psi^*(\mathbf{r})\exp(\mathrm{i}\omega t + \mathrm{i}kz)] \\ &= \mathrm{Re}[A_2^{\omega}(\mathbf{r})\exp(\mathrm{i}\omega t)], \end{aligned} \tag{10.106a}$$

where

$$A_2^{\omega}(\mathbf{r}) = [A_1^{\omega}(\mathbf{r})]^* \quad \text{for} \quad z < z_0, \tag{10.106b}$$

then the field $E_2(\mathbf{r}, t)$ is called a *field with a conjugate phase* in relation to the field $E_1(\mathbf{r}, t)$. We obtain it by taking the conjugation only of the spatial part of the phase factor of the field $E_1(\mathbf{r}, t)$, leaving the factor $\exp(\mathrm{i}\omega t)$ unchanged. (This is equivalent to leaving without change the spatial part of the phase and changing the sign of time. Thus we can regard the field E_2 as obtained from the field E_1 by "time reversal".)

The primary beam propagating in the direction $+z$ through a medium with dielectric permittivity $\varepsilon(\mathbf{r})$ fulfils the equation

$$\Delta E_1 + k_s^2 E_1 = 0, \tag{10.107}$$

where

$$k_s^2 = \omega^2 \mu_0 \varepsilon(\mathbf{r}).$$

After substituting expression (10.106) into (10.107) and using the estimation

$$\left|\frac{\partial^2 \psi}{\partial z^2}\right| \ll k_s \left|\frac{\partial \psi}{\partial z}\right|, \quad \left|\frac{\partial^2 \psi}{\partial z^2}\right| \ll |k_s^2 \psi|,$$

we arrive at the equation describing the changes of amplitude

$$\Delta_t \psi + (k_s^2 - k^2)\psi - 2ik\frac{\partial \psi}{\partial z} = 0, \tag{10.108}$$

where $\Delta_t = \dfrac{\partial^2}{\partial x^2} + \dfrac{\partial^2}{\partial y^2}$.

The equation conjugate to (10.108) has the form

$$\Delta_t \psi^* + (k_s^2 - k^2)\psi^* + 2ik\frac{\partial \psi^*}{\partial z} = 0, \tag{10.108a}$$

and formally can be treated as the equation for the conjugated amplitude of the wave $E_2(\mathbf{r}, t)$, which propagates in the direction $-z$ and has the spatial part of the phase conjugate in relation to the spatial part of the phase of the wave $E_1(\mathbf{r}, t)$.

10.14. Optical Bistability

Optical bistability is a phenomenon in which a nonlinear medium changes its transmittance according to the intensity of the incident light, revealing at the same time the property of hysteresis.

A system is optically bistable if to the same value of the input

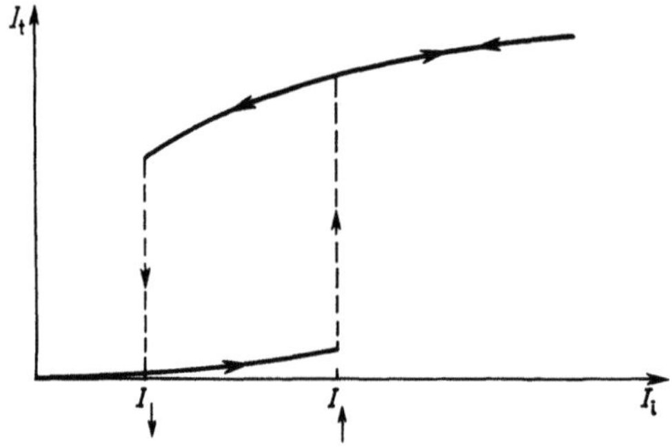

Figure 10.9

intensity from a certain range of its values correspond two different states of the intensity being transmitted.

For each cycle of changes of the incident intensity I_i (Fig. 10.9) the transmittance of the bistable system takes the following course: for a low value of I_i the transmittance is low and its dependence on I_i is represented by the lower segment of the curve in Fig. 10.9. As soon as the intensity I_i has exceeded $I^\uparrow$, a stepwise change of the transmittance value takes places. The transmittance of the system remains on the higher level when I_i decreases until it reaches the critical value $I_\downarrow$ $(<I^\uparrow)$, at which the system goes back to the low transmittance value.

The essential elements of optically bistable systems are:

– feedback, ensuring the influence of the radiation which is being transmitted on the physical state of the medium shined,

– nonlinear electric susceptibility.

The radiation applied ought to be time-coherent (radiation of a continuous-work laser) and space-coherent. Feedback is realized in several different ways, for instance by using the Fabry-Pérot resonance cavity (Fig. 10.10) or a ring cavity (Fig. 10.11).

It is possible to distinguish absorptive optical bistability and dispersive optical bistability. This follows from the fact that optical bistability may be related either to the imaginary part or to the real part of the nonlinear tensor of electric susceptibility

$$\underset{\sim}{\chi}^{(3)} = \underset{\sim}{\chi}_d^{(3)} + \mathrm{i}\underset{\sim}{\chi}_a^{(3)}. \tag{10.109}$$

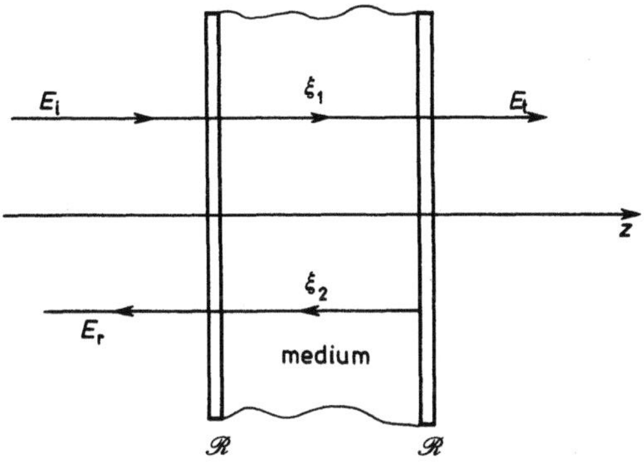

$\mathscr{T} = 1 - \mathscr{R}$

Figure 10.10

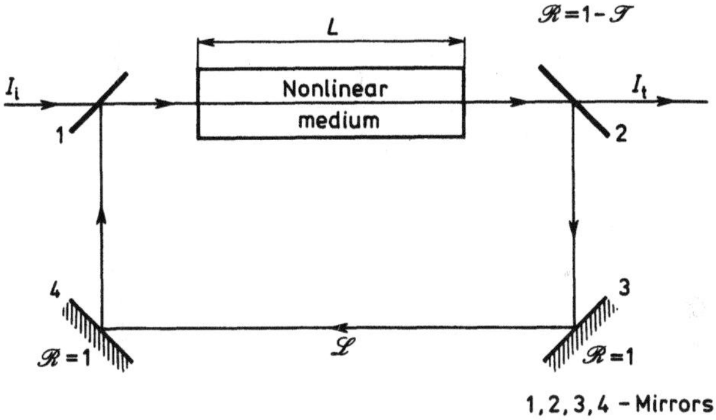

Figure 10.11

An explicit form of the real and imaginary components of the electric susceptibility may be obtained by applying the quantum methods and the model of a medium consisting of two-level atoms [6]. Here is this form:

$$\underset{\sim}{\chi}^{(3)} = \frac{1}{1 + \Delta^2 + \dfrac{I_i}{I_s}} (\alpha\Delta + i\alpha), \tag{10.109a}$$

where

$$\Delta = \omega_a - \omega,$$

$$I_s = \frac{\hbar^2}{|\mu|^2 T_1 T_2},$$

$$\alpha = \frac{N T_2 |\mu|^2}{2\hbar V},$$

ω_a – the frequency of the transition of the electron between the levels,

ω – the field frequency of the wave interacting with the medium,

μ – the dipole moment for the electron transition,

T_1, T_2 – respectively: the longitudinal and the transversal time of relaxation describing the ensembles of atoms composing the system under consideration,

N – the number of atoms enclosed in the volume V of the system of atoms under consideration,

$\hbar = \dfrac{h}{2\pi}$, where h is the Planck constant.

In the case of absorptive bistability occurring for $\Delta = 0$, a nonlinear medium (filling the Fabry–Pérot cavity) plays the role of an absorbent. When the intensity of the incident wave increases, the medium absorbs more and more quanta of energy and undergoes "bleaching". The quality factor of the cavity rises. This in turn causes a further growth of the field inside the cavity and further "bleaching" of the medium. It goes on till the state of absorptive saturation is reached, in which the intensity of the transmitted light I_t becomes equal to the intensity of the incident light I_i. If in this state the intensity of the incident light begins to decrease, the internal field of the cavity will "try" to preserve the state of absorptive saturation and the state of high transmittance of the medium. Only when the decrease of the intensity of the incident light is very significant, "bleaching" begins to diminish and transmittance falls. The course of this phenomenon is illustrated in Fig. 10.9.

The mechanism of dispersive optical bistability is different. At a low value of the radiation incident on the Fabry–Pérot cavity, when the frequency of the incident light is different from the frequency of atom transition $(\omega \neq \omega_a)$, the transmittance of the system is low. In a linear case the Fabry–Pérot cavity has a transmittance which for a constant value of the linear refractive index n_L depends on the thickness of the cavity L. The resonance frequency of the cavity is expressed by the formula

$$\omega_c = \frac{\pi c}{n_L L}, \tag{10.110}$$

and the intensity of the light transmitted through the Fabry–Pérot cavity by the relation

$$I_t = \frac{I_i}{1 + F \sin^2\left(\pi \dfrac{\omega}{\omega_0}\right)}, \tag{10.110a}$$

where F is a quantity determined by the ability of the plates making up the boundaries of the cavity (see Sec. 3.12) to transmit and reflect light.

The intensity of the light I_t transmitted through the cavity reaches (Fig. 10.12) the maximal values when

$$\frac{\omega}{\omega_0} = m \quad (m = 1, 2, 3, \ldots). \tag{10.110b}$$

If the intensity of the incident light increases, we have to take into account the dependence of the refractive index upon the intensity

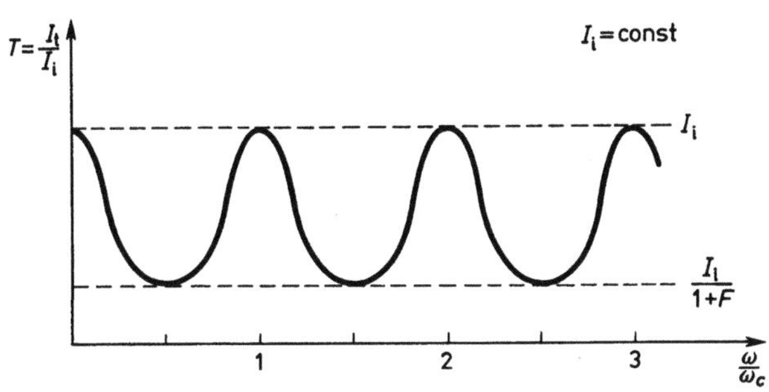

Figure 10.12

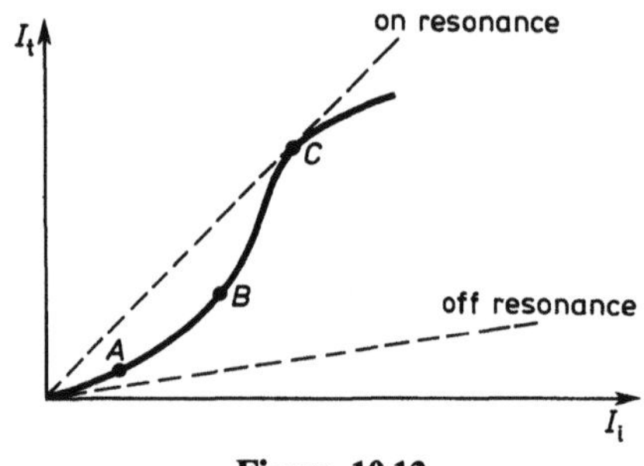

Figure 10.13

$n = n(I)$. This causes the change of the resonance frequencies of the Fabry–Pérot cavity, and consequently the change of its work conditions.

If the primary resonance frequency of the cavity (under low intensities of the incident light) is such that the transmitted light corresponds to point A in Fig. 10.13, then at a small increase of the intensity of the incident light the transmittance of the cavity will increase proportionally. If the relation $n = n(I)$ appears, the resonance frequency of the cavity will change and the resonance curve (Fig. 10.14) will move to the left. The intensity of the transmitted light will now be determined by point B (Fig. 10.13). A further increase of the intensity I_i will be accompanied by an increase of the intensity I_t and at the same time by the growth of the energy density

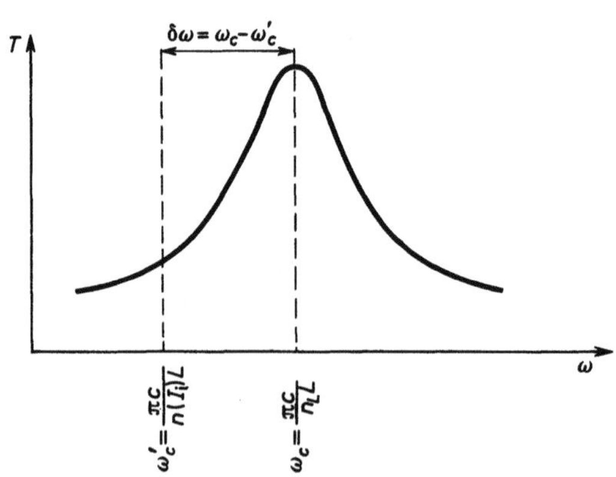

Figure 10.14

of the field inside the cavity. The system attains the state of equilibrium when reaching point C (Fig. 10.13). A further increase of I_i should lead to the mistuning of the cavity and a decrease of I_t. However, this does not happen, since the field energy stored inside the cavity supports the existing state. The value I_t becomes stabilized. The same happens when the value of the intensity I_i decreases. A considerable decrease of I_i is needed to tune the system significantly out of resonance and bring it back to the proportionality region. There appears hysteresis (Fig. 10.9).

Thus the nonlinear and ambiguous dependence of the cavity transmittance on the intensity I_i leads to the appearance of two different transmittance states for a given value of I_i.

Dispersive optical bistability is easier to obtain experimentally than absorptive optical bistability. This results from the difficulty in satisfying the condition $\omega = \omega_a$ during a long period of work of the laser.

In constructing optical systems wide potentialities are shown by thin semiconductor layers. Large values of the linear refractive index $(n_L \approx 4)$ indicate that those layers can be used as the Fabry–Pérot cavities even without additional dielectric cover (the reflectivity of their border-planes is $\approx 35\%$). In addition, they show considerable nonlinearity, which permits a considerable decrease of the order of the intensity I_i indispensable for reaching bistable states. Such semiconductors include, among others, GaAs, InSb, InAs, and CdHgTe.

10.15. Dispersive Optical Bistability

Let us consider the Fabry–Pérot cavity with mirrors positioned perpendicularly to the z-axis (Fig. 10.15) and described by the coefficient of reflectivity $\mathcal{R}$ and transmittance $\mathcal{T}$ $(\mathcal{R} + \mathcal{T} = 1)$. We assume that the medium filling the cavity is a nonlinear isotropic medium described by nonlinear electric susceptibility of the form

$$\chi(|\mathbf{E}|^2) = \mathrm{i}\alpha + \xi|\mathbf{E}|^2, \tag{10.111}$$

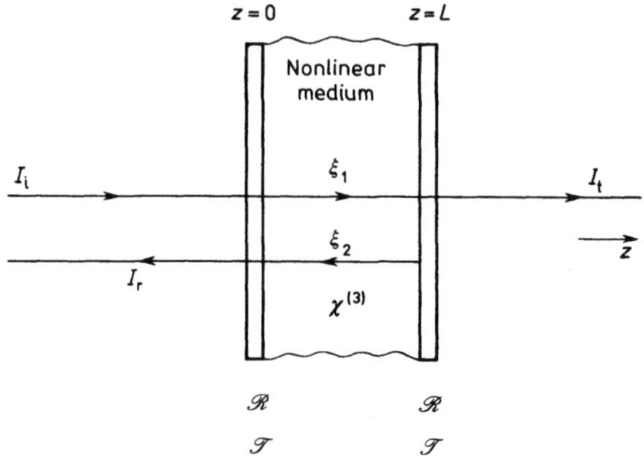

Figure 10.15

where α is the linear absorption coefficient and ξ a real constant which describes nonlinear permittivity.

The magnetic properties of the medium remain linear and are described by the magnetic permeability of the vacuum μ_0.

The electromagnetic waves inside the cavity are described by equation (10.30). Let us suppose that the field incident on the cavity from the side $z < 0$ has the form

$$\mathbf{E}_i = E_0 \mathbf{e}_x \exp[\mathrm{i}(\omega t - k_0 z)], \tag{10.112}$$

and the field inside the cavity may be presented as

$$\mathbf{E}(z,t) = \mathbf{e}_x \{\xi_1(z)\exp[\mathrm{i}(\omega t - kz)] + \xi_2(z)\exp[\mathrm{i}(\omega t + kz)]\}, \tag{10.113}$$

where ξ_1 and ξ_2 are slow-variable functions (10.35) of the variable z. Substitution of (10.113) and (10.112) in equation (10.30) gives, on considering the assumption of the slow-variation of the function $\xi_j(z)$ ($j = 1, 2$), the system of equations

$$\begin{aligned} \frac{\mathrm{d}\xi_1}{\mathrm{d}z} &= \mathrm{i}\gamma[\mathrm{i}\alpha + \zeta(|\xi_1|^2 + 2|\xi_2|^2)]\,\xi_1, \\ \frac{\mathrm{d}\xi_2}{\mathrm{d}z} &= -\mathrm{i}\gamma[\mathrm{i}\alpha + \zeta(2|\xi_1|^2 + |\xi_2|^2)]\,\xi_2, \end{aligned} \tag{10.114}$$

where

$$\gamma = \frac{\omega}{2}\sqrt{\frac{\mu_0}{\varepsilon}}. \tag{10.114a}$$

Each of the amplitudes $\xi_j(z)$ is a complex quantity and may be written in the form

$$\xi_j(z) = \rho_j(z)\exp[\mathrm{i}\varphi_j(z)], \tag{10.115}$$

where $\rho_j(z)$ and $\varphi_j(z)$ are real functions. Substituting (10.115) into equations (10.114) and equating the real and imaginary parts leads to a system of equations which describes the behaviour of the Fabry–Pérot cavity

$$\frac{\mathrm{d}\varphi_1}{\mathrm{d}z} = \gamma\zeta[\rho_1^2 + 2\rho_2^2], \tag{10.116a}$$

$$\frac{\mathrm{d}\varphi_2}{\mathrm{d}z} = -\gamma\zeta[2\rho_1^2 + \rho_2^2], \tag{10.116b}$$

$$\frac{\mathrm{d}\rho_1}{\mathrm{d}z} = -\alpha\gamma\rho_1, \tag{10.116c}$$

$$\frac{\mathrm{d}\rho_2}{\mathrm{d}z} = \alpha\gamma\rho_2. \tag{10.116d}$$

This system of equations implies the following relations (for $\varphi_2(0) = \varphi_1(0) = 0$):

$$\begin{aligned}\rho_1 &= \rho_1(0)\exp[-\alpha\gamma z],\\ \rho_2 &= \rho_2(L)\exp[\alpha\gamma(L-z)],\end{aligned} \tag{10.117a}$$

$$2\Delta\varphi = \varphi_1(L) - \varphi_2(L) = 3\gamma\zeta\int_0^L(\rho_1^2 + \rho_2^2)\mathrm{d}z = 2\kappa I_{\mathrm{eff}}, \tag{10.117b}$$

where L is the thickness of the cavity,

$$\kappa = \sqrt{\frac{\mu_0}{\varepsilon}}\,3\gamma\zeta L.$$

The boundary conditions imply the relations

$$\begin{aligned} \rho_1(0) &= \sqrt{\mathscr{T}}\, E_0 + \sqrt{\mathscr{R}}\, \rho_2(0), \\ \exp[\mathrm{i}\varphi_2(L)]\, \rho_2(L) &= \sqrt{\mathscr{R}}\, \rho_1(L) \exp[\mathrm{i}\varphi_1(L)]. \end{aligned} \tag{10.118}$$

Defining the parameter η which characterizes the absorption of the cavity as

$$\eta = \left| \frac{\rho_1(0)}{\rho_1(L)} \right| = \left| \frac{\rho_2(L)}{\rho_2(0)} \right|, \tag{10.119}$$

we can determine $E_t(L)$ and $E_r(0)$, i.e., the electric field of the wave passing through the cavity

$$E_t(L) = \sqrt{\mathscr{T}}\, \rho_1(L) \exp[\mathrm{i}\varphi_1(L)]. \tag{10.120}$$

The knowledge of the above quantities allows us to calculate the transmittance of the cavity

$$T = \left| \frac{E_t(L)}{E_0} \right|^2 = \frac{\mathscr{T}^2 \eta^2}{(\eta^2 - \mathscr{R})^2 + 4\eta^2 \mathscr{R} \sin^2(\Delta\varphi - kL)}. \tag{10.121}$$

Substituting expressions (10.117a) in (10.117b) and expressing I_{eff} by I_t, we can express $\Delta\varphi$ in the form

$$\Delta\varphi = \nu I_t, \tag{10.122}$$

where

$$\nu = \frac{(n^2 - 1)(\eta^2 + \mathscr{R})}{2\eta^2 \alpha \zeta \mathscr{T} L}. \tag{10.122a}$$

Relation (10.122) enables us to write the transmittance of the cavity in the final form

$$T = \frac{I_t}{I_i} = \frac{\mathscr{T}^2 \eta^2}{(\eta^2 - \mathscr{R})^2 + 4\eta^2 \mathscr{R} \sin^2[\nu I_t - kL]}. \tag{10.123}$$

From (10.123) follows the unambiguous relation $I_i = f(I_t)$, (Fig. 10.16a). The inverse function $I_t = F(I_i)$ (Fig. 10.16b) is a many-valued function, and its diagram has the shape of the letter S. Bistability arises, because to the same value of the incident light

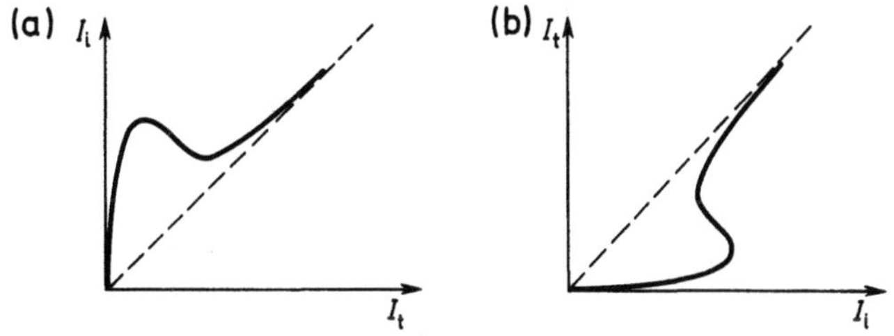

Figure 10.16

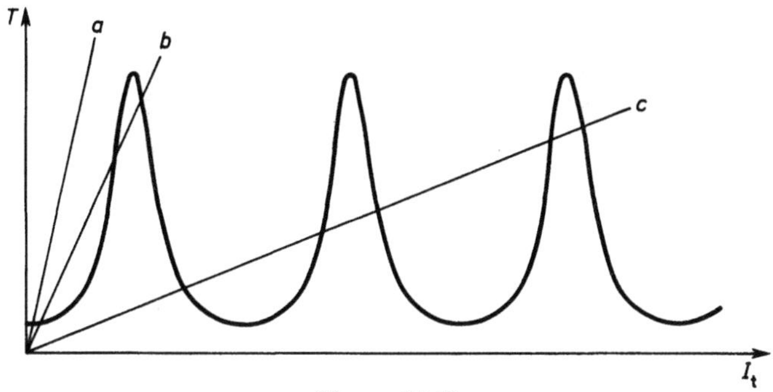

Figure 10.17

intensity can be assigned, in a certain range of I_i, two different values of the transmitted light intensity I_t.

If we make diagram $T = T(I_t)$ by using (10.123), we obtain the curve presented in Fig. 10.17. It is similar to the curve in Fig. 10.12 but it is shifted a distance of νI_t in relation to that curve. Its maxima do not occur for $\omega = \omega_c$ but for $\omega = \omega_c + (1/2\pi)\nu I_t$. Stationary values of the cavity transmittance are obtained from the intersection of curve (10.123) with the straight line

$$T = \frac{I_t}{I_i}. \tag{10.124}$$

The intensity I_i incident on the cavity determines the slope of line (10.124). For small values of I_i (line *a* in Fig. 10.17) there occurs only one point of intersection. For larger values of I_i (line *b*) there occur three points of intersection. For still larger values of I_i (line *c*) we obtain multiple intersection points. For those points the diagram showing the dependene of I_t upon I_i will give the curve from Fig. 10.16b.

It is worthwhile to note the essential role of feedback cases where bistability occurs. In the case of the Fabry–Pérot cavity feedback is realized by reflections from the cavity border-planes. If $\mathscr{T} \to 1$ and $\mathscr{R} \to 0$, curves (10.124) become oblate and bistability does not take place.

Bistabiliy occurs also in the light reflected from the cavity.

10.16. Application of Optical Bistability

Optical bistability can be used in producing various kinds of signal converters. Here are some examples of such applications.

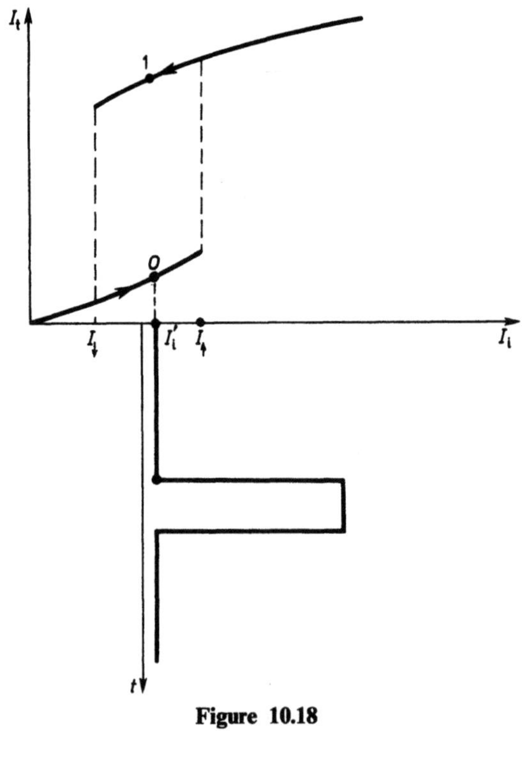

Figure 10.18

A. Bistable Memory Element

A bistable system with hysteresis $I_t = F(I_i)$ (Fig. 10.18) may be used as a memory element. When the incident intensity I_i' remains constant, the system can be in one of the two permissible states of transmittance: the lower or the upper one. Switching from one state to the other may take place in the following way. Transition from a state of high transmittance to one of lower transmittance requires decreasing I_i below $I_\downarrow$. For the reverse transition the intensity I_i should increase above $I^\uparrow$.

B. Optical Transistor

The character of the curves $I_t = F(I_i)$ is determined by the parameters $\mathscr{R}$, $\mathscr{T}$ describing the cavity, the attenuation coefficient

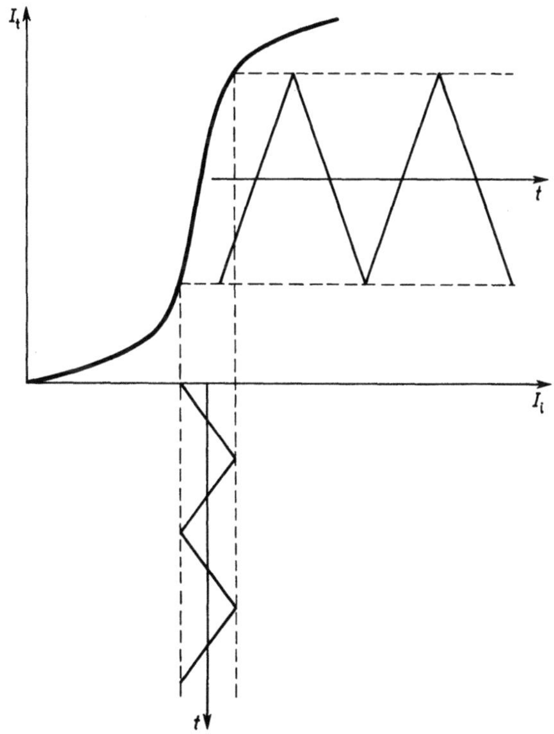

Figure 10.19

η and the cavity length L. Under some conditions [3], the curves $I_t = F(I_i)$ become differential curves (Fig. 10.19) on which

$$\frac{dI_t}{dI_i} > 1.$$

This makes it possible to use these curves for differential amplification of light beams.

If the intensity of the incident light of an average value I_g is modulated, the modulation will be handed over to the transmitted wave. The modulation depth of the transmitted intensity will be greater than that occurring in the incident light (Fig. 10.19). The rate of modulation of the incident light should be comparatively slow, so that the system could move along the fixed state curve.

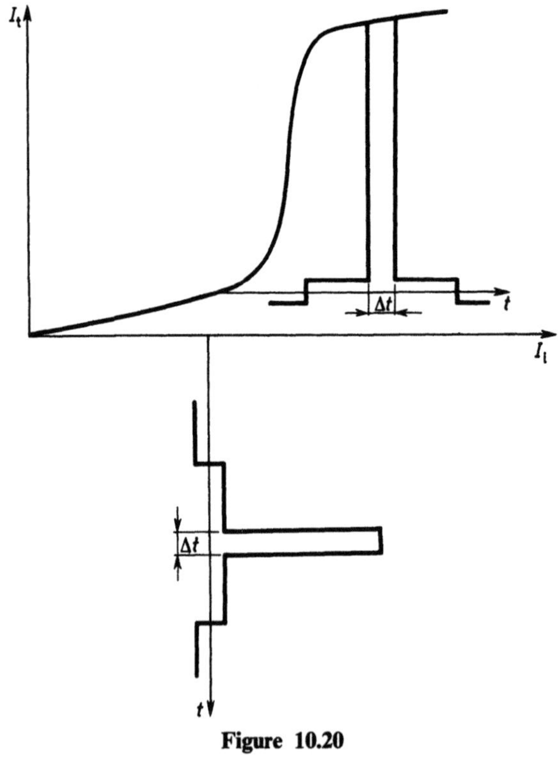

Figure 10.20

C. Forming and Limiting Light Impulses

The transmission characteristic presented in Fig. 10.20 may be applied in forming the shape of a light impulse. The components of the input signal which are smaller than $I_{\downarrow}$ decrease passing through the system described by the curve in Fig. 10.20. Those impulse components which are greater than $I^{\uparrow}$ are transmitted with an insignificant decrease.

To restrict the height of the impulses we make use of the fact that the upper branch of the transmission characteristic (Fig. 10.20) is almost a flat curve. Owing to this, on reaching the state of "high transmission" the impulse power emerging from the system remains constant for a wide range of values of the input flux power (Fig. 10.21).

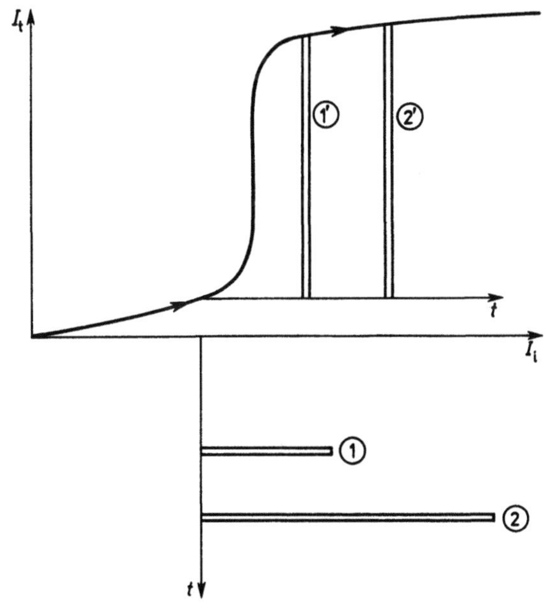

Figure 10.21

10.17. Nonlinear Reflective Phenomena

A. Generation of the Second Harmonic at the Boundary of a Nonlinear Medium

We consider a situation where a nonlinear medium with square nonlinearity fills up the half-space $z > 0$. Upon the border of this medium $z = 0$ falls from the region $z < 0$ a plane wave with a wave vector in the plane $y = 0$ and the plane polarization

$$\mathbf{E}_{\mathrm{i}} = \mathbf{e}_y E_0 \exp\left[\mathrm{i}(\omega t - k_0 x \sin\alpha - k_0 z \cos\alpha)\right]. \tag{10.125}$$

Suppose that the nonlinear medium is within a linear range an anisotropic uniaxial medium whose optical axis coincides with the z-axis. In this situation the wave refracted in the crystal will be the ordinary wave. Its wave vector $\mathbf{k}_1^T$ is determined by the Snell law, and the amplitude $\mathbf{E}_1^T$ is calculated from the Fresnel formulas. The refracted wave has the frequency of the incident wave ω. It produces in the medium (for sufficiently high intensity) nonlinear polarization of the form

$$\mathbf{P}^{(\mathrm{NL})} = \chi^{(2)}(2\omega) : \mathbf{E}_1^T \mathbf{E}_1^T \exp\left[-\mathrm{i}(2\mathbf{k}_1^T \mathbf{r} - 2\omega t)\right]. \tag{10.126}$$

This nonlinear polarization becomes the source of the field of the second harmonic, which can be found by means of equation (10.29).

The general solution of this equation, which allows for (10.126), is the sum of the general solution of the homogeneous equation and a particular solution of the non-homogeneous equation. It has the form

$$\begin{aligned} \mathbf{E}_2^T &= \mathbf{e}_T \xi_2^T \exp\left[\mathrm{i}(\mathbf{k}_2^T \cdot \mathbf{r} - 2\omega t)\right] + \\ &+ \frac{\mu_0 P_0^{(\mathrm{NL})} 4\omega^2 \left[\mathbf{p} - 4(k_2^T)^{-2} \mathbf{k}_1^T (\mathbf{p} \cdot \mathbf{k}_1^T)\right]}{4(k_1^T)^2 - (k_2^T)^2} \exp\left[\mathrm{i}2(\mathbf{k}_1^T \cdot \mathbf{r} - \omega t)\right], \end{aligned} \tag{10.127}$$

where $k_1^T = \omega\sqrt{\mu_0 \varepsilon(\omega)}$, $k_2^T = 2\omega\sqrt{\mu_0 \varepsilon(2\omega)}$, and $\mathbf{p}$ is the versor which determines the direction of the vector $\mathbf{P}^{(\mathrm{NL})}$, $P_0^{(\mathrm{NL})} = \left|\mathbf{P}^{(\mathrm{NL})}\right|$.

(The occurrence of the relations $\varepsilon(\omega)$, $\varepsilon(2\omega)$ is accounted by the dispersive properties of the medium.)

The magnetic field of the second harmonic conjugate with (10.127) has the form

$$\mathbf{H}_2^T = \frac{1}{2\omega\mu_0}(\mathbf{k}_2^T \times \mathbf{e}_T)\xi_2^T \exp\left[\mathrm{i}(\mathbf{k}_2^T \cdot \mathbf{r} - 2\omega t)\right] + \\ + \frac{4\omega P_0^{(\mathrm{NL})}}{4(k_1^T)^2 - (k_2^T)^2}(\mathbf{k}_1^T \times \mathbf{p})\exp\left[\mathrm{i}2(\mathbf{k}_1^T \cdot \mathbf{r} - \omega t)\right]. \tag{10.128}$$

In the region $z < 0$ equation (10.29) has a solution of the form

$$\mathbf{E}_2^R = \mathbf{e}_R \xi_2^R \exp\left[\mathrm{i}2(\mathbf{k}_1^R \cdot \mathbf{r} - \omega t)\right], \\ \mathbf{H}_2^R = \frac{1}{\omega\mu_0}(\mathbf{k}_1^R \times \mathbf{e}_R)\xi_2^R \exp\left[\mathrm{i}2(\mathbf{k}_1^R \cdot \mathbf{r} - \omega t)\right]. \tag{10.129}$$

The directions of the wave vectors $2\mathbf{k}_1^{R,T}$, $\mathbf{k}_2^{R,T}$, the polarization vectors $\mathbf{e}_R$, $\mathbf{e}_T$, and also the values of the amplitudes ξ_2^R and ξ_2^T are determined from the boundary conditions. (The tangent components of the vectors **E** and **H** must be continuous at the boundary of the medium at any place and in any time.)

The boundary conditions are fulfilled when the following relations hold:

$$k_0 \sin\alpha = k_{1x}^R = k_{1x}^T \quad \text{for} \quad \omega, \\ 2k_{1x}^T = k_{2x}^R = k_x^T \quad \text{for} \quad 2\omega. \tag{10.130}$$

Relations (10.130) enable us to find the angles which the appropriate wave vectors form with the normal to the boundary plane $z = 0$:

$$\sin\theta_2^R = \sin\alpha, \tag{10.131a}$$

$$\sin\theta_2^T = \frac{\sin\alpha}{\sqrt{\varepsilon_r(2\omega)}}, \tag{10.131b}$$

$$\sin\theta_2^p = \frac{\sin\alpha}{\sqrt{\varepsilon_r(\omega)}}, \tag{10.131c}$$

where θ_2^T is the angle of the second harmonic homogeneous wave, θ_2^R the angle of the reflected second harmonic, and θ_2^p the angle for a wave induced by nonlinear polarization.

As can be seen from the above, a wave produced by nonlinear polarization propagates in the same direction as the refracted wave with frequency ω. The homogeneous wave with frequency 2ω has in general a different direction of propagation. The wave vectors for a wave with nonlinear polarization and for a homogeneous wave are parallel only when $\varepsilon(\omega) = \varepsilon(2\omega)$, or when the external wave falls perpendicularly upon the plane $z = 0$. The situation is illustrated in Fig. 10.22.

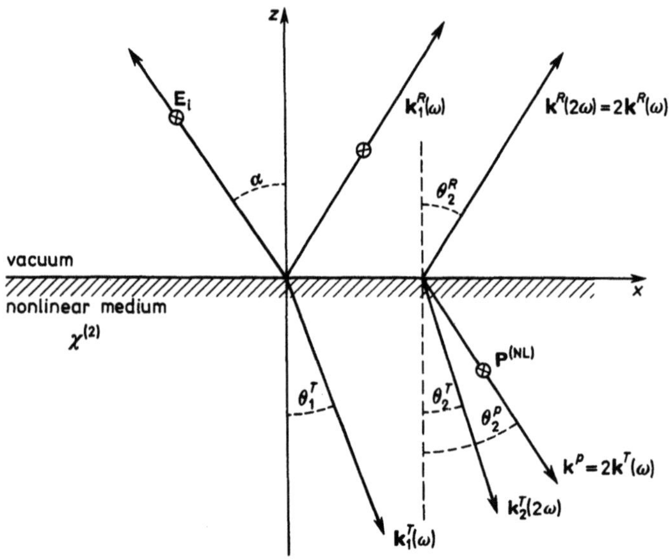

Figure 10.22

If a nonlinear medium belongs to the trigonal system (C_3), then for the incident wave (10.125) nonlinear polarization has the versor $\mathbf{p} = \mathbf{e}_y$. The continuity conditions for the tangent components E_{2y} and H_{2x} in the plane $z = 0$ will be fulfilled if we assume $\mathbf{e}_T = \mathbf{e}_R = \mathbf{e}_y$.

Then the continuity of the E_{2y} component gives the equality

$$\xi_2^T + \frac{P_0^{(\mathrm{NL})}}{\varepsilon(\omega) - \varepsilon(2\omega)} = \xi_2^R, \tag{10.132a}$$

and the continuity of the component H_{2x} gives

$$\xi_2^T \sqrt{\varepsilon_r(2\omega)} \cos\theta_2^T + \frac{P_0^{(\mathrm{NL})}}{\varepsilon(\omega) - \varepsilon(2\omega)} \cos\theta_2^g = -\xi_2^R \cos\theta_2^R. \tag{10.132b}$$

From (10.132) we obtain

$$\xi_2^R = \frac{P_0^{(\mathrm{NL})}\left[\sqrt{\varepsilon_r(2\omega)}\cos\theta_2^T - \sqrt{\varepsilon_r(\omega)}\cos\theta_2^g\right]}{\left[\varepsilon(2\omega) - \varepsilon(\omega)\right]\left[\sqrt{\varepsilon_r(2\omega)}\cos\theta_2^T + \cos\theta_2^R\right]}, \tag{10.133a}$$

$$\xi_2^T = \frac{-P_0^{(\mathrm{NL})}}{\varepsilon(2\omega) - \varepsilon(\omega)} + \xi_2^R. \tag{10.133b}$$

In order to find the dependence of the reflected wave with the frequency 2ω upon the incident angle of the primary wave with frequency ω we have to find $P_0^{(\mathrm{NL})}$. This can be done by using the Fresnel formula for a linear medium in order to find the amplitude of the refracted wave with primary frequency. However, we should remember that results (10.127) together with (10.133) remain in force only for that range of the variable z in which the power of the wave with frequency 2ω remains small. When this power increases, we must take into account a reverse process — an energy conversion from the wave with frequency 2ω to the wave with frequency ω, and then the constant field approximation is no longer applicable. Thus the results obtained concern only the primary process of generating a higher harmonic.

B. *Reflection and Refraction of a Wave on the Surface of a Medium with Cubic Nonlinearity*

Suppose that the half-space $z < 0$ is filled up with an ordinary linear dielectric with the refractive index n_0, while the positive half-space ($z > 0$) with a nonlinear material whose refractive index has the linear component $n_L = n_0 - \Delta$ ($\Delta > 0$); for high intensities this refractive index is described by the relation

$$n = n_0 - \Delta + n_2|E|^2, \tag{10.134}$$

where E is the intensity of the electric field of the wave in that medium.

If from the region $z < 0$ there falls a plane wave upon the boundary plane at an angle $\theta > \arcsin(1 - \Delta/n_0)$, then in the case of low intensity there will be total internal reflection. There is no refracted wave in the area $z > 0$, but there is a field decaying exponentially with the increase of the distance from the plane $z = 0$. For $n_2 > 0$ the field increases the initial value of the refractive index in the area $z > 0$ together with increasing the intensity of the incident wave. Then the difference between the refractive indices on both sides of the boundary plane becomes smaller, increasing the effective angle of the total internal reflection, so that it approaches the incident angle θ. This in turn will cause the growth of the field in the area $z > 0$ and the appearence of feedback. It may be expected that for a certain critical intensity of the incident wave there will be a sudden transition from a state of total internal reflection to a state in which part of the energy of the field will begin to propagate inside the nonlinear area (Fig. 10.23). For this to happen the following condition must be fulfilled:

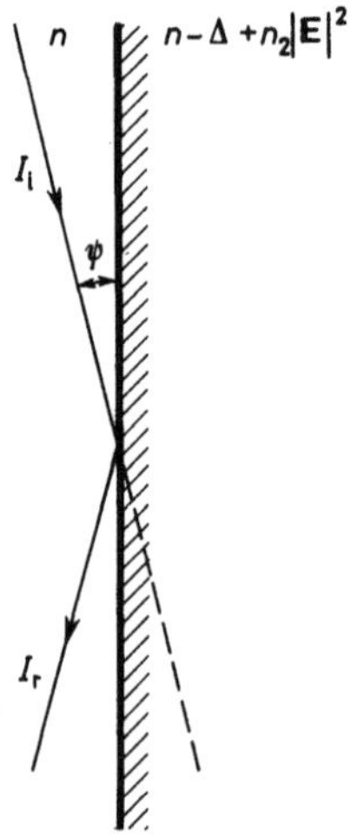

Figure 10.23

$$-\Delta + n_2|E|^2 = 0. \tag{10.135}$$

We can determine the critical intensity I_i^{cr} of the incident wave under which for a given incident angle θ there is a transition from the state of total reflection to the state of partial transmission of the wave into the region $z > 0$.

To this end we introduce, for the sake of convenience, the angle $\psi = \pi/2 - \theta$. In the case of ordinary reflection, where in the medium filling the region $z > 0$ the refractive wave has the nature of a wave propagating at an angle ψ_1 in relation to the boundary plane, the Snell law can be written in the form

$$\frac{\cos\psi}{\cos\psi_1} = \frac{n_0 - \Delta + n_2|E|^2}{n_0}. \tag{10.136}$$

On the assumptions

$$\begin{aligned} &\left|-\Delta + n_2|E|^2\right| \ll n_0, \\ &\psi \ll \frac{\pi}{2}, \end{aligned} \tag{10.136a}$$

for the coefficients of transmittance τ and reflection r, we can write (from Fresnel's formulas) for both states of wave polarization the expressions

$$\tau = \frac{E}{E_0} = \frac{2\psi}{\psi + \psi_1}, \tag{10.137a}$$

$$r = \frac{-\psi_1 + \psi}{\psi_1 + \psi}, \tag{10.137b}$$

where E_0 is the intensity of the incident wave and E – of the refracted wave.

From (10.137a) we obtain

$$E = E_0 \frac{2\psi}{\psi + \psi_1}. \tag{10.138}$$

Substituting in formula (10.136) expression (10.138) for E and replacing the cosines in it by the expressions

$$\cos\psi = 1 - \tfrac{1}{2}\psi^2, \tag{10.139}$$

we arrive at the relation

$$(\psi + \psi_1)^2\left[\tfrac{1}{2}(\psi_1^2 - \psi^2) + \frac{\Delta}{n_0}(1 - \tfrac{1}{2}\psi_1^2)\right] + \\ + \frac{n_2}{n_0}4|E_0|^2\psi^2(1 - \tfrac{1}{2}\psi_1^2) = 0. \tag{10.140}$$

Hence, disregarding the lower order quantities, we obtain the "nonlinear Snell law"

$$(\psi + \psi_1)^2\left[(\psi_1^2 - \psi^2) + \frac{2\Delta}{n_0}\right] - 8\frac{n_2}{n_0}|E_0|^2\psi^2 = 0. \tag{10.141}$$

From formula (10.136), by $\psi_1 = 0$, replacing $\cos\psi$ by (10.139), and disregarding $n_2|E|^2$, we obtain for $\psi_{cr} = -\pi/2 - \theta_{lim}$ the expression

$$\psi_{cr} = \sqrt{\frac{2\Delta}{n_0}}. \tag{10.142}$$

From (10.141), setting $\psi_1 = 0$ and taking into account (10.142), we can find the critical wave intensity E_0^{cr}, which for a given angle of incidence ψ produces a transition from the state of total reflection to the state of partial transmittance. This intensity is

$$\frac{n_2|E_0^{cr}|^2}{\Delta} = \frac{1}{4}\left[1 - \left(\frac{\psi}{\psi_{cr}}\right)^2\right]. \tag{10.143}$$

Applying (10.137b), we can rewrite equality (10.141) in the form

$$r\psi^2 - (1 + r)^2\frac{\Delta}{2n_2} + \frac{n_2}{2n_0}|E_0|^2(1 + r)^4 = 0, \tag{10.144}$$

which can finally be written as

$$\frac{n_2|E_0|^2}{\Delta} = \frac{1}{(1 + \mathrm{r})^2}\frac{n_0}{n_2}\left[1 - \frac{4\frac{n_2}{n_0}r}{(1 + r)^2}\left(\frac{\psi}{\psi_{cr}}\right)^2\right]. \tag{10.145}$$

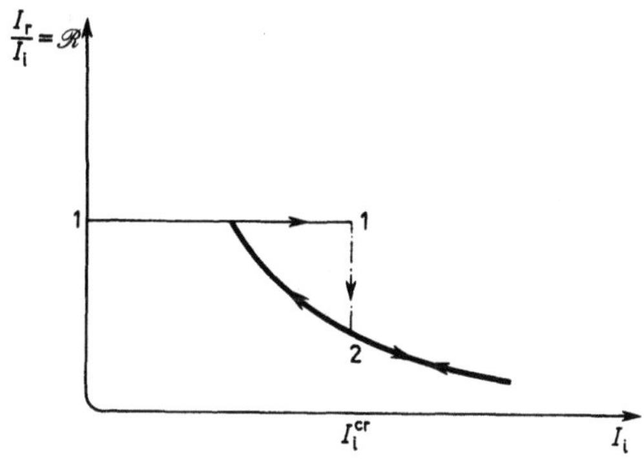

Figure 10.24

Figure 10.24 shows the dependence of the calculated coefficient of reflection $\mathscr{R} = r^2$ on the intensity of the incident light I_i. We can conclude that there is a limited range of values I_i for which, the angle ψ with respect to the boundary plane being given, there is total internal reflection. Then comes a transitory region, in which a transmission wave is present (vertical segment 1–2 in the diagram). As I_i decreases, the system will return to the state of total internal reflection along curve 2, reaching that primary state when $I_i < I_i^{cr}$. There appears a hysteresis curve, which suggest the possibility of the occurrence of bistability. This effect, however, is possible only for an incident plane wave. If the light incident on the boundary plane is a Gaussian beam, then the return to the state of total internal reflection may be different from that shown in Fig. 10.24.

The phenomenon discussed above makes it possible to construct a directional switch of light controlled by its intensity.

Bibliography

1. Bloembergen, N.: (1965), 'Nonlinear Optics', W.A. Benjamin, New York.
2. Fisher, R.A.: (1983), 'Optical Phase Conjugation', Academic Press, New York.
3. Gibbs, H.M.: (1985), 'Optical Bistability: Controlling Light with Light', Academic Press, New York.
4. Hellwarth, R.W.: (1977), 'Progress in Quantum Electronics', Vol. 5, pp. 1–68, Pergamon Press, Oxford.

5. Kaplan, A.E.: (1977), *Sev. Phys. JETP* **45**, 896.
6. Lugiate, L.A.: (1984), 'Theory of Bistability', in: 'Progress in Optics', Vol. XXI, p. 69, Wolf, E. (ed.), North-Holland, Amsterdam.
7. Miller, D.A.B.: (1981), *IEEE Quantum Electronics* **QE–17**, p. 306.
8. Perina, J.: (1984), 'Quantum Statistics of Linear and Nonlinear Optical Phenomena', D. Reidel Publishing Company, Dordrecht.
9. Schubert, M. and Wilhelmi, B.: (1971), 'Einführung in die Nichtlineare Optik', G. Teubner Verlag, Leipzig.
10. Shen, Y.R.: (1986), 'Principles of Nonlinear Optics', John Wiley, New York.
11. Yariv, A.: (1973), 'Quantum Electronics', John Wiley, New York.

11. Waveguide Optics

11.1. Waveguide Properties of a Medium whose Refractive Index Changes Axially

We consider a non-homogeneous medium whose refractive index varies according to the relation

$$n^2 = n_0^2\left(1 - \frac{\rho^2}{a}\right) + 1, \tag{11.1}$$

where ρ is the distance from the axis of non-homogeneity of the medium, while n_0 and a are constant values. The change of the refractive index described by (11.1) takes place in the region where $\rho \leq \sqrt{a}$, and for greater values of ρ the refractive index is constant and equals unity.

It will be shown below that in a medium of this kind light waves, which are solutions of Maxwell's equations, have the form of Gaussian beams in which there is a periodical dependence of the crosswise distribution of the amplitude on the coordinate z, this being the direction in which the waves propagate. The field of these waves converges about the axis of non-homogeneity and for $\rho > \sqrt{a}$ becomes practically zero.

The possibility of such spatial restriction of the region in which electromagnetic waves propagate, conditioned by the spatial distribution of the dielectric properties of the medium, form the basis of so called *waveguide optics*. Dielectric waveguides may have different cross-sections: there are cylindrical waveguides used as transmission lines, there are also planar waveguides, and stripped waveguides, which serve as elements of coding and decoding information carried by light waves. Waveguide optics describes a number of interesting physical effects specific only for that branch of physics.

To determine the field of a monochromatic wave in a medium described by the refractive index (11.1) we use the Helmholtz equation

$$\nabla^2\mathbf{E} + k_0^2 n^2 \mathbf{E} = \mathbf{0}, \tag{11.2}$$

where $k_0 = \omega c^{-1}$.

We seek a solution of this equation in the form

$$E_j = A_j(\rho, z)\exp(-\mathrm{i}k_0 z), \tag{11.3}$$

where j (= 1, 2, 3) stands for the Cartesian component of the field.

Let us suppose that the functions $A_j(\rho, z)$ are slow-variable functions of variable z.

Putting (11.3) in equation (11.2) given in cylindrical coordinates, and disregarding $\dfrac{\partial^2 A_j}{\partial z^2}$ (with respect to $k_0 \dfrac{\partial A_j}{\partial z}$ and $k_0^2 A_j$), we obtain the simplified equation

$$\frac{1}{\rho}\frac{\partial A_j}{\partial \rho} + \frac{\partial^2 A_j}{\partial \rho^2} - 2\mathrm{i}k_0 \frac{\partial A_j}{\partial z} + k_0^2 n_0^2\left(2 - \frac{\rho^2}{a}\right) A_j = 0. \tag{11.4}$$

The solution of (11.4) is sought in the form

$$A_j(\rho, z) = \exp\left[\mathrm{i}P(z) - \tfrac{1}{2}\mathrm{i}\rho^2 Q(z)\right], \tag{11.5}$$

where the functions $P(z)$ and $Q(z)$ remain unknown.

Substituting (11.5) in (11.4) and comparing the coefficients of the same powers of ρ, we obtain two equations for the unknown functions $P(z)$ and $Q(z)$

$$Q^2 + k_0 \frac{\mathrm{d}Q}{\mathrm{d}z} + n_0^2 \frac{1}{a} k_0^2 = 0, \tag{11.6a}$$

$$\mathrm{i}Q - k_0 \frac{\mathrm{d}P}{\mathrm{d}z} - n_0^2 k_0^2 = 0. \tag{11.6b}$$

By substituting

$$Q(z) = \frac{\mathrm{i}k_0 n}{a} + \frac{1}{u}, \tag{11.7}$$

we reduce equation (11.6a) to the form

$$\frac{\mathrm{d}u}{\mathrm{d}z} - \frac{2in_0}{\sqrt{a}}u - \frac{1}{k_0} = 0. \tag{11.8}$$

The solution of (11.8) is the function

$$u = C_0 \exp\left(2in_0 \frac{z}{\sqrt{a}} + \mathrm{i}\beta\right) + \frac{\mathrm{i}\sqrt{a}}{2k_0 n_0}, \tag{11.9}$$

where C_0 and β are the constants of integration.

Returning to the function $Q(z)$ (11.7), we obtain

$$Q(z) = \left[\frac{4C_0k_0n_0\sqrt{a}\cos\left(2n_0\frac{z}{\sqrt{a}} + \beta\right) + \mathrm{i}(4n_0^2k_0^2C_0^2 - a)}{4C_0k_0n_0\sqrt{a}\sin\left(2n_0\frac{z}{\sqrt{a}} + \beta\right) + (4n_0^2k_0^2C_0^2 + a)}\right]\frac{k_0n_0}{\sqrt{a}}. \tag{11.10}$$

It is a complex quantity, which can be written as

$$Q(z) = Q^r(z) + \mathrm{i}Q^i(z). \tag{11.11}$$

Since the required form of the solution is a Gaussian beam, the functions $Q^r(z)$ and $Q^i(z)$ should determine the curvature of the wave front $R(z)$

$$R(z) = \frac{k_0}{Q^r(z)}, \tag{11.12}$$

and the square of the width of the beam $w^2(z)$

$$w^2(z) = -\frac{2}{Q^i(z)}. \tag{11.13}$$

If we assume that for $z = 0$ the wave front is a plane (i.e. $R(0) = \infty$), we should equate β with the right angle: $\beta = \pi/2$. Putting $z = 0$ in (11.13), we obtain the relation between the constant C_0 and the given primary width of the beam $w^2(0)$:

$$C_0 = \frac{w^2(0)k_0n_0 - 2\sqrt{a}}{2a + k_0n_0w^2(0)}\frac{\sqrt{a}}{2k_0n_0}. \tag{11.14}$$

All this enables us to write an explicit and unambigous form of the function $Q(z)$

$$Q(z) = \left[\frac{[4a - w^4(0)k_0^2n_0^2]\sin(\gamma z) - i4w^2(0)k_0n_0\sqrt{a}}{[w^4(0)k_0^2n_0^2 - 4a]\cos(\gamma z) + [w^4(0)k_0^2n_0^2 + 4a]}\right]\frac{k_0n_0}{\sqrt{a}}, \tag{11.15}$$

where $\gamma = 2n/\sqrt{a}$.

The function $P(z)$ can be found from equation (11.6b) after using (11.15) and the boundary condition $P(0) = 0$. This gives

$$\begin{aligned} P(z) = & -k_0n_0^2z + \\ & + i\tfrac{1}{2}\ln\left|[w^4(0)k_0^2n_0^2 - 4a]\cos(\gamma z) + [w^4(0)k_0^2n_0^2 + 4a]\right| + \\ & + \arctan\left[\frac{2\sqrt{a}\tan(\frac{1}{2}\gamma z)}{w^2(0)k_0n_0}\right] - i\tfrac{1}{2}\ln\left|2[w^4(0)k_0^2n_0^2]\right|. \end{aligned} \tag{11.16}$$

As follows from (11.5), (11.15) and (11.16) the explicit form of $A_j(\rho, z)$ may be written as

$$\begin{aligned} A_j(\rho, z) = & \exp\left\{-i\left[n_0^2k_0z - \arctan\left(\frac{2\sqrt{a}\tan(\frac{1}{2}\gamma z)}{w^2(0)k_0n_0}\right) + \frac{\rho^2}{2}Q^r(z)\right]\right\} \times \\ & \times \exp\left[\frac{\rho^2}{2}Q^i(z)\right] \times \\ & \times \left\{\frac{2[w^4(0)k_0^2n_0^2]}{[w^4(0)k_0^2n_0^2 - 4a]\cos\gamma z + [w^4(0)k_0^2n_0^2 + 4a]}\right\}^{1/2}. \end{aligned} \tag{11.17}$$

From the form $A_j(\rho, z)$ we can find:

(1) the width of "throat" of the beam:

$$w^2(z) = \frac{[w^4(0)k_0^2n_0^2 - 4a]\cos(\gamma z) + [w^4(0)k_0^2n_0^2 + 4a]}{2w^2(0)k_0^2n_0^2}, \tag{11.18}$$

(2) the radius of curvature of the wave front

$$R(z) = \frac{\sqrt{a}\{[w^4(0)k_0^2n_0^2 - 4a]\cos(\gamma z) + [w^4(0)k_0^2n_0^2 + 4a]\}}{n_0[4a - w^4(0)k_0^2n_0^2]\sin(\gamma z)}. \tag{11.19}$$

From (11.18) it follows that the width of the "throat" of the beam attains its boundary value at points of coordinates $z = 2\pi m/\gamma$ $(m = 0, 1, 2...)$. In view of (11.19) the radius of the curvature of the wave front becomes infinite for $z = \pi m/\gamma$.

The amplitude of the wave $A_j(\rho, z)$ is also subjected to periodical changes, which accompany the changes of the width of the beam.

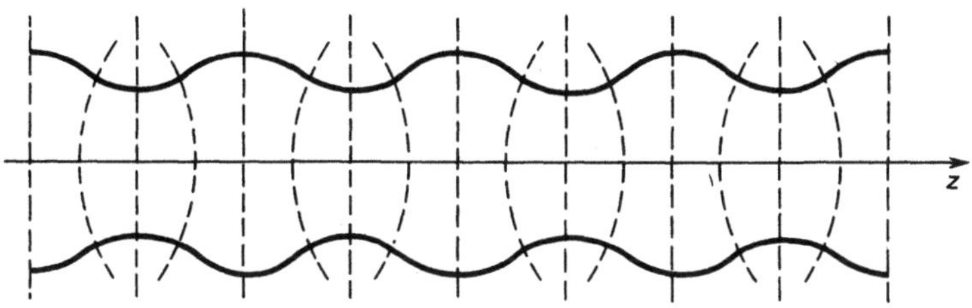

Figure 11.1

The above analysis shows that an electromagnetic wave which propagates in a medium with the refractive index (11.1) undergoes variations and periodically reconstructs its original form. The situation is presented in Fig. 11.1, where the continuous lines show the change of the width of the beam, and the dashed lines – the surface of the constant phase. The direction of propagation of the beam coincides with the positive direction of the z-axis.

In multilayer structures parallel to the direction of propagation of the wave, such as waveguides, the wave front undergoes a delay, which is proportional to the local value of the refractive index. This leads to a deformation of the front and to a tendency of the light to focus in the region of the greatest refractive index. This enables us to formulate an important rule for the qualitative analysis of waveguide systems, the so-called *law of the greatest refractive index*:

In a multilayer structure, such as a waveguide, parallel to the direction of propagation of the wave, light propagates in the region where the refractive index is the greatest or, in other words, where light velocity is the lowest.

11.2. Spatial Structure of the Electromagnetic Field in a Planar Dielectric Waveguide

We assume that the waveguide structure under consideration consists of three areas separated by the planes $x = 0$ an $x = W$. The material media filling these regions have magnetic permeability equal to that of the vacuum μ_0, but they differ in their electric permittivity. The dielectric constant is ε_c in the region $x > W$, ε_f in the region $0 < x < W$, and ε_s in the region $x < 0$. We assume that the above quantities are bound by the relation

$$\varepsilon_c < \varepsilon_s < \varepsilon_f. \tag{11.20}$$

Assuming the isotropy of all the three media, we choose the direction of the axes of the Cartesian coordinate system in such a way that the wave vectors of the wave fields in question remain perpendicular to the y-axis. We shall consider only the fields dependent on the spatial variables x, z and monochromatically variable in time.

Under these assumptions Maxwell's equations in the regions free of conduction currents and of charges can be divided into two independent subsystems. The first of them connects only the components E_y with B_x and B_y, and H_x with H_y and D_y:

$$\frac{\partial E_y}{\partial z} = \dot{B}_x, \qquad \frac{\partial E_y}{\partial x} = -\dot{B}_z, \qquad \frac{\partial H_x}{\partial z} - \frac{\partial H_z}{\partial x} = \dot{D}_y. \tag{11.21}$$

The second subsystem relates the components E_x and E_z with B_y, and D_x and D_z with H_y:

$$\frac{\partial E_x}{\partial z} - \frac{\partial E_z}{\partial x} = -\dot{B}_y, \quad \frac{\partial H_y}{\partial z} = -\dot{D}_x, \quad \frac{\partial H_y}{\partial x} = \dot{D}_z. \tag{11.22}$$

These two subsystems of equations are not related with each other, which enables us to state that when the electromagnetic field is independent of one of the spatial variables, it can be divided for two independent fields. The first of the subsystem (11.21) contains only the y-component of the electric field; the vector $\mathbf{E}$ is perpendicular to the z-axis. A field of this type is called a *TE field.* The second subsystem (11.22) contains only the y-th component of the magnetic field. The magnetic vector of the field described by the subsystem (11.22) is perpendicular to the z-axis. This field is called a *TM field.*

The eigenfunctions of the appropriate subsystems (11.21) and (11.22) which fulfil the conditions arising from the requirement of the continuity of the tangent components of the electric field and the magnetic field at the interface of different dielectric media, are called *TE* or *TM modes.* All the *TE* and *TM* modes form a complete set of eigenfuctions for a field dependent of two of the spatial variables. An arbitrary field in a waveguide structure may be presented as a superposition of mode fields.

A. TE Modes

TE modes are the eigenfunctions of system (11.21) or of the equation

$$\frac{\partial^2 E_y}{\partial x^2} + \frac{\partial^2 E_y}{\partial z^2} + \omega^2 \mu_0 \varepsilon(x) E_y = 0, \tag{11.23}$$

where

$$\varepsilon(x) = \begin{cases} \varepsilon_c & x > W, \\ \varepsilon_f & 0 < x < W, \\ \varepsilon_s & x < 0. \end{cases} \tag{11.23a}$$

In order to find the TE modes for a plane waveguide it is necessary to find the solutions of equation (11.23) for each of three adjoining media and to interrelate them through the boundary conditions at their interfaces, i.e., in the planes $x = 0$ and $x = W$.

We can divide TE modes into subgroups.

– The first of the subgroups comprises the waves confined in the layer $0 < x < W$. These modes are called *discrete* or *guided* modes. They are eigenfunctions of Maxwell's equations, and for $|x| \to \infty$ they show an exponential decay of the field vectors outside the layer (Fig. 11.2a).

– The second subgroup comprises the waves which can propagate across the planes confining the waveguide layer. They are the modes of emission to the ground ($x < 0$) and to the covering ($x > W$) (Fig. 11.2b, c).

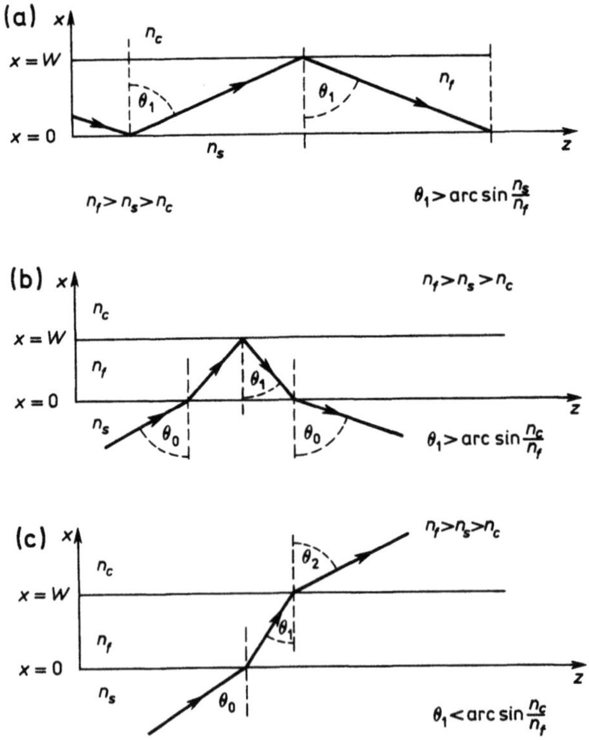

Figure 11.2

TE discrete modes. Discrete modes or waveguide modes are waves imprisoned within the layer, which happens as the result of their total internal reflections from the boundary planes. In the case of total internal reflection the electromagnetic field in regions with a lower refractive index is not equal to zero but decreases exponentially with the increase of the distance from the interface.

The solution of equation (11.23) satisfying this demand has the form

$$E_y(x,z) = \begin{cases} A\exp(\delta x)\exp(-\mathrm{i}\beta z) & \text{for } x<0, \\ [A\cos(\kappa x) + B\sin(\kappa x)]\exp(-\mathrm{i}\beta z) & \text{for } 0<x<W, \\ [A\cos(\kappa W) + B\sin(\kappa W)]\exp[\gamma(W-x)]\exp(-\mathrm{i}\beta z) & \\ & \text{for } x>W, \end{cases} \qquad (11.24)$$

where the parameters δ, γ and κ are defined by the relations

$$\delta = \sqrt{\beta^2 - \omega^2\mu_0\varepsilon_s}, \qquad (11.24a)$$

$$\gamma = \sqrt{\beta^2 - \omega^2\mu_0\varepsilon_c}, \qquad (11.24b)$$

$$\kappa = \sqrt{\omega^2\mu_0\varepsilon_f - \beta^2}. \qquad (11.24c)$$

The parameter β is called the *propagation constant* and physically it is the z-th component of the wave vector of the imprisoned wave. The quantity $\omega^2\mu_0\varepsilon_l$ $(l = c, f, s)$ is the square of the wave number of the wave in the l-th medium.

The knowledge of E_y enables us, after applying the system (11.21), to determine the z-th component of the magnetic field

$$H_z = \begin{cases} \dfrac{\mathrm{i}}{\omega\mu_0}\delta A\exp(\delta x)\exp(-\mathrm{i}\beta z) & \text{for } x<0, \\ \dfrac{-\mathrm{i}}{\omega\mu_0}\kappa[A\sin(\kappa x) - B\cos(\kappa x)]\exp(-\mathrm{i}\beta z) & \\ & \text{for } < x < W, \\ \dfrac{-\mathrm{i}}{\omega\mu_0}\gamma[A\cos(\kappa W) + & \\ \quad + B\sin(\kappa W)]\exp[\gamma(W-x)]\exp(-\mathrm{i}\beta z) & \\ & \text{for } x>W. \end{cases} \qquad (11.25)$$

The condition of continuity for the component H_z in the planes $x = 0$ and $x = W$ leads to the following system of equations:

$$\begin{gathered}\delta A + \kappa B = 0, \\ [\kappa \sin(\kappa W) - \gamma \cos(\kappa W)]A + [\kappa \cos(\kappa W) + \gamma \sin(\kappa W)]B = 0.\end{gathered} \tag{11.26}$$

This homogeneous system of equations has nonzero solutions for A and B only if its determinant vanishes. This condition implies the equation for the permissible values of propagation constants β. It is the so-called *dispersion equation for a planar waveguide.* Its form is

$$\tan(\kappa W) = \frac{\kappa(\delta + \gamma)}{\kappa^2 - \delta\gamma}. \tag{11.27}$$

All values of the parameter β which satisfy this equality determine the value of the z-th component of the wave vector of those waves for which there is total internal reflection from the planes bounding the layer.

Indeed, we can show that it is possible to interpret the wave field of the guided mode as a superposition of two plane waves.

To this aim we use the identities

$$\begin{gathered}\sin(\kappa x) = \frac{1}{2\mathrm{i}}[\exp(\mathrm{i}\kappa x) - \exp(-\mathrm{i}\kappa x)], \\ \cos(\kappa x) = \frac{1}{2}[\exp(\mathrm{i}\kappa x) + \exp(-\mathrm{i}\kappa x)],\end{gathered} \tag{11.28}$$

with the help of which the electromagnetic waves inside the layer $0 < x < W$ (see 11.23) can be presented as

$$\begin{aligned}E_y = {} & \frac{A - \mathrm{i}B}{2}\exp(\mathrm{i}\kappa x - \mathrm{i}\beta x) + \\ & + \frac{A + \mathrm{i}B}{2}\exp(-\mathrm{i}\kappa x - \mathrm{i}\beta x).\end{aligned} \tag{11.29}$$

This fact is the basis for the "zigzag" model describing the behaviour of the field in the layer in the geometric optics approximation. A plane wave can be associated with a parallel beam of rays perpendicular to its phase surface. The internal reflections of the plane wave from the upper and the lower planes of the layer can be presented graphically by following the path of a ray which has the shape of a broken "zigzag" line (Fig. 11.2a).

An analytical solution of the dispersive equation (11.27) cannot be obtained. Some approximate information is obtainable graphically. To this end we make diagrams of the functions

$$\tan(\kappa W), \qquad F(\kappa W) = \frac{W\kappa(\delta W + \gamma W)}{(\kappa W)^2 - W^2\delta\gamma},$$

in relation to the arguments κW. Figure 11.3 shows a diagram of the functions $\tan(\kappa W)$ (continuous lines) and the function $F(\kappa W)$ (dashed line). The points of intersection of these curves give the values κW, which are the roots of equation (11.27). Each of the values obtained in this way gives one of the values β, which describes a waveguide mode of *TE* type.

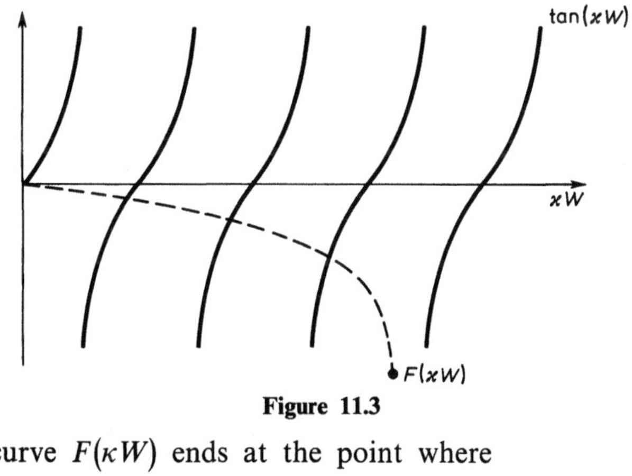

Figure 11.3

The curve $F(\kappa W)$ ends at the point where

$$(\kappa W)^2 - W^2\omega^2\mu_0(\varepsilon_f - \varepsilon_s) = 0 \quad (\text{for } \varepsilon_c < \varepsilon_s). \tag{11.30}$$

This is the so-called *cut-off point*. Its position depends on the refractive coefficients of the media forming the waveguide structure and on the thickness of the layer. The cut-off point determines the number of guided modes arising in the structure.

Guided modes exists if $\beta/\omega > \sqrt{\mu_0 \varepsilon_s}$ $(\varepsilon_s > \varepsilon_c)$. Then the parameters δ and γ are real quantities and the field outside the layer decays exponentially. If β takes the values from the interval

$$\omega\sqrt{\mu_0 \varepsilon_c} < \beta < \omega\sqrt{\mu_0 \varepsilon_s}, \tag{11.31a}$$

the parameter γ remains real and the parameter δ is an imaginary quantity. This means that inside the region $x < 0$ the field becomes a simple wave propagating in the direction of the wave vector $\mathbf{k} = [|\delta|, 0, \beta]$. In the region $x > W$ there is exponential decay. The structure of the field corresponding to this situation is called the *mode of radiation to the ground* (Fig. 11.2b).

If the propagation constant fulfils the condition

$$|\beta| < \omega\sqrt{\mu_0 \varepsilon_c}, \tag{11.31b}$$

then the parameters δ and γ are both imaginary. In this case the field assumes in all three areas the character of a propagating wave which can now pass from the ground, through the layer, to the covering, and vice versa. Such a structure of the field is called the *modes of radiation to the ground and to the covering* (Fig. 11.2c). The propagation constants now take (under restriction (10.31b)) all possible values, changing continuously.

Inequalities (11.31) allow us to determine in the plane (ω, β) the regions of the occurrence of different kinds of modes. The situation is illustrated in Fig. 11.4. The region of the occurrence of discrete

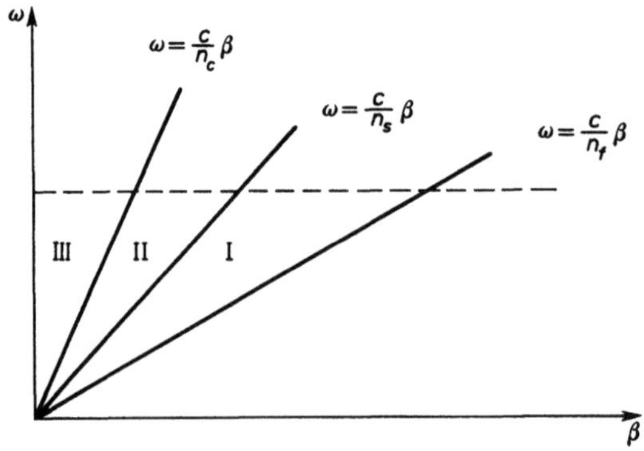

Figure 11.4

modes is marked as region I. For a fixed field frequency ω this region is bounded from the above. Region II, determined by inequality (11.31a), applies to the modes radiating to the ground, region III to the modes radiating both to the ground and the covering. The modes of radiation are responsible for the escape of energy from the waveguide to the ground and the covering.

B. TM Modes

TM modes are the eigenfunctions of the field described by the components E_x, E_z, H_y. In a plane waveguide with the structure and geometry described above *TM* modes satisfy equations (11.21). From these equations we obtain the second order equation for the component H_y

$$\left[\frac{\partial^2}{\partial x^2} + \frac{\partial^2}{\partial z^2} + \omega^2 \mu_0 \varepsilon(x)\right] H_y = 0, \tag{11.32}$$

where $\varepsilon(x)$ is defined by (11.23a).

Seeking the solutions of the equation in the form of homogeneous plane waves which propagate along the z-axis and undergo total internal reflections at the boundary planes $x = 0$ and $x = W$, we obtain the following dispersive equation:

$$\tan(\alpha W) = \frac{\kappa\left[\delta\left(\frac{n_f}{n_s}\right)^2 + \gamma\left(\frac{n_f}{n_c}\right)^2\right]}{\kappa^2 - \delta\gamma\left(\frac{n_f}{n_s}\right)^2\left(\frac{n_f}{n_c}\right)^2}, \qquad n = \sqrt{\frac{\varepsilon}{\varepsilon_0}}. \tag{11.33}$$

We solve this equation graphically in a way similar to that followed in the case of *TE* modes.

In addition to guided discrete modes we can also discuss radiating modes for *TM* fields. The discussion of the full set of *TM* modes is a repetition of what was said in the case of *TE* modes.

11.3. Tunnel Effect and the Introduction of Light into Planar Waveguides

A. Tunnel Effect in Optics

We consider a medium which fills the space $x > S$ and has the refractive index n_3. Below the plane $x = S$ (Fig. 11.5) there is a medium with the refractive index n_2 $(< n_3)$ bounded from the bottom by the plane $x = 0$. The region $x < 0$ is filled with a medium

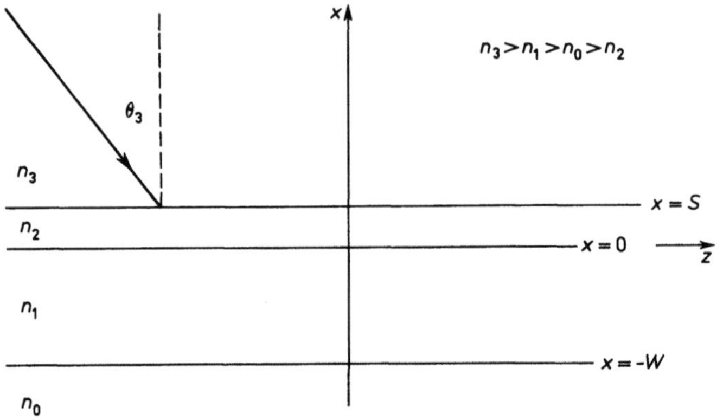

Figure 11.5

with the refractive index n_1 $(> n_2)$. Suppose that in the medium with the refractive index n_3 propagates an electromagnetic wave *TE* described by the relation

$$E_y = E_{0y}\left(-\mathrm{i}\frac{\omega}{c}n_3 x\cos\theta_3 + \mathrm{i}\frac{\omega}{c}n_3 z\sin\theta_3\right). \tag{11.34}$$

The direction of the wave vector forms with the x-axis an angle $\pi - \theta_3$, where $\theta_3 > \arcsin(n_2/n_3)$.

From the boundary conditions imposed on the field on the boundary planes it follows that the z-th component of the wave vector of the required field should be the same in all three media. This means that

$$\beta = \frac{\omega}{c}n_3\sin\theta_3 = \frac{\omega}{c}n_1\sin\theta_1 = k_z^{(2)} > \frac{\omega}{c}n_2. \tag{11.35}$$

This condition allows us to seek the field E_y in the media with the refractive indices n_1 and n_2 in the form

$$E_y^{(j)} = E_y^{(j)}(x)\exp(\mathrm{i}\beta z) \quad (j = 1, 2, 3), \tag{11.36}$$

where $E_y^{(j)}(x)$ are the solutions of the equations

$$\frac{\mathrm{d}^2 E_y^{(j)}(x)}{\mathrm{d}x^2} + \left[\frac{\omega^2}{c^2} n_j^2 - \beta^2\right] E_y^{(j)}(x) = 0. \tag{11.37}$$

Introducing the notation

$$\begin{aligned} \kappa_1 &= \sqrt{\frac{\omega^2}{c^2} n_1^2 - \beta^2}, \\ \kappa_3 &= \sqrt{\frac{\omega^2}{c^2} n_3^2 - \beta^2}, \\ \kappa_2 &= \sqrt{\beta^2 - \frac{\omega^2}{c^2} n_2^2}, \end{aligned} \tag{11.38}$$

we can write the solutions of equations (11.37) as follows:

$$\begin{aligned} E_y^{(1)} &= C_1 \exp(-\mathrm{i}\kappa_1 x) && \text{for} \quad x < 0, \\ E_y^{(2)} &= C_2 \exp(-\kappa_2 x) + B_2 \exp(\kappa_2 x) && \text{for} \quad 0 < x < S, \\ E_y^{(3)} &= E_{0y} \exp\left[-\mathrm{i}\left(\frac{\omega}{c} n_3 \cos\theta_3\right) x\right] + \\ &\quad + C_3 \exp\left[\mathrm{i}\left(\frac{\omega}{c} n_3 \cos\theta_3\right) x\right] && \text{for} \quad x > S. \end{aligned} \tag{11.39}$$

The conditions of the continuity of the components of the field on the boundary planes $x = 0$ and $x = S$ make it possible to determine the integration constants C_1, C_2, B_2, C_3.

The explicit form of the field in the area $x < 0$ will be expressed in the following way:

$$E_y^{(1)}(x) = \frac{4\kappa_2\kappa_3 E_{0y} \exp(-\mathrm{i}\kappa_1 x)\exp(-\mathrm{i}\kappa_3 S)\exp(-\kappa_2 S)}{(\kappa_2 + \kappa_3)^2 \exp(-2\kappa_2 S) - (\kappa_2 - \kappa_3)^2}. \tag{11.40}$$

For large values of $\kappa_2 S$, i.e., for great thickness of the layer with the refractive index n_2, we can simplify the expression (11.40) and write it in the form

$$E_y^{(1)}(x) = \frac{4\kappa_2\kappa_3}{(\kappa_2 - \kappa_3)^2} E_{0y} \exp(-i\kappa_1 x)\exp(-i\kappa_2 S)\exp(-\kappa_2 S). \tag{11.41}$$

The calculations above show that the presence of a "third" medium radically changes the situation of total internal reflection. Even for $\theta_3 > \arcsin(n_2/n_3)$ the wave enters the medium with the refractive index n_1. The smaller the thickness of the medium with the refractive index n_2, the greater the amplitude of that wave (11.41). The field $E_y^{(2)}(x)$ in the region $0 < x < S$ is described by two exponential functions (11.39): increasing and decreasing, and can propagate only in a direction parallel to the confining planes. Such behaviour of the wave is called the *tunnel effect*, since it is completely analogous to the tunnel effect in quantum mechanics occurring when a particle passes through the potential barrier.

B. Introducing Light into a Planar Waveguide

The tunnel effect can be used to introduce light into a dielectric planar waveguide. By placing over the waveguide layer a prism with the refractive index $n_p > n_f > n_c$ we form the structure illustrated in Fig. 11.6. We assume that from the region $x > S$ a plane wave falls upon the base of the prism at an angle $\theta_p > \arcsin(n_c/n_p)$. As a result of the tunnel effect, part of the energy of that wave penetrates in the form of a mode wave into the region of the layer $0 > x > -W$. This happens when the direction of the wave vector of the wave incident on the base of the prism is properly orientated, i.e. if we have the equality

$$\beta = \frac{\omega}{c} n_p \sin\theta_p, \tag{11.42}$$

where β is the propagation constant of the waveguide mode of the layer. This is the so-called *synchronism condition.*

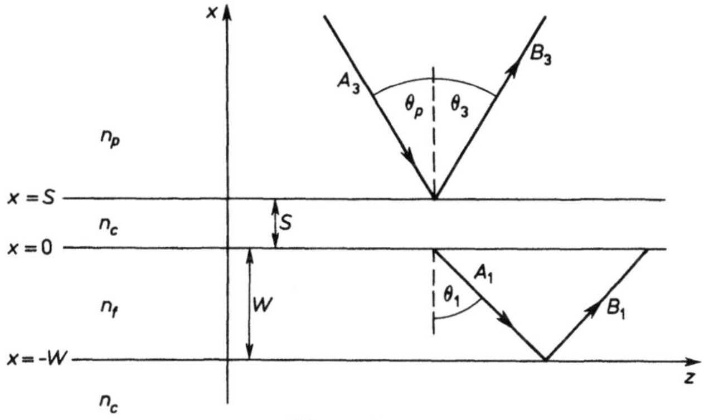

Figure 11.6

The amplitude of the wave incident upon the base of the prism is related through the boundary conditions on the planes $x = -W$, and $x = 0$ and $x = S$ to the amplitude of the wave excited in the layer.

The process of introducing of light into the waveguide layer can be explained in a simplified way without the necessity of providing an exact electromagnetic description. The total wave field is divided in two groups of waves, and each of them is examined independently. The complex amplitudes of the incident wave and of the wave reflected at the base of the prism are denoted by A_p and B_p respectively, and the amplitudes of the waves within the layer by A_1 and B_1. The amplitude A_1 relates to the wave proceeding from the upper to the lower boundary plane, and the amplitude B_1 to the wave going in the opposite direction (Fig. 11.7).

The wave A_p incident on the base of the prism in the neighbourhood of point z_m excites a wave A_1'' in the layer. This wave is summed up with the wave A_1', produced from the reflection of the wave B_1, which in turn is a derivative of the wave A_1 generated at point z_{m-1}. Thus the resultant wave at point z_m equals

$$A_1(z_m) = A_1'(z_m) + A_1''(z_m). \tag{11.43}$$

Similarly, the wave B_p reflected from the prism at point z_m will be the sum of B_p' and B_p''

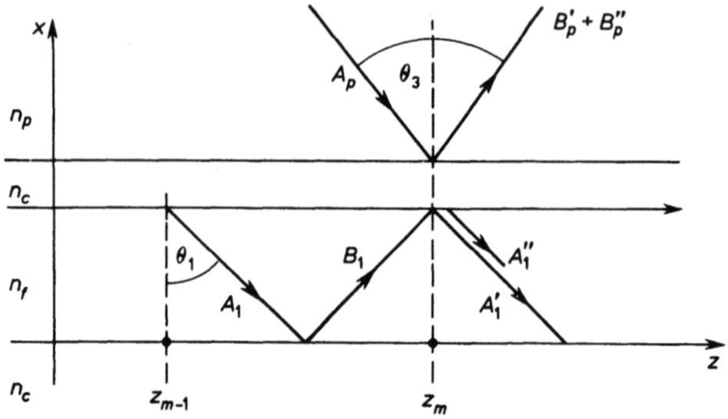

Figure 11.7

$$B_p(z_m) = B'_p(z_m) + B''_p(z_m), \tag{11.44}$$

i.e. the sum of the wave reflected at point z_m and the wave transmitted from the layer to the prism at point z_m.

Using the coefficient of reflection and the coefficient of transmission, which can both be calculated on the ground of electromagnetic considerations [1] for the multilayer structures

$$\begin{aligned} R_3 &= r_{3210}, \quad & R_1 &= r_{123}, \\ T_3 &= t_{123}, \quad & T_1 &= t_{321}, \end{aligned} \tag{11.45}$$

we can write

$$B_p(z_m) = R_3 A_p(z_m) + T_3 B_1(z_m), \tag{11.46a}$$

$$A_1(z_m) = T_1 A_p(z_m) + R_1 B_1(z_m). \tag{11.46b}$$

The complex amplitude of the wave $B_1(z_m)$ is connected with the amplitude $A_1(z_{m-1})$ through the relation

$$B_1(z_m) = A_1(z_{m-1}) \exp(\mathrm{i}2\kappa W - \mathrm{i}2\Phi_{10}), \tag{11.47}$$

where $2\Phi_{10}$ is the phase factor of the coefficient of total reflectivity at the border of the media with the refractive indices n_1 and n_c, and κW describes the change of phase connected with wave propagation in the direction of the x-axis.

The coefficient of reflection R_1 is a complex quantity, since it describes the tunnel effect as the wave passes from the layer to the prism. It can be written as

$$R_1 = |R_1|\exp(-i2\Phi_{123}). \tag{11.48}$$

Taking this into account, we can present (11.46) in the form

$$A_1(z_m) - A_1(z_{m-1}) = T_1 A_p(z_m) - (1 - |R_1|)A_1(z_{m-1}). \tag{11.49}$$

We have used here the equality

$$2\kappa W - 2\Phi_{10} - 2\Phi_{123} = 2m\pi. \tag{11.50}$$

This equality is an equivalent of the equation of dispersion (11.27) written in the form

$$\kappa W = \arctan\frac{\delta}{\kappa} + \arctan\frac{\gamma}{\kappa} + m\pi, \tag{11.51}$$

in the case of a three-layer structure (Fig. 11.6). Dividing both sides of (11.48) by the distance between points z_{m-1} and z_m, which are the successive points of total reflections of the wave from the boundary planes of the layer:

$$z_m - z_{m-1} = 2W\tan\theta_1$$

(θ_1 being the angle of reflection of the wave from the planes bordering the layer), we obtain the equation

$$\frac{dA_1}{dz} = \frac{1}{2W\tan\theta_1}\left[T_1 A_p(z) - (1 - |R_1|)A_1(z)\right]. \tag{11.52}$$

Equation (11.52) describes the change of the amplitude of the wave excited in the layer as the result of the tunnel effect. Analysing it we can see that:

(1) wherever $A_p = 0$, the amplitude $A_1(z)$ undergoes exponential decay,

(2) if the term $(1 - |R_1|)A_1(z)$ can be disregarded, $A_1(z)$ will increase with the increase of the variable z.

Thus the first term in equation (11.52) describes the increase of the amplitude $A_1(z)$ and is related with the flow of energy to the layer, whereas the second term describes the escape of energy back to the prism.

If for $z = 0$ we have $A_1(0) = 0$, then initially there is an increase of $A_1(z)$, but then the second term becomes significant, and for a certain value of z the following equality will be satisfied:

$$\frac{dA_1(z)}{dz} = 0, \tag{11.53a}$$

which holds for

$$A_1(z) = \frac{T_1}{1 - |R_1|} A_p(z). \tag{11.53b}$$

If A_p is not equal to 0 and remains constant for $0 < z < L$, the solution (11.52) takes the form

$$A_1(z) = \frac{T_1 A_p}{1 - |R_1|} \left\{ 1 - \exp\left[-\left(\frac{1 - |R_1|}{2W \tan\theta_1} \right) z \right] \right\} \quad \text{for } z \leq L, \tag{11.54a}$$

$$A_1(z) = \frac{T_1 A_p}{1 - |R_1|} \left\{ 1 - \exp\left[-\left(\frac{1 - |R_1|}{2W \tan\theta_1} \right) L \right] \right\} \times$$
$$\times \exp\left[-\left(\frac{1 - |R_1|}{2W \tan\theta_1} \right) z \right] \quad \text{for } z \geq L. \tag{11.54b}$$

From the form of the solution (11.54) we can draw a practical conclusion: in introducing light into a layer with the use of a prism, the illuminated region of the base of the prism should be located close to the end of the prism. Otherwise there will be a significantly growing reverse process involving an escape of energy from the layer.

11.4. Periodic Structures in Planar Waveguides and Their Application

The simplest description of the periodic structure of a planar waveguide (Fig. 11.8) is provided by the so-called *Rayleigh method*.

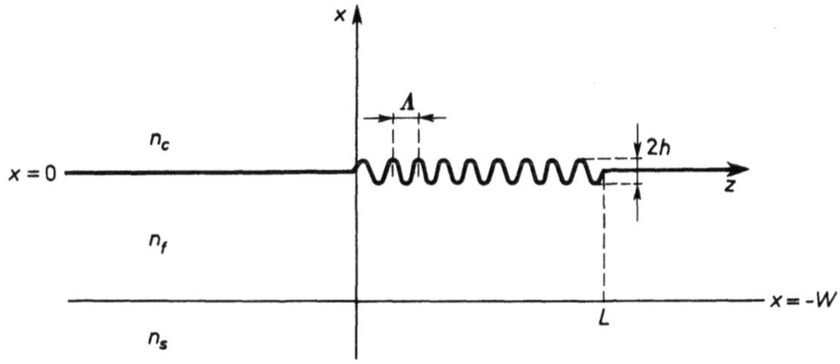

Figure 11.8

In this method we assume that the diffractional field induced by the periodic structure may be treated as a superposition of the whole set of homogeneous and non-homogeneous plane waves. If we confine ourselves to this solution, the problem is considerably simplified, but only the qualitative character of the phenomenon in question is then taken into account, resulting in a model which is far from reality.

Let us consider a diffraction phase grating, formed by the local fold of one of the boundary planes of a plane waveguide. Let this folding, which appears on the upper surface of the layer (Fig. 11.8) given by the relation $x = 0$, be described by

$$x = h\sin\left(\frac{2\pi}{\Lambda}z\right), \tag{11.55}$$

where h and Λ are, respectively, the amplitude and spatial phase of the folding.

A plane light wave incident on the grating at the angle θ and described by the equation

$$E_y(x, z) = A\exp[\mathrm{i}(\omega t + k_c z\sin\theta - k_c x\cos\theta)], \tag{11.56}$$

where $k_c = \omega n_c/c$ (n_c is the refractive index of the covering), having passed through the region of non-homogeneity caused by the folds, changes the phase according to the relation

$$\psi = M\sin\left(\frac{2\pi}{\Lambda}z\right), \tag{11.57}$$

where M is given by the relation

$$M = \frac{\omega h}{c}(n_f - n_c) \tag{11.58}$$

(n_f is the refractive index of the material of the layer). On passing through the grating and reaching the surface of the layer $x = 0$, the wave has the phase factor

$$\exp\left\{\mathrm{i}\left[M\sin\left(\frac{2\pi}{\Lambda}z\right) + \frac{\omega}{c}zn_c\sin\theta\right]\right\}, \tag{11.59}$$

and may be presented in the form

$$E_y(0,z) = A\sum_m J_m(M)\exp\left[\mathrm{i}\left(\frac{\omega}{c}zn_c\sin\theta + m\frac{2\pi}{\Lambda}z\right)\right]. \tag{11.60}$$

To obtain this we use the relation

$$\exp\left[\mathrm{i}M\sin\left(\frac{2\pi}{\Lambda}z\right)\right] = \sum_m J_m(M)\exp\left[im\frac{2\pi}{\Lambda}z\right], \tag{11.61}$$

where $J_m(M)$ is the Bessel function.

Expression (11.60) shows that the diffractional field behind the grating is a superposition of divergent plane waves, each of which has the z-th component of the wave vector of the form

$$k_z^{(d)} = \frac{\omega}{c}n_c\sin\theta + m\frac{2\pi}{\Lambda}, \tag{11.62}$$

where $m = 0,\ \pm 1,\ \pm 2, \ldots$

The component $k_z^{(d)}$ has two terms. The first $((\omega/c)n_c\sin\theta)$ is the z-th component of the wave vector of the incident wave, the second $(m2\pi/\Lambda)$ occurs as the result of interaction with the grating. It is a multiple of the grating vector $2\pi/\Lambda$. This vector lies in the plane of the grating and remains perpendicular to the lines of the projections. If $k_z^{(d)}$ is equal to one of the propagation constants β_j of the guided mode, then the diffracted wave corresponding to it will propagate in the waveguide as a guided mode. This condition can be written as

$$\beta_j = \frac{\omega}{c} n_c \sin\theta + \frac{2\pi}{\Lambda} m. \tag{11.63}$$

Thus the grating at the surface of the plane waveguide may serve as a device which introduces light into the waveguide. If condition (11.63) is not fulfilled, the appropriate diffracted wave becomes a radiating mode and passes to the ground of the layer.

Thus the folded surface is the source of the wave vector $|\mathbf{G}| = 2\pi/\Lambda$, which lies in the plane of the layer. Denoting the wave vector of the incident wave by $\boldsymbol{\beta}_i$ and the wave vector of the propagation constant of the mode by $\boldsymbol{\beta}_j$, we can rewrite (11.63) in the vector form

$$\boldsymbol{\beta}_j = \boldsymbol{\beta}_i + \mathbf{G}, \tag{11.63a}$$

which is called the *Bragg condition.*

Since the propagation constant of the guided mode is larger than the length of the wave vector in the ground,

$$\beta_j > \frac{\omega}{c} n_s \quad (n_s > n_c), \tag{11.64}$$

in order to transform part of the energy of the incident wave into a guided mode, we need a factor which will change the z-th component of the wave vector and that factor is the grating which supplies the grating vector or its multiple.

The variation of the wave vector which occurs when the incident wave interacts with the grating placed on the surface of the waveguide may be presented graphically. To this end it is necessary to make use of the plane of the coordinates of the wave vector k_x, k_z. The k_z-axis constitutes here the axis of abscissae, and the k_x-axis the axis of ordinates (Fig. 11.9). In the upper half-plane of the plane in question, starting from the origin of the coordinate system, we draw a semicircle of radius $\omega n_c/c$, and in the lower half-plane a semicircle of radius $\omega n_f/c$ $(n_f > n_s)$. By means of this diagram we can illustrate the law of reflection. Vector $\overrightarrow{0M}$ represents the z-th component of the vector of the incident wave. We carry out the following operation: we shift vector $\overrightarrow{0M}$ to the right till its end-point

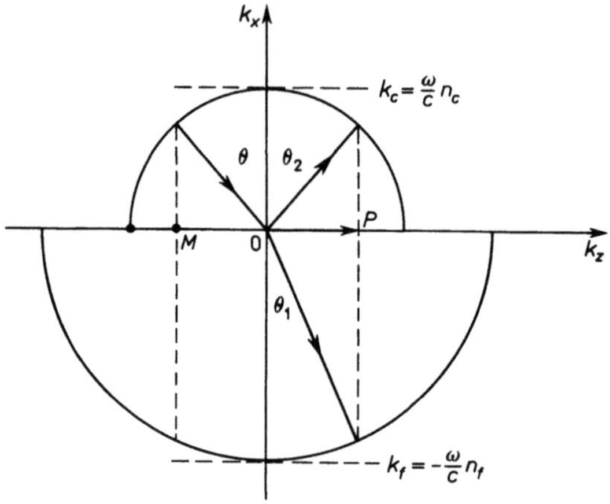

Figure 11.9

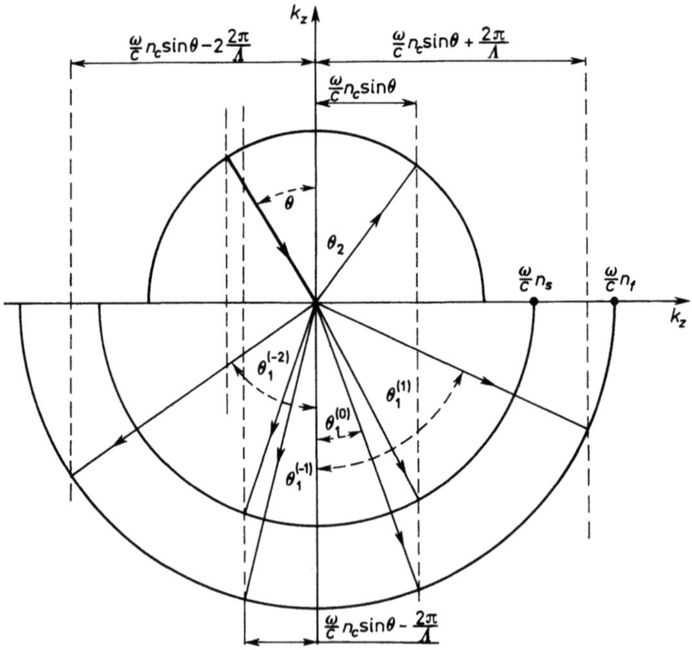

Figure 11.10

M coincides with the origin of the coordinte system. In this way we obtain vector $\overrightarrow{0P}$. Drawing through point P a perpendicular to the k_z-axis, we obtain two points of its intersection with the semicircles whose radii are $\omega n_c/c$ and $\omega n_f/c$. Joining these points with the origin of the system, we obtain the directions of the refracted ray θ_1 and of the reflected ray θ_2.

By means of a similar construction (Fig. 11.10) we can illustrate the condition of forming a guided mode but we must remember of condition (11.64). There are two possibilities:

1. If the lattice vector $G = 2\pi/\Lambda$ is such that

$$\frac{\omega}{c} n_s < \left| \frac{\omega}{c} n_c \sin\theta + \frac{2\pi}{\Lambda} \right| < \frac{\omega}{c} n_f, \tag{11.65}$$

then laying off the segment $|\omega n_c \sin\theta/c + 2\pi/\Lambda|$ on the k_z-axis from the origin of the coordinate system, and drawing through the origin a straight line perpendicular to the k_z-axis, we obtain one point of intersection with the semicircle of radius $\omega n_f/c$. Joining this point with the origin of the coordinate system, we can determine the angle $\theta_1^{(1)}$, which defines the direction of the wave vector of the waveguide mode with respect to the axis perpendicular to the planes bounding the layer.

2. It is also possible to form a guided mode for negative values of m. Then it is required that

$$\frac{\omega}{c} n_s < \left| \frac{\omega}{c} n_c \sin\theta + m\frac{2\pi}{\Lambda} \right| < \frac{n}{c} n_f. \tag{11.66}$$

This situation is reflected in Fig. 11.10 by the direction marked as $\theta_1^{(-2)}$. As can be seen from Fig. 11.10, when the vector $2\pi/\Lambda$ is sufficiently large it may happen that the modes guided in the positive direction of the z-axis cannot be excited at all. A greater chance have the modes guided in the direction $z < 0$.

The interaction of the surface grating with the guided modes may be determined by using the theorem about reversing the path of the ray. A grating which operates on a waveguide mode with a propagation constant β_s generates diffracted waves for which the z-th component of the wave vector is

$$k_z = \beta_s + \frac{2\pi}{\Lambda} m. \tag{11.67}$$

In order for a wave of the radiating mode to appear among these diffracted waves, the following condition must fulfilled:

$$\left| \beta_s + \frac{2\pi}{\Lambda} m \right| < \frac{\omega}{c} n_c. \tag{11.68}$$

In view of inequality (11.64), condition (11.68) may be satisfied only for negative m. Then inverse radiating modes will dominate (Fig. 11.11).

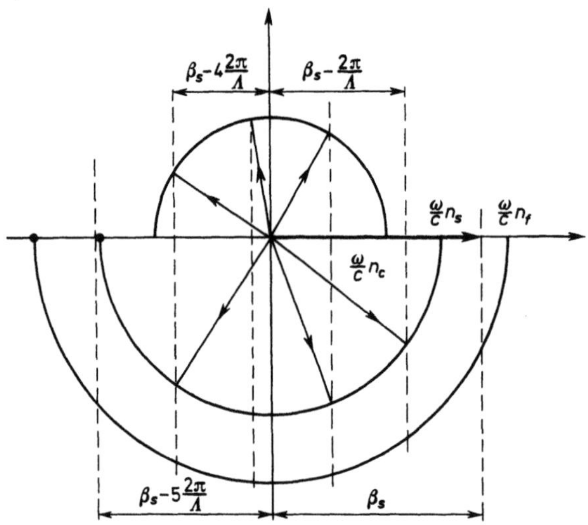

Figure 11.11

If condition (11.68) is not fulfilled for any value of m $(= 0, \pm 1, \pm 2, \ldots)$, the grating will only bring about a spatial reconstruction of the electromagnetic field: part of the energy carried by the mode coming over the periodic structure of the waveguide will pass to the other permissible modes guided in the layer, proceeding both along the positive and the negative direction of the z-axis.

Of particul interest is the case where one of the excited diffracted waves has the z-th component of the wave vector equal to the negative value of the propagation constant of the appropriate mode, i.e. when

$$k_z^{(d)} = -\beta_s. \tag{11.69}$$

Then the diffracted wave in question is simply a guided mode reflected in the negative direction of the z-axis. This may occur if the grating vector **G** fulfils the condition

$$m|\mathbf{G}| = -2\beta_s. \tag{11.70}$$

Hence, taking into account the equality $\beta_s \approx 2\pi/\lambda_f$, we can obtain

$$\Lambda = -\frac{1}{2}\lambda_f. \tag{11.71}$$

Thus, if the spatial period of the grating is equal to half the length of the wave in the medium of the waveguide layer, the diffracted wave of order $m = -1$ will be the induced mode reflected backwards.

A folded planar waveguide can be used as a Bragg deflector (mirror). From the physical point of view its action can be explained as follows: a wave going along the waveguide passes by a great many grooves of the folded surface. Each groove reflects part of the energy of the wave. If the phase of the wave reflected from the first groove is taken as a reference point, the phase of the wave reflected from the second groove differs by 2π in the case of (11.71). These two partial waves sum up in phase. Largely the same holds for the wave reflected from the third, and the fourth groove, etc.

Bibliography

1. Born, M. and Wolf, E.: (1965), 'Principles of Optics', Pergamon Press, Oxford–London.
2. Kapany, N.S. and Burke, J.J.: (1972), 'Optical Waveguides', Academic Press, New York.
3. Marcuse, D.: (1974), 'Theory of Dielectric Optical Waveguides', Academic Press, New York.
4. Tamir, T.: (1975), 'Integrated Optics', Springer Verlag, Berlin.
5. Tien, P.K.: (1977), *Review of Modern Physics*, **49**, 361.
6. Tien, P.K, Smolinsky, G. and Martin, R.J.: (1975), *IEEE Trans. Microwave Theory Tech.* **MTT–23**, 79.

12. Geometric Optics Approximation

12.1. Eikonal Equation

Geometric optics is a limiting case of the wave optics, when the wavelength tends to zero.

The basic equations of geometric optics are derived from the Helmholtz's equations

$$\Delta U + k^2 U = 0. \tag{12.1}$$

To this end we postulate the solution in the form

$$U = A \exp(ik_0 S), \tag{12.2}$$

where A is the amplitude of a monochromatic wave, which in general is a function of position, S is the phase factor, which is also a function of a point, k_0 is the wave number of the monochromatic wave in the vacuum, while k is the wave number of the same wave in the medium (wheares $k = k_0 n$, n – the refractive index of the medium).

When $k_0 \to \infty$, the function $U(\mathbf{r})$ is a quick-variable function of the coordinates. It is connected with the exponential factor in expression (12.2), despite the fact that the functions $A(\mathbf{r})$ and $S(\mathbf{r})$ remain slow-variable functions.

In order to introduce the basic equations of geometric optics, we calculate the Laplacian for field (12.2). To this aim we use the equations

$$\begin{aligned}
&\Delta\left(A e^{ik_0 S}\right) = e^{ik_0 S} \Delta A + A \Delta\left(e^{ik_0 S}\right) + 2(\nabla A)\left(\nabla\left(e^{ik_0 S}\right)\right),\\
&\Delta\left(e^{ik_0 S}\right) = -k_0^2 e^{ik_0 S} (\operatorname{grad} S)^2 + ik_0 e^{ik_0 S} \Delta S, \\
&\operatorname{grad}\left(e^{ik_0 S}\right) = ik_0 e^{ik_0 S} \operatorname{grad} S.
\end{aligned} \tag{12.3}$$

This enables us to obtain from equation (12.1) for field (12.2) the expressions

$$[\Delta A + k^2 A - A k_0^2 (\operatorname{grad} S)^2 + i k_0 (A \Delta S + 2(\nabla A)(\nabla S)] = 0. \tag{12.4}$$

Comparing the real and the imaginary part of expression (12.4) to zero, we obtain

$$(\operatorname{grad} S)^2 - n^2 = \frac{\Delta A}{A k_0^2}, \tag{12.5}$$

$$2(\operatorname{grad} A)(\operatorname{grad} S) + A \Delta S = 0. \tag{12.6}$$

In the limiting case where $\lambda \to 0$ $(k_0 \to \infty)$, from (12.5) it follows

$$(\operatorname{grad} S)^2 = n^2, \tag{12.7}$$

the so-called *eikonal equation*, which is the primary equation in geometric optics and it determines the direction of the rays. The surface $S(\mathbf{r}) = \text{const}$ is called the *geometric wave front.*

If the quantity $\Delta A/A$ is comparable with $(1/k_0)^2$, equation (12.7) is no longer valid. This situation will take place in the vicinity of the focus, where we can expect significant deviations from what geometric optics expects.

Equation (12.6) may be written in the form

$$\operatorname{grad}(\ln A) \cdot (\operatorname{grad} S) = -\tfrac{1}{2} \Delta S, \tag{12.8}$$

which enables us to find the variation of $\ln A$ in the direction of the gradient S. Equation (12.8) does not provide any information about the value of gradient $\ln A$ in the direction perpendicular to grad S.

12.2. Light Rays

Light rays are defined as lines perpendicular to the geometric wave fronts $S(x, y, z) = \text{const}$.

If $\mathbf{r}$ is the position vector of a current point on the radius, while

s is the length of the ray measured from a fixed point lying on it, then equation (12.7) may take the form

$$n \frac{\mathrm{d}\mathbf{r}}{\mathrm{d}s} = \operatorname{grad} S. \tag{12.9}$$

Differentiating this equation with respect to the length of the arc of the ray s, we obtain

$$\frac{\mathrm{d}}{\mathrm{d}s}\left(n \frac{\mathrm{d}\mathbf{r}}{\mathrm{d}s}\right) = \frac{\mathrm{d}}{\mathrm{d}s} \operatorname{grad} S = \frac{\mathrm{d}\mathbf{r}}{\mathrm{d}s} \operatorname{grad}(\operatorname{grad} S). \tag{12.10}$$

Taking (12.9) into account, we can give to equality (12.10) the form

$$\frac{\mathrm{d}}{\mathrm{d}s}\left(n \frac{\mathrm{d}\mathbf{r}}{\mathrm{d}s}\right) = \frac{1}{n}(\operatorname{grad} S)\operatorname{grad}(\operatorname{grad} S), \tag{12.11}$$

which, after taking into account the eikonal equation, leads to the equation

$$\frac{\mathrm{d}}{\mathrm{d}s}\left(n \frac{\mathrm{d}\mathbf{r}}{\mathrm{d}s}\right) = \frac{1}{2n} \operatorname{grad} n^2, \tag{12.12}$$

or finally to

$$\frac{\mathrm{d}}{\mathrm{d}s}\left(n \frac{\mathrm{d}\mathbf{r}}{\mathrm{d}s}\right) = \operatorname{grad} n. \tag{12.13}$$

It is a vector form of a differential equation for light rays. For a homogeneous medium ($n = \text{const}$) this equation reduces to

$$\frac{\mathrm{d}^2\mathbf{r}}{\mathrm{d}s^2} = 0. \tag{12.14}$$

The solution of (12.14) is

$$\mathbf{r} = s\mathbf{a} + \mathbf{b}, \tag{12.15}$$

where $\mathbf{a}$ and $\mathbf{b}$ are constant vectors. It is a vector equation of a straight line of the direction of the vector $\mathbf{a}$ passing through the point $\mathbf{r}_0 = \mathbf{b}$. Thus, light rays are straight lines in a homogeneous medium.

In the case of a non-homogeneous medium the light ray is a curve with the vector of curvature defined as

$$\mathbf{K} = \frac{\mathrm{d}\mathbf{s}}{\mathrm{d}s} = \frac{1}{\rho}\mathbf{v}, \tag{12.16}$$

where ρ is the radius of curvature and $\mathbf{v}$ a versor normal to the curvature.

For a non-homogeneous medium with $n = n(x, y, z)$ this vector is found from the equation

$$n\mathbf{K} = \operatorname{grad} n - \frac{\mathrm{d}n}{\mathrm{d}s}\mathbf{s}, \tag{12.17}$$

where $\mathbf{s}$ is a versor tangent to the curve and $\mathrm{d}s$ a length element of the arc of the curve.

Equality (12.17) shows that $\operatorname{grad} n$ lies in the osculating plane at a point under consideration.

Multiplying scalarly both sides of (12.17) by $\mathbf{K} = (1/\rho)\mathbf{v}$, we obtain

$$n|\mathbf{K}|^2 = \frac{1}{\rho}(\mathbf{v}\cdot\operatorname{grad})n, \tag{12.18}$$

and hence

$$|\mathbf{K}| = \frac{1}{\rho} = \frac{\partial}{\partial v}(\ln n). \tag{12.19}$$

The radius of curvature ρ is always positive, which means that while moving along the principal normal the refractive index increases, i.e. the ray deflects towards the region with a higher refractive index.

From (12.9) it follows that

$$\operatorname{curl}\left(n\frac{\mathrm{d}\mathbf{r}}{\mathrm{d}s}\right) = \operatorname{curl}(n\mathbf{s}) = \mathbf{0}, \tag{12.20}$$

and hence, on the ground of the Stokes theorem, the integral of the refractive index n taken over a closed curve L equals zero

$$\oint_L n\,\mathrm{d}l = 0. \tag{12.21}$$

It is the so-called *Lagrange integral invariant*, from which it follows that the integral taken over an arbitrary curve joining points P_1 and P_2 at which $n = n(x, y, z)$ is independent of the shape of the integration path

$$\int_{P_1}^{P_2} n\mathbf{s}\,d\mathbf{r} = \text{invariant}. \tag{12.22}$$

As a starting point in geometric optics we can take the so-called Fermat principle (the principle of the shortest optical path) instead of the eikonal equation.

12.3. Fermat's Principle

The relation between gradient of S and the refractive index $n = n(x, y, z)$ enables us to establish the so-called *Fermat principle* (the principle of the shortest optical path). To this end we use the quantity

$$\oint_L \mathbf{s}n\,d\mathbf{l} = 0, \tag{12.23}$$

where L is a closed integration contour.

The integration contour is chosen is such a way that a part of it coincides with the real light ray passing through the points A and B, and the other part ACB is an arbitrary curve which does not coincides with the ray (Fig. 12.1).

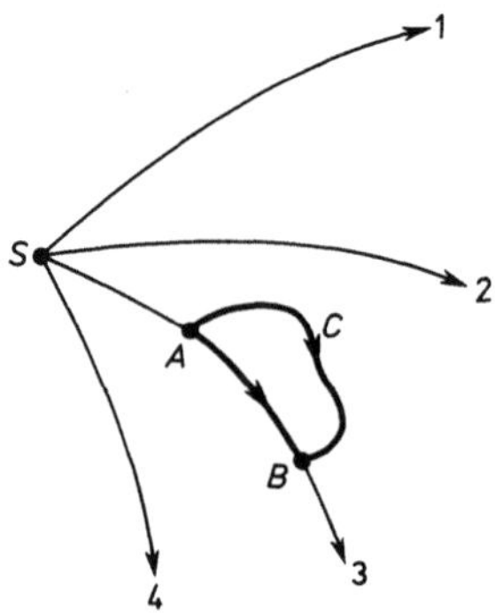

Figure 12.1

Formula (12.23) can be rewritten in the form

$$\int_{ACB} n\mathbf{s}\,\mathrm{d}\mathbf{l} + \int_{BA} n\,\mathrm{d}l = 0. \tag{12.24}$$

But as the inequality

$$\left|\int_{ACB} n(\mathbf{s}\,\mathrm{d}\mathbf{l})\right| < \int_{ACB} n\,\mathrm{d}l$$

holds, we obtain from (12.24) the inequality

$$\int_{AB} n\,\mathrm{d}s \leq \int_{ACB} n\,\mathrm{d}l. \tag{12.25}$$

Hence, integrating over the radius we obtain in fact

$$\int_{AB} n\,\mathrm{d}s = \min. \tag{12.26}$$

This means either that the optical path

$$\int_{P_1}^{P_2} n\,\mathrm{d}s \tag{12.27}$$

along the ray passing through the points P_1 and P_2 is the shortest of all possible calculated among all other curves joining these two points and lying in the neighbourhood, or that

$$\delta\left(\int_{P_1}^{P_2} n\,\mathrm{d}s\right) = 0. \tag{12.28}$$

The equivalence of condition (12.28) with equation (12.13), which is a direct consequence of the eikonal equation, results immediately from the general principles of the variational calculus. Integral (12.28) may be written as

$$\delta\left(\int_{P_1}^{P_2} n(x, y, z)\sqrt{\left(\frac{\mathrm{d}x}{\mathrm{d}s}\right)^2 + \left(\frac{\mathrm{d}y}{\mathrm{d}s}\right)^2 + \left(\frac{\mathrm{d}z}{\mathrm{d}s}\right)^2}\,\mathrm{d}s\right) = 0, \tag{12.29}$$

where $\mathrm{d}s = \sqrt{(\mathrm{d}x)^2 + (\mathrm{d}y)^2 + (\mathrm{d}z)^2}$ is a length element of the arc of the curve in question. For the above variation the Euler equations take the form

$$\frac{\partial}{\partial x}\left(n\sqrt{x'^2+y'^2+z'^2}\right)-\frac{\mathrm{d}}{\mathrm{d}s}\left[\frac{\partial}{\partial x'}\left(n\sqrt{x'^2+y'^2+z'^2}\right)\right]=0,$$

$$\frac{\partial}{\partial y}\left(n\sqrt{x'^2+y'^2+z'^2}\right)-\frac{\mathrm{d}}{\mathrm{d}s}\left[\frac{\partial}{\partial y'}\left(n\sqrt{x'^2+y'^2+z'^2}\right)\right]=0,$$

$$\frac{\partial}{\partial z}\left(n\sqrt{x'^2+y'^2+z'^2}\right)-\frac{\mathrm{d}}{\mathrm{d}s}\left[\frac{\partial}{\partial z'}\left(n\sqrt{x'^2+y'^2+z'^2}\right)\right]=0.$$

These equations may be written in a vector form

$$\operatorname{grad} n=\frac{1}{\sqrt{x'^2+y'^2+z'^2}}\frac{\mathrm{d}}{\mathrm{d}s}\left(n\frac{\mathrm{d}\mathbf{r}/\mathrm{d}s}{\sqrt{x'^2+y'^2+z'^2}}\right). \quad (12.30)$$

Taking into account that

$$\sqrt{x'^2+y'^2+z'^2}=\frac{1}{\mathrm{d}s}\sqrt{(\mathrm{d}x)^2+(\mathrm{d}y)^2+(\mathrm{d}z)^2}=1,$$

we indeed obtain equation (12.13) from (12.28), which is the desired conclusion.

Using the Fermat principle, we can establish the behaviour of the light ray in each definite case.

12.4. Law of Intensities in Geometric Optics

Multiplying equation (12.6) by A we get

$$[\operatorname{grad}(A^2)]\cdot\operatorname{grad} S+A^2\Delta S=0. \quad (12.31)$$

The obtained equality may be written in the form

$$\operatorname{div}(A^2\nabla S)=0. \quad (12.32)$$

Now we take into consideration a region V (Fig. 12.2), which has a shape of a truncated cone with the lateral surface C determined by the directions of the curves for which their tangents overlap at every

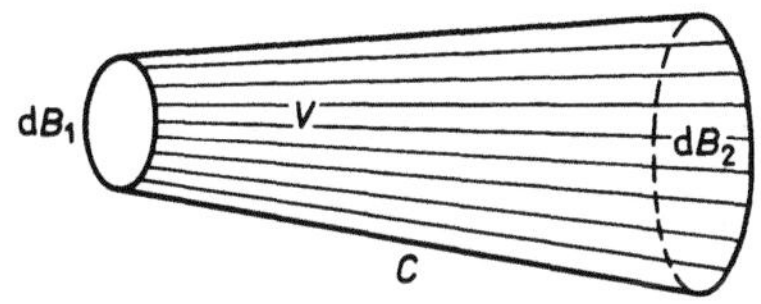

Figure 12.2

point with ∇S, and the bottoms dB_1 and dB_2, perpendicular to ∇S, in the case $n = \text{const}$ are mutually parallel. Integrating (12.32) over the region V and applying the Gauss theorem, we arrive at

$$0 = \iiint_V \operatorname{div}(A^2 \nabla S)\, d\tau = \iint \mathbf{n}(A^2 \nabla S)\, df. \tag{12.33}$$

In view of the structure of the surface the above relation resolves itself into the equality

$$\mathbf{n}_1 \cdot (A^2 \nabla S) dB_1 + \mathbf{n}_2 \cdot (A^2 \nabla S) dB_2 = 0, \tag{12.34}$$

where $\mathbf{n}_1$ and $\mathbf{n}_2$ are the normals external to the surface elements in question. Since $\mathbf{n}_1 = -\mathbf{n}_2 = \mathbf{n}$, we may rewrite expression (12.34) in the form

$$\mathbf{n} \cdot (A^2 \nabla S)_1 dB_1 = \mathbf{n} \cdot (A^2 \nabla S)_1 dB_2. \tag{12.35}$$

The quantity $A^2 \nabla S$ occurring in formulas (12.34) and (12.35) may be interpreted as the vector of energy flux for the wave field $U = A e^{ik_0 S}$. The absolute value of this expression $I = A^2 |\nabla S|$ is the intensity of the wave in question.

In view of this, we may interpret (12.35) as the law of intensities in geometric optics, which says:

The product of the intensity of the wave and the area of the cross-section of the cone formed by the rays drawn by an arbitrary spatial curve is constant:

$$I_1 dB_1 = I_2 dB_2 = \text{const}. \tag{12.36}$$

The energy of the wave may propagate only within the cone created by the rays.

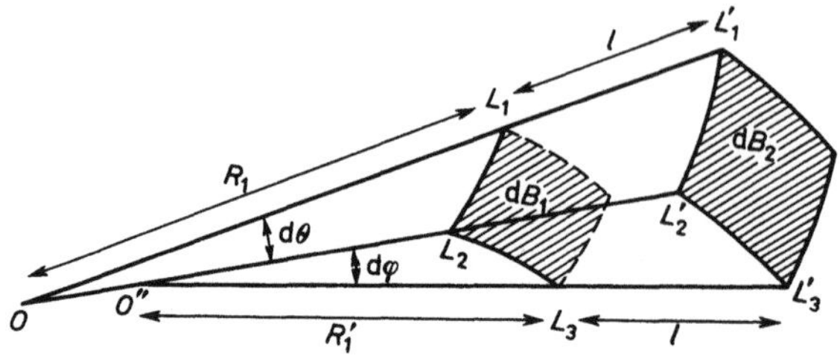

Figure 12.3

If a medium is homogeneous ($n = \text{const}$), the rays are straight lines and the law of intensities may be expressed in a slightly different form. The surface elements dB_1 and dB_2 are confined by the segments of the curves (Fig. 12.3) L_1L_2 and L_2L_3. Taking into account their radii of curvature, the lengths of these segments can be written as

$$L_1L_2 = R_1 d\theta, \quad L_2L_3 = R'_1 d\varphi, \tag{12.37}$$

where $d\theta$ and $d\varphi$ are the angles at which the segments L_1L_2 and L_2L_3 are visible from the middle of their curvatures.

Then

$$dB_1 = (L_1L_2)\cdot(L_2L_3) = R_1R'_1 d\theta d\varphi, \tag{12.38a}$$

$$dB_2 = (L'_1L'_2)\cdot(L'_2L'_3) = R_2R'_2 d\theta d\varphi, \tag{12.38b}$$

where

$$R_2 = R_1 + l, \quad R'_2 = R'_1 + l, \tag{12.39}$$

where l is the distance between points L_1 and L'_1, and between L_3 and L'_3.

On the ground of the above expressions for the surface elements the law of intensities (12.36) may be written as

$$\frac{I_2}{I_1} = \frac{dB_1}{dB_2} = \frac{R_1R'_1}{(R_1 + l)(R'_1 + l)}. \tag{12.40}$$

Hence for $R_1 \ll l$ and $R'_1 \ll l$ we get the equality

$$\frac{I_2}{I_1} = \frac{R_1R'_1}{l^2}. \tag{12.41}$$

The quantity $1/R_1R_1'$, the inverse of the product of two principal radii of curvature, is called the Gauss curvature.

Relation (12.41) may be read out in the following way:

Intensity at an arbitrary point of the rectilinear radius is proportional to the Gauss curvature of the wave front which passes through a given point.

In the particular case where $R_1 = R_1'$ and $R_2 = R_2'$, i.e. where the wave front has the shape of a sphere, the law of intensities takes the form

$$I = \frac{\text{const}}{R^2}. \tag{12.42}$$

12.5. Light Rays in a Non-homogeneous Medium

Below we discuss a few examples of how geometric optics can be used for describing the propagation of light in laminar non-homogeneous media, i.e. in media with the refractive index being a function of one spatial variable.

In a plane-laminar medium the refractive index depends only on one Cartesian coordinate. Let variable z be this coordinate. Then we can write $n = n(z)$. Only the z-th component of $\operatorname{grad} n$ is not equal to zero, and the equations of ray (12.13) take the form

$$\frac{\mathrm{d}}{\mathrm{d}s}\left(n\frac{\mathrm{d}x}{\mathrm{d}s}\right) = 0, \tag{12.43a}$$

$$\frac{\mathrm{d}}{\mathrm{d}s}\left(n\frac{\mathrm{d}y}{\mathrm{d}s}\right) = 0, \tag{12.43b}$$

$$\frac{\mathrm{d}}{\mathrm{d}s}\left(n\frac{\mathrm{d}z}{\mathrm{d}s}\right) = \frac{\mathrm{d}n}{\mathrm{d}z}. \tag{12.43c}$$

Assuming that the components of the tangent vector $\mathbf{s} = [\mathrm{d}x/\mathrm{d}s, \mathrm{d}y/\mathrm{d}s, \mathrm{d}z/\mathrm{d}s]$ are the directional cosines of this vector with respect to the axis of the coordinate system: $\mathrm{d}x/\mathrm{d}s = \cos\theta_1$, $\mathrm{d}y/\mathrm{d}s = \cos\theta_2$,

$dz/ds = \cos\theta_3$ (since its length is unitary), we obtain from (12.43a) and (12.43b)

$$n(z)\cos\theta_1 = n_0\cos\theta_1^0,$$
$$n(z)\cos\theta_2 = n_0\cos\theta_2^0, \tag{12.44}$$

where the constants n_0, θ_1^0, θ_2^0 are found from the boundary conditions. Assuming that $\theta_2^0 = \pi/2$, we obtain from (12.44) that the angle $\theta_2 = \pi/2$ is constant and

$$\frac{dx}{ds} = \frac{n_0}{n(z)}\cos\theta_1^0. \tag{12.45}$$

This and the equality $|\mathbf{s}| = 1$ imply

$$\frac{dz}{ds} = \sqrt{n^2(z) - n_0^2\cos^2\theta_1^0}\cdot\frac{1}{n(z)}. \tag{12.45a}$$

Equation (12.45) describes the behaviour of the ray in the xz-plane. From it follows the equation:

$$\frac{dz}{dx} = \sqrt{\frac{n^2(z) - n_0^2\cos^2\theta_1^0}{n_0^2\cos^2\theta_1^0}}. \tag{12.46}$$

From (12.45a) results that for a decreasing function $n(z)$ the angle of inclination of the ray with respect to the perpendicular (the z-axis, Fig. 12.4) decreases, while for an increasing function $n(z)$ this angle increases.

If the refractive index increases monotonously with the altitude, $\operatorname{grad} n(z) > 0$, and if the ray comes from a certain point P of a non-

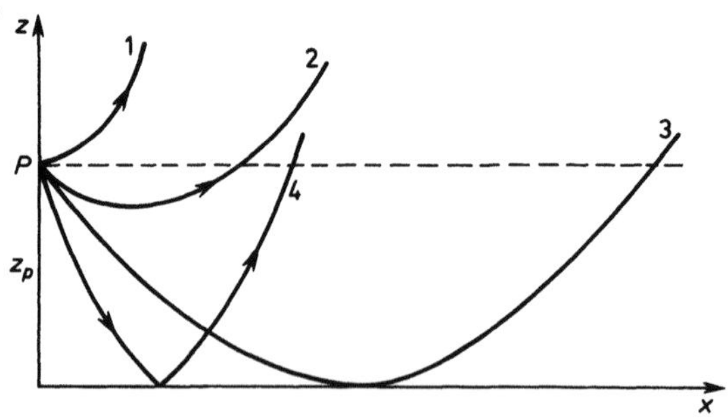

Figure 12.4

homogeneous medium situated at the altitude z_p over a perfectly reflecting surface and makes the angle $\theta_1^0 > 0$ with the level, the ray undergoes curving towards the region with higher value of the refractive index, and for large values of z it tends asymptotically to the perpendicular (ray 1 in Fig. 12.4).

If $\theta_1^0 < 0$, the ray coming from the point P is directed towards the plane $x = 0$. The ray propagates now in the region of decreasing $n(z)$, therefore it deflects more and more towards the perpendicular as the altitude decreases. Two possibilities exist:

a. The ray may turn back towards the region with the increasing $n(z)$ (ray 2 in Fig. 12.4). The ray reaches the turning point at the altitude at which holds the equality

$$n(z) = n_0 \cos\theta_1^0. \tag{12.47}$$

Then $\theta_1^0 = 0$ and the ray becomes parallel to the x-axis. If condition (12.47) takes place for $z = 0$ (i.e. if the turning point of the ray lies on the x-axis), the trajectory of the ray (ray 3 in Fig. 12.4) is a border of the region reached by light coming from the point P and the non-illuminated region (the region of shadow).

b. When the angle θ_1^0 has a sufficiently large negative value, the ray coming from point P will reach the straight line x under the angle $\theta_1 < 0$, and if the plane $x = 0$ is a perfectly reflecting plane, the ray will propagate after reflection towards the increasing values $n(z)$, drawing trajectory 4 in Fig. 12.4.

If $\operatorname{grad} n(z) < 0$ and the altitude increases $(\theta_1^0 > 0)$, the ray coming from point $(0, 0)$ deflects more and more from the perpendicular, and at the altitude at which condition (12.47) holds it becomes parallel to the x-axis. Here we have the change of the sense of the ray, and the ray approaches again the x-axis. If the plane $z = 0$ reflects perfectly, reflection happens on it and the ray will repeat periodically its first cycle of the course (Fig. 12.5).

In a medium whose refractive index $n(z)$ varies symmetrically, e.g.

$$n^2(z) = \begin{cases} n_0^2\left(2 - \dfrac{z^2}{z_s^2}\right), & |z| < |z_s|, \\ n_0^2, & |z| > |z_s|, \end{cases} \tag{12.48}$$

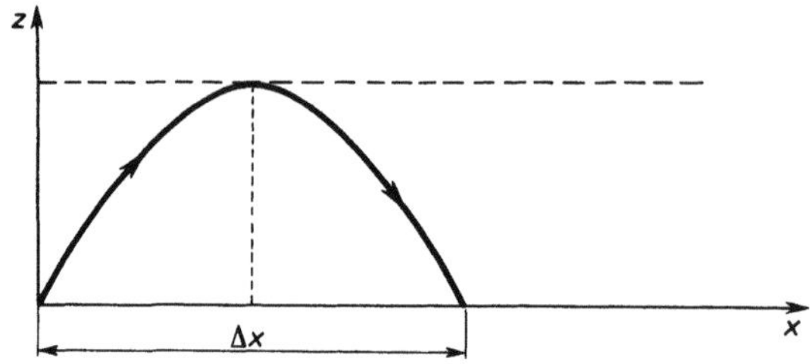

Figure 12.5

the point of emerging of the ray is $(0, z_0)$, while the angle formed by this ray and the x-axis is θ_1^0 and fulfils the condition

$$\cos^2\theta_1^0 \geq \frac{z_s^2}{2z_s^2 - z_0^2}. \tag{12.49}$$

Rays satisfying condition (12.49) in a medium with the refractive index (12.48) have a limited region of propagation. They propagate oscillating in the area confined by the planes $z = \pm z_s$. In this case in a non-homogeneous medium arises a waveguide channel of width $2z_s$. The justification for this is given below.

For a medium with the refractive index (12.48) the equation of ray (12.46) takes the form

$$\frac{dz}{dx} = \frac{\sqrt{\left[2 - \left(\frac{z}{z_s}\right)^2\right] - \left[2 - \left(\frac{z_0}{z_s}\right)^2\right]\cos^2\theta_1^0}}{\sqrt{2 - \left(\frac{z_0}{z_s}\right)^2}\cos\theta_1^0}. \tag{12.50}$$

This equation has the solution of the form

$$z = z_s\sqrt{2 - \left[2 - \left(\frac{z_0}{z_s}\right)^2\right]\cos^2\theta_1^0}\,\sin\left[\frac{x}{z_s\sqrt{2 - \left(\frac{z_0}{z_s}\right)^2}\cos\theta_1^0} + \right.$$

$$\left. + \arcsin\frac{\frac{z_0}{z_s}}{\sqrt{2 - \left[2 - \left(\frac{z_0}{z_s}\right)^2\right]\cos^2\theta_1^0}}\right]. \tag{12.51}$$

The extremal distances of the rays from the x-axis are

$$z_l = (-1)^l z_s \sqrt{2 - \frac{2z_s^2 - z_0^2}{z_s^2} \cos^2 \theta_1^0} \quad (l = 1, 2, \ldots). \quad (12.52)$$

The values (12.52) show evidently the restricted permissible region for light rays. The points of intersection of the ray with the x-axis are given by

$$x_m = z_s \left[\sqrt{2 - \left(\frac{z_0}{z_s} \right)^2} \cos \theta_1^0 \right] \times$$

$$\times \left[\pi m - \arcsin \frac{\dfrac{z_0}{z_s}}{\sqrt{2 - \left[2 - \left(\dfrac{z_0}{z_s} \right)^2 \right] \cos^2 \theta_1^0}} \right], \quad (12.53)$$

$m = 1, 2, \ldots$ The distance between two successive points of intersection of the ray with the x-axis is

$$\Delta x = \pi z_s \sqrt{2 - \left(\frac{z_0}{z_s} \right)^2} \cos \theta_1^0. \quad (12.54)$$

As can be seen from the above the region of propagation of light is limited, and its width does not exceed $2|z_s|$, i.e. it is closely connected with the distribution of the refractive index (12.48) of the medium in question. This phenomenon is called the *waveguide effect*, and the region of propagation of light forms an optical waveguide. The trajectories of the rays are shown in Fig. 12.6.

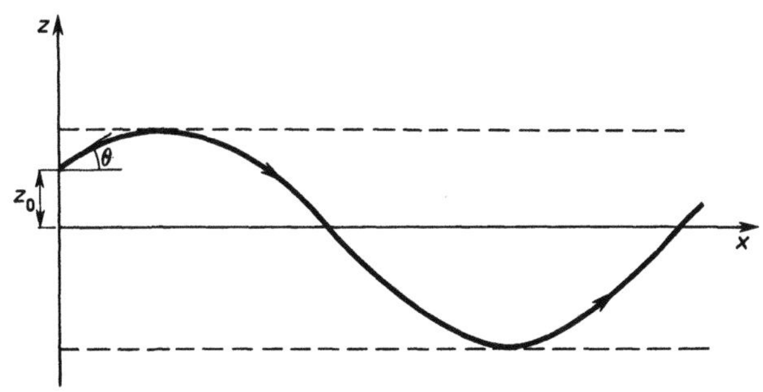

Figure 12.6

If condition (12.49) is not fulfilled, the rays emerging from the point $(0, z_0)$ under the angle θ_1^0 will not reach the turning point in the region $|z| \leq |z_s|$ and will pass into the region $|z| > |z_s|$, i.e. they will not be subject to the waveguide effect. This is a proof that not only the properties of a medium have the influence on the effect of waveguide propagation of the rays but also the spatial orientation of the rays themselves.

12.6. Propagation of Waves in a Non-homogeneous Medium

Propagation of monochromatic wave in a non-homogeneous medium is described by the equation

$$\nabla^2 u + k^2(x, y, z)u = 0, \tag{12.55}$$

where $u = (x, y, z)$ is a wave function, while $k(x, y, z)$ the wave number dependent on spatial variables. In a general case the solution of equation (12.55) cannot be expressed by elementary functions. Therefore approximate methods of finding the wave fields in non-homogeneous media are of great significance. One of them is, for arbitrary but sufficiently slow spatial variations of the medium properties, a geometric method. This method is most useful if the properties of the medium vary only along one of the axes of the rectangular coordinate system, for example along the z-axis, remaining constant in the planes perpendicular to that axis. To such media we can number, in the first approximation. Earth's atmosphere (disregarding the curvature of the Earth) and water of the seas.

We assume that a non-homogeneous medium fills the space $z > 0$ and that its refractive index is $n = n(z)$. A wave dependent on variables x and z and upon the boundary plane $z = 0$ from the side $z < 0$ is being considered.

In the case under consideration the eikonal equation (12.7) takes the form

$$0 = \left(\frac{\partial S}{\partial x}\right)^2 + \left(\frac{\partial S}{\partial z}\right)^2 - n^2(z) = \Phi\left(z, \frac{\partial S}{\partial x}, \frac{\partial S}{\partial z}\right). \tag{12.56}$$

It is a nonlinear partial differential equation of the first order. The system of ordinary equations characteristic for equation (12.56) is expressed in the following way:

$$\frac{\mathrm{d}x}{P_x} = \frac{\mathrm{d}z}{P_z} = \frac{\mathrm{d}S}{p_x P_x + p_z P_z} = -\frac{\mathrm{d}p_x}{X_x} = -\frac{\mathrm{d}p_z}{X_z}, \tag{12.57}$$

where

$$p_{x_j} = \frac{\partial S}{\partial x_j}, \quad P_{x_j} = \frac{\partial \Phi}{\partial p_{x_j}}, \quad X_{x_j} = -\frac{\partial \Phi}{\partial x_j} \tag{12.57a}$$

(above we have used the symbols $x = x_1$, $y = x_2$, $z = x_3$), $\mathrm{d}S$ is an element of the phase surface, while n the refractive index.

On introducing the vector $\mathbf{s} = \frac{1}{n}\nabla S$ and taking into account (12.56), equations (12.57) may be written in a vector form

$$\frac{\mathrm{d}(n\mathbf{s})}{\mathrm{d}s} = \nabla n, \quad \frac{\mathrm{d}\mathbf{r}}{\mathrm{d}s} = \mathbf{s}, \tag{12.57b}$$

or in the form of one scalar equation

$$\frac{\mathrm{d}S}{\mathrm{d}s} = \pm n. \tag{12.57c}$$

The above equations, together with the boundary condition which determines the direction of the ray for $\mathbf{r} = \mathbf{r}_0$, define the trajectory of the ray $\mathbf{r} = \mathbf{r}(s)$. If the trajectory is known, the phase (eikonal S) may be expressed as

$$S(\mathbf{r}) = \int_{M_0}^{M} n[\mathbf{r}(s)]\,\mathrm{d}s. \tag{12.58}$$

Often it is more convenient to use coordinates connected with the ray; the so-called ray coordinates. Let the wave front be defined by the equation $S = S_0(\mathbf{r}) = \text{const}$, and let parameters α and β determine the location of point M_0 on this surface. From each point $M_0(\alpha, \beta)$ we draw a radius perpendicular to the phase surface at this point. Then the points on this radius may be described by the quantity

$$S = S_0 \pm \int_{M_0}^{M} n\,\mathrm{d}s. \tag{12.59}$$

Thereby the parameters α and β describe the direction of the ray at point M_0, and eikonal S determines a point on the radius. Each point $M(\mathbf{r})$ of the ray is a univocal function α, β and S, i.e. $\mathbf{r} = (\alpha, \beta, S)$.

In the case in question, equations (12.57) and (12.57a) imply

$$\frac{\partial S}{\partial x} = p_x = \text{const.} \tag{12.60}$$

If for $z = 0$ the direction of the ray makes the angle θ_1 with the x-axis and $n(0) = n_0$, then for $z = 0$ is $p_x = n_0 \cos\theta_1$. From equation (12.56) it follows:

$$p_z = \frac{\partial S}{\partial z} = \pm\sqrt{n^2(z) - n_0^2 \cos^2\theta_s}. \tag{12.61}$$

Hence

$$S = \pm\int\sqrt{n^2(z) - n_0^2\cos^2\theta_1}\,\mathrm{d}z + S_1(x). \tag{12.62}$$

The boundary condition $p_x = n_0\cos\theta_1 = \text{const}$ implies

$$S_1(x) = x n_0 \cos\theta_1. \tag{12.62a}$$

Finally, for equation (12.56) the eikonal $S(x, z)$ has the form

$$S(x, z) = x n_0 \cos\theta_1 \pm \int_0^z \sqrt{n^2(z) - n_0^2\cos^2\theta_1}\,\mathrm{d}z. \tag{12.63}$$

To determine the amplitude of the discussed wave in a non-homogeneous medium, knowing its eikonal (12.63), we have to use equation (12.6) or (12.32). This leads to the following equation for the amplitude of the wave $A(x, z)$ in a medium with $n = n(z)$:

$$2\frac{\partial A}{\partial x} n_0 \cos\theta_1 \pm 2\sqrt{n^2(z) - n_0^2\cos^2\theta_1}\,\frac{\partial A}{\partial z} \pm$$

$$\pm A\frac{\mathrm{d}}{\mathrm{d}z}\sqrt{n^2(z) - n_0^2\cos^2\theta_1} = 0, \tag{12.64}$$

with the boundary conditions

$$\begin{aligned} &z = 0, \\ &A(x,0) = A_0(x). \end{aligned} \tag{12.64a}$$

On introducing the function $\psi = \ln A$ and the implicit function $\Omega(x,z,\psi) = 0$, equation (12.64) assumes the form

$$\begin{aligned} 2n_0\cos\theta_1 \frac{\partial\Omega}{\partial x} &\pm 2\sqrt{n^2(z) - n_0^2\cos^2\theta_1}\,\frac{\partial\Omega}{\partial z} \pm \\ &\pm \frac{\mathrm{d}}{\mathrm{d}z}\left(\sqrt{n^2(z) - n_0^2\cos^2\theta_1}\right)\frac{\partial\Omega}{\partial\psi} = 0. \end{aligned} \tag{12.65}$$

The characteristic equations for (12.65) are as follows:

$$\begin{aligned} \frac{\mathrm{d}x}{2n_0\cos\theta_1} &= \frac{\pm\mathrm{d}z}{2\sqrt{n^2(z) - n_0^2\cos^2\theta_1}} \\ &= \frac{\mp\mathrm{d}\psi}{\frac{\mathrm{d}}{\mathrm{d}z}\sqrt{n^2(z) - n_0^2\cos^2\theta_1}}. \end{aligned} \tag{12.66}$$

They have the solutions of the form

$$\begin{aligned} &\frac{x}{2n_0\cos\theta_1} \pm \tfrac{1}{2}\int\frac{\mathrm{d}z}{\sqrt{n^2(z) - n_0^2\cos^2\theta_1}} = C_1, \\ &\tfrac{1}{2}\ln\sqrt{n^2(z) - n_0^2\cos^2\theta_1} + \psi = C_2, \end{aligned} \tag{12.67}$$

where C_1 and C_2 are constants. From now on we use the notation

$$N(z) = \int\frac{\mathrm{d}z}{\sqrt{n^2(z) - n_0^2\cos^2\theta_1}}. \tag{12.67a}$$

The general solution of equation (12.65) will be the surface defined by the function $F(C_1, C_2) = \Omega(x,z,\psi)$. Its explicit form we obtain from the condition that the surface $F(C_1, C_2) = 0$ should pass through the curve determined by boundary conditions (12.64a). This leads to the relation

$$\psi\left(2n_0\cos\theta_1[C_1 \pm N(0)]\right) = \tfrac{1}{2}\ln\sqrt{n^2(0) - n_0^2\cos^2\theta_1} - C_2. \tag{12.68}$$

Hence, taking into account (12.67), we obtain an explicit form of the equation $F(C_1, C_2) = 0$

$$\psi\left(x \mp [N(z) - N(0)]\,2n_0\cos\theta_1\right) + \tfrac{1}{2}\ln\sqrt{n^2(0) - n_0^2\cos^2\theta_1} - \\ - \tfrac{1}{2}\ln\sqrt{n^2(z) - n_0^2\cos^2\theta_1} - \psi(x, z) = 0. \tag{12.69}$$

Thus, having taken into account the relation $\psi(x, z) = \ln A(x, z)$, we obtain the sought form of the wave amplitude $A(x, z)$ in a non-homogeneous medium

$$A(x, z) = A\left(x \mp [N(z) - N(0)]\,2n_0\cos\theta_1\right)\sqrt[4]{\frac{n^2(0) - n_0^2\cos^2\theta_1}{n^2(z) - n_0^2\cos^2\theta_1}}. \tag{12.70}$$

Having amplitude (12.70) and phase (12.63) of the wave, we can give the form of the wave in a laminar non-homogeneous medium in the geometric optics approximation

$$u(x, z) \\ = A\left(x - [N(z) - N(0)]\,2n_0\cos\theta_1\right)\sqrt[4]{\frac{n^2(0) - n_0^2\cos^2\theta_1}{n^2(z) - n_0^2\cos^2\theta_1}} \times \\ \times \exp\left[\mathrm{i}k_0\left(n_0\cos\theta_1 x + \int_0^z \sqrt{n^2(z) - n_0^2\cos^2\theta\,\mathrm{d}z}\right)\right] + \\ + A\left(x + [N(z) - N(0)]\,2n_0\cos\theta_1\right)\sqrt[4]{\frac{n^2(0) - n_0^2\cos^2\theta_1}{n^2(z) - n_0^2\cos^2\theta_1}} \times \\ \times \exp\left[\mathrm{i}k_0\left(n_0 x\cos\theta_1 - \int_0^z \sqrt{n^2(z) - n_0^2\cos^2\theta_1}\,\mathrm{d}z\right)\right]. \tag{12.71}$$

Solution (12.71) is the sum of two components. Each of them can be presented in the form

$$U_j(x, z) = \exp[\mathrm{i}k_0 n_0\cos\theta_1 x]\,f_j(z) \quad (j = 1, 2), \tag{12.72}$$

where $f_j(z)$ is an approximate solution of the equation

$$\frac{d^2 f}{dz^2} + k_0^2[n^2(z) - n_0^2 \cos^2\theta_1] f(z) = 0 \tag{12.73}$$

and is expressed by the formula

$$f_j(z) = A\big(x + (-1)^j [N(z) - N(0)] 2n_0 \cos\theta_1\big) \times$$

$$\times \sqrt[4]{\frac{n^2(0) - n_0^2\cos^2\theta_1}{n^2(z) - n_0^2\cos^2\theta_1}} \times$$

$$\times \exp\Big[ik_0(-1)^{j+1} \int_0^z \sqrt{n^2(z) - n_0^2\cos^2\theta_1}\, dz\Big]. \tag{12.73a}$$

The solution (12.73a) is the *approximate solution* obtained by means of the WKB (Wentzel, Kramers, Brillouin) method.

Although (12.71) are the solutions of the boundary problem (12.64a), they do not include waves reflected from the non-homogenous medium. This follows from the approximation of the equations of geometric optics. Reflection can take place only in such areas where the conditions for application of geometric optics are broken. Deriving equations (12.7) we neglect a term which in some cases may have an essential influence on the form of the solution. It seems that a more correct method is neglecting small terms in the obtained solutions of the equations instead of neglecting them in the equations themselves. For this purpose we seek the solutions of equation (12.55) in the form of the asymptotic extension

$$u(\mathbf{r}) = [A_0(\mathbf{r}) + k_0^{-1}A_1(\mathbf{r}) + k_0^{-2}(\mathbf{r}) + \ldots] \exp[ik_0 S(\mathbf{r})], \tag{12.74}$$

where $\mathbf{r}$ is the position vector of the considered point of space. Substituting (12.74) into equation (12.55) and comparing the coefficients by different powers of k_0 to zero, we obtain the following system of equations:

$$[n^2(\mathbf{r}) - (\nabla S)^2] A_0 = 0,$$

$$A_0 \nabla^2 S + 2\nabla S \cdot \nabla A_0 = 0,$$

$$A_1\nabla^2 S + 2\nabla S\cdot\nabla A_0 = 0, \qquad (12.75)$$
$$A_1\nabla^2 S + 2\nabla S\cdot\nabla A_1 = \mathrm{i}\nabla^2 A_0,$$
$$\cdots\cdots\cdots\cdots\cdots\cdots\cdots\cdots\cdots$$

The first two equations (12.75) agree with equations (12.6) and (12.7). Finding from them $S(\mathbf{r})$ an $A(\mathbf{r})$ we can get, with the use of the other equations (12.75), the successive corrections for the obtained values.

The results given above are true for an arbitrary dependence of the refractive index on variable z. For a given dependence $n(z)$ only the function $N(z)$ (12.67a) will take a definite form. Of course the rate of the variations of the properties of a non-homogeneous medium should stay in agreement with the assumptions of geometric optics.

12.7. Optical Mapping

From point M_0 (at which a source may be situated) placed in a medium with the refractive index $n(x, y, z)$ may emerge an infinite number of rays. Through another arbitrary point M of this medium may generally pass only a finite number of rays. In particular cases if there is such a point M_1 that the number of rays coming from point M_0 and going through M_1 is infinite, then point M_1 is called a sharp (stigmatic) mapping of point M_0.

An optical system is formed by consecutive refracting and reflecting surfaces limiting different homogeneous media. An optical system is perfect if for each point M_0 of a three-dimensional region called an *object space* it gives a stigmatic mapping into a certain point M_1 of a three-dimensional region called an *image space.*

If point M_0 draws curve C_0 in the object space, point M_1 draws in the image space a curve C_1 conjugated to C_0. Both curves are not necessarily geometrically similar to each other. If curve C_0 in the object space is geometrically similar to its image C_1, we speak of perfect mapping. An optical system which gives stigmatic mapping of a three-dimensional object is called an *absolute system*. For an absolute system holds Maxwell's theorem which states:

In the case of an absolute optical system the optical length of an arbitrary curve C_0 lying in the object space is equal to the optical length of its mapping C_1

$$\int_{C_0} n_0 \mathrm{d}s_0 = \int_{C_1} n_1 \mathrm{d}s_1. \tag{12.76}$$

From Maxwell's theorem it follows that an infinitesimal triangle with the sides $\mathrm{d}s_0^{(1)}$, $\mathrm{d}s_0^{(2)}$, $\mathrm{d}s_0^{(3)}$ in the object space, where the refractive index is n_0, is transformed in a triangle with the sides $\mathrm{d}s_1^{(1)}$, $\mathrm{d}s_1^{(2)}$ and $\mathrm{d}s_1^{(3)}$ in the image space, where the refractive index is n_1.

In view of Maxwell's theorem take place the following equalities:

$$n_0 \mathrm{d}s_0^{(j)} = n_1 \mathrm{d}s_1^{(j)} \quad (j = 1, 2, 3). \tag{12.77}$$

When the values n_0 and n_1 are constant, from (12.77) results that the triangles in the object space and the image space are similar, their appropriate angles are equal. Thus the mapping obtained with an absolute optical instrument is a conformal transformation. Since the only conformal transformation possible in a three-dimensional space is projective transformation or inversion, the next theorem follows:

Conformal transformation performed by an absolute system is a projective transformation, an inversion, or the combination of these two transformations.

Projective transformation of point $M_0(x, y, z)$ from the object space into point $M_1(x', y', z')$ from the image space is realized through the relations

$$x'_j = \frac{a'_j x + b'_j y + c'_j z + d'_j}{a'_0 x + b'_0 y + c'_0 z + d'_0} = \frac{F_j}{F_0}, \tag{12.78a}$$

$$x_k = \frac{a_j x' + b_j y' + c_j z' + d_j}{a_0 x' + b_0 y' + c_0 z' + d_0} = \frac{F'_j}{F'_0}, \tag{12.78b}$$

where we use the notation: $x_1 = x$, $x_2 = y$, $x_3 = z$ $(j, k = 1, 2, 3)$.

The image of point M_0 lying in the plane $F_0 = 0$ is point M_1 lying in ifinity. Analogically, the image M_1 lying in the plane $F'_0 = 0$ has its object point M_0 in infinity. The plane $F_0 = 0$ is called the *focal plane* in the object space, while $F'_0 = 0$ the focal plane in the image space.

Most of the optical systems contain rotational surfaces with the same symmetry axis, e.g. the z-axis of the coordinate system is the symmetry axis of the optical system. These are the so-called *centred optical systems.* In the case of centred optical systems projective transformations (12.78) become simplified. Mapping of an arbitrary point M_0 lying in the plane passing through the symmetry axis (this plane is taken as the plane $x = 0$) is described by the projective transformation of the form

$$\begin{aligned} y' &= \frac{b'_2 y}{c'_0 z + d'_0}, \qquad z' = \frac{c'_3 z + d'_3}{c'_0 z + d'_0}, \\ y &= \frac{c_0 d'_3 - c_3 d'_0}{b_2} \frac{y'}{c_0 z' - c_3}, \qquad z = \frac{-d_0 z' + d_3}{c_0 z' - c_3}, \end{aligned} \tag{12.79}$$

Thus, in the presence of the symmetry axis of the optical system the projective transformation is determined by five independent parameters. In this case the focal planes are given by the equations

$$c'_0 z + d'_0 = 0, \qquad c_0 z' - c_3 = 0. \tag{12.80}$$

These planes intersect the symmetry axis at the points

$$z = -\frac{d'_0}{c'_0} \quad \text{and} \quad z' = \frac{c_3}{c_0}. \tag{12.81}$$

They are called *principal foci.*

Both in the object and in the image space we can introduce own reference systems, placing their origins in appropriate principal foci, i.e. putting

$$\begin{aligned} y &= Y, \quad c'_0 z + d'_0 = c'_0 Z, \\ y' &= Y', \quad c_0 z' - c_3 = c_0 Z'. \end{aligned} \tag{12.82}$$

Then, introducing the symbols

$$f = \frac{b'_2}{c'_0} \quad \text{and} \quad f' = \frac{c'_0 d'_3 - c'_3 d'_0}{c'_0}, \tag{12.83}$$

we can rewrite relations (12.79) in the form

$$\frac{Y'}{Y} = \frac{f}{Z} = \frac{Z'}{f'}. \tag{12.84}$$

f is called the *focal length* in the object space, f' the focal length in the image space.

Expression (12.84) shows that transverse magnification Y'/Y depends only on Z, and does not depend on Y. Transverse magnification is equal to a unity if $Z = f$ and $Z' = f'$. The planes

$$Z = f \quad \text{and} \quad Z' = f' \tag{12.85}$$

are called *principal planes* of the system and they are denoted by letters H and H'. Intersection of these planes with the symmetry axis of the system determines the principal points of the optical system.

Two points of principal foci F and F' and the principal points of the optical system H and H' fully determine the properties of optical systems within the framework of the so-called *Gauss optics*. It is that part of geometric optics which describes propagation of rays in infinitesimaly small volume concentrated around the symmetry axis of the optical system. It is called also *paraxial optics*.

12.8. Matrix Description of Gauss' Optics

Let us consider transition of light through the system which is a set of different homogeneous media with precisely defined boundary surfaces. These systems separate regions with refractive indices n_1, n_2, n_3...

Thus, the system under consideration makes up a set of transparent media separated by spherical surfaces whose centres lie on a straight line called the *optical axis* of the discussed system. Of all possible light rays we take into consideration only these which passing through the system remain close to the optical axis and make small angles with respect to this axis. It is, therefore, a case of the so-called *paraxial transformations*, often called *Gauss' optics*.

Under these assumptions, sines and tangents of arbitrary angles made by the rays with the optical axis may be replaced by the values of angles themselves given in arc measure.

To describe the behaviour of a ray in paraxial approximation we introduce a dextrorotatory system of rectangular coordinates with the z-axis covering the optical axis of the system and with the sense the same as the direction of propagation of light. The y-axis is chosen as lying in the plane of the figure and pointed upwards, whereas the x-axis is perpendicular to this plane.

On the assumption of rotational symmetry around the z-axis, it is enough to consider rays which lie in the yz-plane. The paths of these rays crossing different refractive surfaces of the system consist of segments of straight lines. The position of each of them is defined by the coordinte y of one of its points and by the angle made by the ray with the z-axis. Therefore, with respect to the arbitrary plane $z = \text{const}$, a light ray is described by two parameters: the height y, on which it crosses the plane, and the angle α, which it makes with the z-axis. This angle is positive if going from the z-axis to the direction of the ray we move counter-clockwise.

According to the law of refraction, if a ray crosses the boundary between two media, we have the values $n \sin\alpha = \text{const}$. Within the framework of Gauss' optics this quantity may be replaced by $n\alpha = V$, since the condition $\sin\alpha = \alpha$ is fulfilled.

As the ray crosses the system of refracting spherical surfaces separating media of different refractive indices and of different thickness, for the description of the behaviour of light we have to consider closely two basic processes: (a) the displacement of the ray along the optical path between the successive refracting surfaces and (b) the refraction of the ray at the boundary surface of two regions with different refractive indices.

In order to examine the trajectory of the light ray in the considered optical system, we introduce a number of reference planes

$z = \text{const}$. In each of these planes the parameters of the ray (its height y and the directional parameter $V = n\alpha$) may be different. And so, in the reference plane OP_1 the ray may have the parameters: y_1 and V_1, while in the other plane OP_2 – y_2 and V_2. Thus in each reference plane the behaviour of the ray.is described by a vector of the form

$$\begin{bmatrix} y_j \\ V_j \end{bmatrix}.$$

Two vectors describing the behaviour of the ray in different reference planes are related with each other through linear dependencies. They can be written in the form of matrix equation

$$\begin{bmatrix} y_2 \\ V_2 \end{bmatrix} = \begin{bmatrix} A & B \\ C & D \end{bmatrix} \begin{bmatrix} y_1 \\ V_1 \end{bmatrix}, \tag{12.86}$$

in which the matrix of transformation has the determinant equal to unity.

As we can choose the reference planes arbitrarily, we can assign to each element of the optical system in question an appropriate matrix of transformation of parameters of the ray. Within the framework of Gauss' optics all transformations of parameters of the ray can be resolved into displacements and refractions. Multiplying in an appropriate order matrices of these two kinds of elementary transformations, we obtain the matrix of transformation for the whole optical system.

12.9. The Matrix of Displacement

In order to find the matrix of displacement, we take into consideration two reference planes OP_1 and OP_2, which confine a homogeneous medium with the refractive index n and of the thickness d.

If in the plane OP_1 the ray had parameters y_1, V_1, because of the homogeneity of the medium the parameters of the ray in the plane OP_2 will be respectively

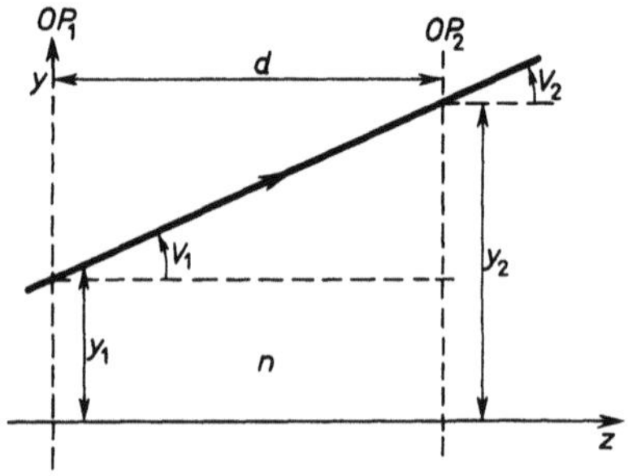

Figure 12.7

$$y_2 = y_1 + d\frac{V_1}{n}, \qquad V_2 = V_1. \tag{12.87}$$

These relations are obvious, they are illustrated in Fig. 12.7. Equations (12.87) written in a matrix form are

$$\begin{bmatrix} y_2 \\ V_2 \end{bmatrix} = \begin{bmatrix} 1 & T \\ 0 & 1 \end{bmatrix} \begin{bmatrix} y_1 \\ V_1 \end{bmatrix}, \tag{12.88}$$

where the matrix element $T = d/n$.

Thus the displacement of the ray in the direction of the z-axis is described by the matrix

$$\check{T} = \begin{bmatrix} 1 & T \\ 0 & 1 \end{bmatrix}. \tag{12.89}$$

12.10. The Matrix of Refraction

The matrix of refraction describes the change of the parameters of the ray caused by the spherical surface which separates two media with the refractive indices n_1 and n_2.

If the centre of the curvature of this surface is situated right from this surface, the curvature is positive. The behaviour of the ray in this surface separating two media is shown in Fig. 12.8. On both sides of the refracting surface both parameters of the ray y and V ($V = n\alpha$) have positive values. The angles shown in the figure are

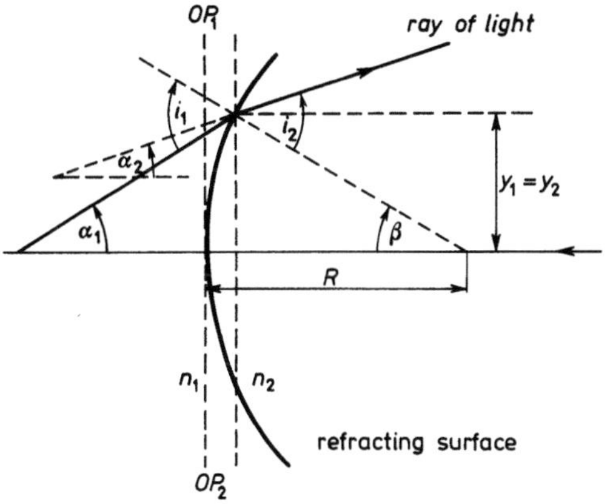

Figure 12.8

magnified exaggeratedly. In reality, according to the assumption of Gauss' optics, these angles are small. This is the reason why the distance from the refractive plane OP_1 led by central point of the refracting surface to the reference plane OP_2 passing through the point of intersection of the ray with the refracting surface is to neglect. In the general case the distance between these two surfaces is $r(1 - \cos\alpha)$; for paraxial rays, because of small value of the angle, we can assume $\cos\alpha \cong 1$. Therefore, in fact, for paraxial rays the distance between the reference planes chosen in a way described above is negligible.

As a consequence, it is possible to write the equality

$$y_1 = y_2. \tag{12.90}$$

In the case of paraxial rays the law of refraction

$$n_1 \sin i_1 = n_2 \sin i_2$$

for small angles takes the form

$$n_1 i_1 = n_2 i_2. \tag{12.91}$$

From the theorem on the external angles of a triangle (Fig. 12.8) result the relations

$$i_1 = \alpha_1 + \beta = \alpha_1 + \frac{y_1}{r},$$
$$i_2 = \alpha_2 + \beta = \alpha_2 + \frac{y_2}{r}. \tag{12.92}$$

Taking into account relations (12.92) in equality (12.91), we arrive at the relationship

$$n_1\left(\alpha_1 + \frac{y_1}{r}\right) = n_2\left(\alpha_2 + \frac{y_1}{r}\right). \tag{12.93}$$

Equality (12.90) together with (12.93) may be written as one matrix equation

$$\begin{bmatrix} y_2 \\ n_2\alpha_2 \end{bmatrix} = \begin{bmatrix} 1 & 0 \\ -\dfrac{n_2 - n_1}{r} & 1 \end{bmatrix} \begin{bmatrix} y_1 \\ n_1\alpha_1 \end{bmatrix}. \tag{12.94}$$

Thus the matrix of refraction of the ray on the spherical surface in the paraxial case is of the form

$$\check{R} = \begin{bmatrix} 1 & 0 \\ -\dfrac{n_2 - n_1}{r} & 1 \end{bmatrix}. \tag{12.95}$$

12.11. The Matrix of Transformation of a Ray for an Optical System

If an optical system is composed of l refracting surfaces separating $l - 1$ media, then in order to find the matrix of transformation of the ray for the system, we have to follow the description below.

As the input reference plane OP_1 we choose the plane situated left from the first refracting surface. The next reference planes OP_2 and OP_3 are placed directly on the left and right side of the first refracting surface (their location with respect to the refracting surface is such as described in Sec. 12.10). The succeeding reference planes OP_4 and OP_5 are chosen in much the same way on the both sides of the second refracting surface, and so on. The last output reference plane OP_{2l+2} we choose in the distance d_b from the last refracting surface.

In order to get the general matrix $\check{M}$ of transformation of a ray of the system, which transforms directly the vector of the input ray $\begin{bmatrix} y_1 \\ n_1\alpha_1 \end{bmatrix}$ into the vector of the output ray $\begin{bmatrix} y_2 \\ n_2\alpha_2 \end{bmatrix}$, we should act in the following way.

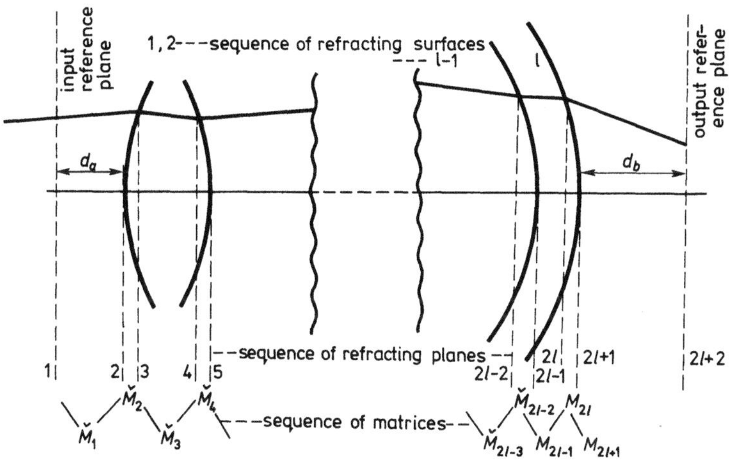

Figure 12.9

To each element of the optical system we assign an appropriate matrix of transformation with the ordinal number the same as the number of the reference plane situated left from the nearest element of the system (see Fig. 12.9). The matrix of transformation of the ray by the optical system may be presented as a product of the matrices assigned to the individual elements of the system, i.e.

$$\check{M} = \check{M}_{2l+1}\check{M}_{2l}\dots\check{M}_2\check{M}_1. \tag{12.96}$$

It is worth noting that the matrices of the component elements are arranged in the product according to the order reverse to that taken in their numeration. It is a consequence of the succeeding changes of the vector of the ray caused by the following elements of the complex optical system.

12.12. Physical Sense of the Elements of the Matrix of Transformation of a Ray

The knowledge of the elements of the matrix

$$\check{M} = \begin{bmatrix} A & B \\ C & D \end{bmatrix} \tag{12.97}$$

of the optical system in question allows us to associate the vector of a ray at the input with the vector of a ray at the output of the optical system. They have a defined physical sense, which is determined by the analysis of particular cases.

Thus, if $D = 0$, we have $V_2 = Cy_1$. It means that all rays coming from the same point of the input reference plane have the same angle with respect to the optical axis of the system at the points of the output reference plane. From this it follows that the input reference plane OP_1 is in this case the front focal plane of the system (Fig. 12.10).

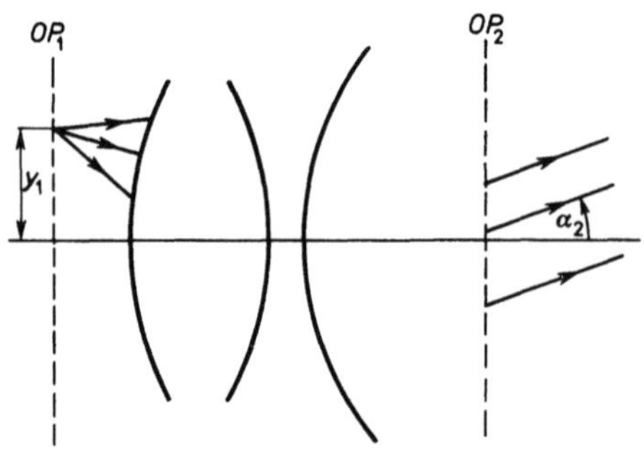

Figure 12.10

If in the matrix of transformation of a ray the element $B = 0$, then holds the relationship

$$y_2 = Ay_1, \tag{12.98}$$

which informs that all rays leaving the point y_1 of the input plane of the optical system go through the same point y_2 lying in the output plane OP_2. It means that the point y_2 is the image of the point y_1. OP_1 and OP_2 are the *conjugate planes* of the considered optical system. The element of the matrix of this system $A = y_2/y_1$ is a measure of linear magnification occuring in this case.

An optical system whose matrix has the element $C = 0$ generates the relation

$$n_2\alpha_2 = Dn_1\alpha_1. \tag{12.99}$$

Such an optical system transforms a beam of parallel rays into a parallel beam, in general the change of direction of their propagation may take place. Angular magnification given by the optical system in question is

$$\frac{\alpha_2}{\alpha_1} = D\frac{n_1}{n_2}. \tag{12.100}$$

In the case of an optical system with $A = 0$, we have the relation

$$y_2 = Bn_1\alpha_1. \tag{12.101}$$

It denotes that rays entering the system under the same angle are transformed into a convergent beam focused in the point y_2 of the output plane OP_2. This plane is the *back focal plane of the system* (Fig. 12.11).

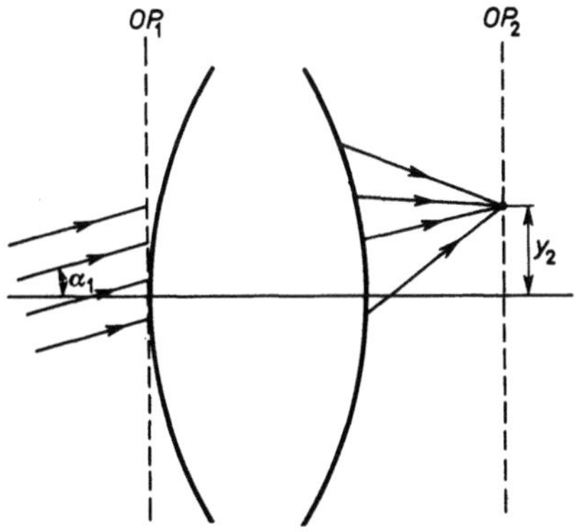

Figure 12.11

12.13. Cardinal Points of an Optical System

The knowledge of the elements of the matrix of transformation of a ray for an optical system allows us to find the position of its cardinal points with respect to the input and output reference planes.

To the *cardinal points* of the optical system belong:

(a) the front and back focus of the system: F and F';

(b) the principal points of the system: H and H';

(c) the nodal points of the system: K and K'.

The *principal points* of the optical system are the points of intersection of the principal planes of an optical system with the optical axis. The principal planes of the optical system have such a property that an object placed in one of them gives its image in the other, the linear magnification being then a unity.

The *foci* of the optical system are the points on which focus the rays of the beam parallel to the optical axis of the system after emerging from the system.

The *nodal points* of an optical system have such a property that a light ray falling on the system and going through one of these points emerges the system in the direction parallel to the incident ray, and the straight line determining this direction goes through the other nodal point of the system.

The knowledge of the cardinal points of the system allows us to determine the image of the object produced by the system as a whole, without considering the work of its component parts.

And so, to obtain the image of point P (Fig. 12.12) we can choose

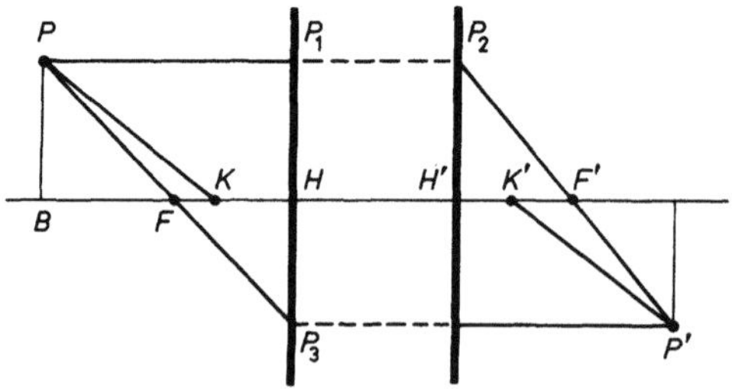

Figure 12.12

one of three pairs of the rays PP_1, PP_3; PP_1, PK; PP_3, PK. The ray PP_1 running parallel to the optical axis crosses the front principal plane of the system and runs further to the back focus of the system F'. The ray PP_3 which passes through the front focus of the system F reaches the first principal plane and then becomes parallel to the optical axis. Finally, the ray PK passing through the first nodal point of the system goes out from the second nodal point K' in the same direction as it initially had, giving the ray $K'P'$. All these rays intersect at a point P', which is the image of the point P.

If for the optical system in question we know the matrix of transformation of a ray

$$\check{M} = \begin{bmatrix} A & B \\ C & D \end{bmatrix}, \tag{12.102}$$

which connects the parameters of the ray in the chosen input reference plane OP_1 with its parameters in the output plane OP_2, then we can find the position of the cardinal points of the system and determine the distance from these points to the reference planes.

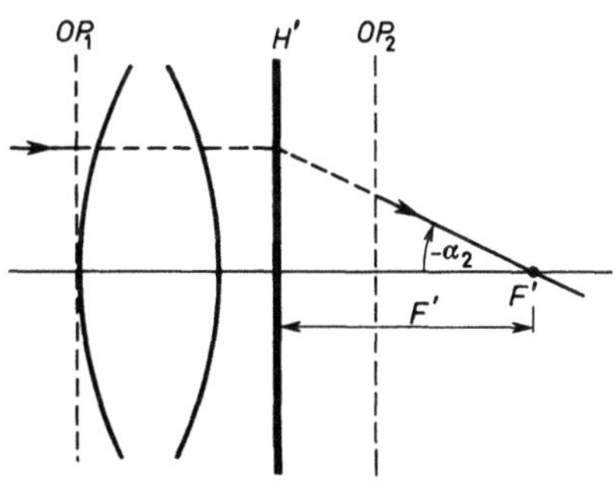

Figure 12.13

a. *The position of the foci and of the principal points of the system.* The ray which is parallel to the optical axis and which has in the input reference plane the parameters y_1, 0 (Fig. 12.13) will be at the same height in the back principal plane, behind which it changes its direction and then goes through the focus F'. According to the matrix equation, the parameters of the ray in the output plane OP_2 are

$$y_2 = Ay_1, \quad \alpha_2 = -C\frac{y_1}{n_2}. \tag{12.103}$$

From the analysis of Fig. 12.13 it follows that the distance from the focus F' to the plane OP_2 is

$$s_2 = -n_2\frac{A}{C}, \tag{12.104}$$

while its distance to the plane H' is given by

$$F' = -\frac{y_1}{\alpha_2} = -\frac{n_2}{C}. \tag{12.105}$$

Hence, the back principal plane of the system H' is distant from OP_2 by the value

$$d_2 = s_2 - F' = n_2\frac{1-A}{C}. \tag{12.106}$$

If the ray going through the focus F enters the optical system under the angle α_1, then having crossed the front principal plane H, it will be parallel to the optical axis, reaching the output plane OP_2 at the angle $\alpha_2 = 0$. This implies (Fig. 12.14) the equality $Cy_1 + Dn_1\alpha_1 = 0$ or

$$y_1 = -\frac{D}{C}n_1\alpha_1. \tag{12.107}$$

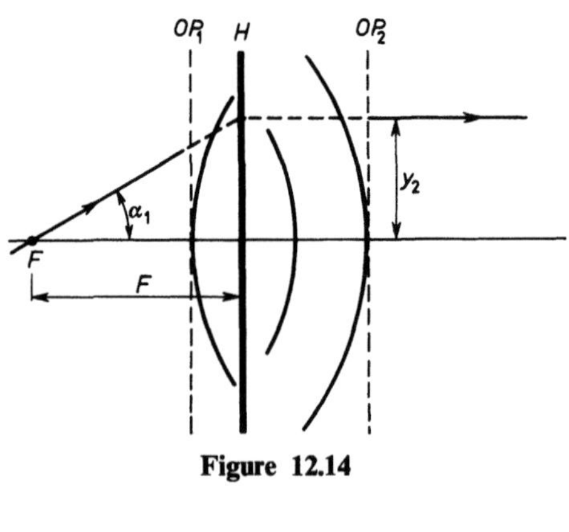

Figure 12.14

The distance from the focus F of the plane OP_1 (Fig. 12.14) is

$$s_1 = -\frac{y_1}{\alpha_1} = n_1 \frac{D}{C}. \tag{12.108}$$

Since in the front principal plane H the coordinate of the ray is $y = y_2$, and at the same time, as a result from the matrix equation, is $y_2 = Ay_1 + Bn_1\alpha_1$, we have

$$F = \frac{y_2}{\alpha_1} = -\frac{DA}{C} n_1 + Bn_1 = -\frac{n_1}{C}, \tag{12.109}$$

as $AD - BC = 1$.

From the above (Fig. 12.14) it follows also that the distance from the front principal plane to OP_1 is

$$d_1 = s_1 + F = n_1 \frac{D-1}{C}. \tag{12.110}$$

b. There are still to be determined the *distances from the nodal points to the input plane and output plane of the optical system.* To this end we use the situation shown in Fig. 12.15.

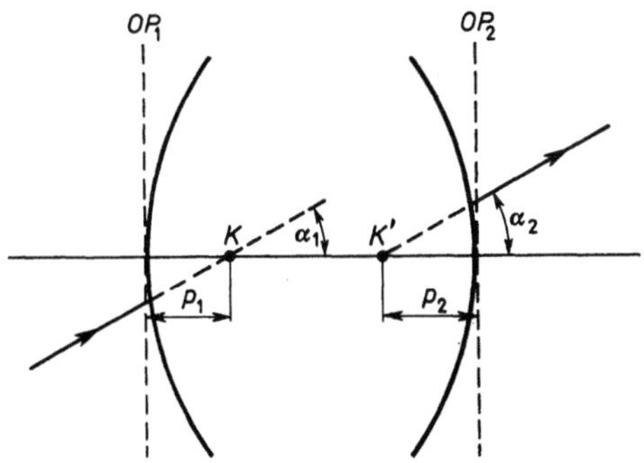

Figure 12.15

Denoting the distances from K and K' to the planes OP_1 and OP_2 by p_1 and p_2 respectively, we can write the matrix of transformation relating the parameters of the ray at the points K and K'. It has the form of the product

$$\Phi = \begin{bmatrix} 1 & \dfrac{p_2}{n_2} \\ 0 & 1 \end{bmatrix} \begin{bmatrix} A & B \\ C & D \end{bmatrix} \begin{bmatrix} 1 & -\dfrac{p_1}{n_1} \\ 0 & 1 \end{bmatrix}. \tag{12.111}$$

As to the nodal points correspond the parameters $y_1 = y_2 = 0$ and $\alpha_1 = \alpha_2$, the elements of the matrix are $\Phi_{12} = 0$ and $\Phi_{22}(n_1/n_2) = 1$, which leads to the equality

$$p_1 = (Dn_1 - n_2)\frac{1}{C}, \quad p_2 = (n_1 - An_2)\frac{1}{C}. \tag{12.112}$$

The carried out analysis allows us to make up a table given below, in which the distances of particular cardinal points of the optical system are expressed by the elements of the transformation matrix of a ray. As can be seen from the table the knowledge of the matrix for the optical system in question makes it possible, within the framework of Gauss' optics, to find the position of all its cardinal points, and consequently – to draw an image obtained by means of the system.

Let us take as an example an optical system composed of two refracting surfaces with the radii of curvatures R_1 and $-R_2$ respectively at a distance of d far from each other, and separating media with the refractive indices n_1, n_2 and n_3 (Fig. 12.16). As the reference planes (the input reference plane OP_1 and the output reference plane OP_2) of the system we can choose the planes tangent to the refracting surfaces at the points of their intersection with the optical axis.

The matrix of the system can be represented as a product of three matrices: the refracting matrix of the first surface, the matrix of displacement of a ray in the medium with the refractive index n_2, and the matrix of the second refracting surface, arranged in the order given in Sec. 12.11.

Table

Parameter of system	Measured from	Measured to	Function of matrix elements
Front focus	OP_1 input reference plane	F_1 focus	$\frac{n_1 D}{C}$
Front focal length	F_1 front focus	H_1 front principal point	$-\frac{n_1}{C}$
Front principal point	OP_1 input reference plane	H_1 front principal point	$\frac{n_1(D-1)}{C}$
Front nodal point	OP_1 input reference plane	L_1 front nodal point	$\frac{Dn_1 - n_2}{C}$
Back focus	F_2 back focus	OP_2 output reference plane	$-n_2 \frac{A}{C}$
Back focal length	H_2 back principal point	F_2 back focus	$-\frac{n_2}{C}$
Back principal point	OP_2 output reference plane	H_2 back principal point	$\frac{n_2(1-A)}{C}$
Back nodal point	OP_2 output reference plane	L_2 back nodal point	$\frac{n_1 - An_2}{C}$

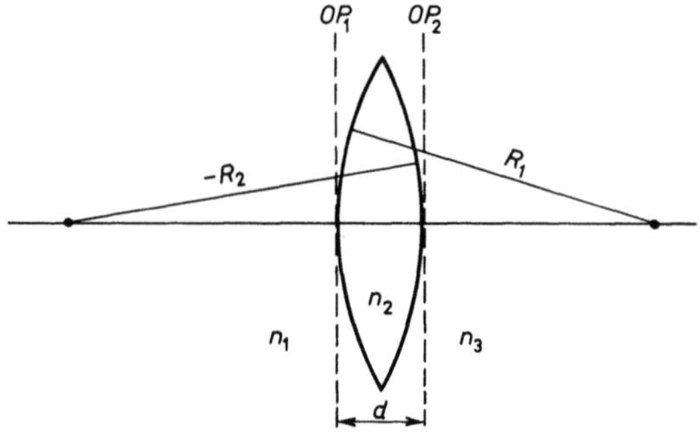

Figure 12.16

Thus this matrix is expressed as shown below

$$\check{M} = \begin{bmatrix} 1 & 0 \\ Q_2 & 1 \end{bmatrix} \begin{bmatrix} 1 & \dfrac{d}{n_2} \\ 0 & 1 \end{bmatrix} \begin{bmatrix} 1 & 0 \\ -Q_1 & 1 \end{bmatrix}, \tag{12.113}$$

where $Q_1 = (n_2 - n_1)/R_1$, $Q_2 = (n_3 - n_2)/R_2$. On multiplying the matrices this matrix takes the final form

$$\check{M} = \begin{bmatrix} 1 - Q_1 \dfrac{d}{n_2} & \dfrac{d}{n_2} \\ -Q_1 + Q_2 - Q_1 Q_2 \dfrac{d}{n_2} & 1 + Q_2 \dfrac{d}{n_2} \end{bmatrix}. \tag{12.114}$$

It is the matrix of transformation of a ray for a thick lens. If the thickness d of the lens is small in comparison with the radii of the curvatures of the refracting surfaces, then putting $d = 0$, we obtain a matrix for a thin lens:

$$\check{M}_1 = \begin{bmatrix} 1 & 0 \\ -(Q_1 - Q_2) & 1 \end{bmatrix}. \tag{12.115}$$

In this case, as follows from the table, the principal points of the system are situated in the middle of the lens. If the medium before and behind the lens is identical, the nodal points will also be in the middle of the lens, and both focal lengths will be equal.

Bibliography

1. Born, M. and Wolf, E.: (1965), 'Principles of Optics', Pergamon Press, Oxford–London.
2. Gerrard, A. and Bruch, J.M.: (1975), 'Introduction to Matrix Methods in Optics', John Wiley, London.
3. Kravchenko, I.T.: (1985), 'Theory of Wave Processes' (in Russian), Vysshaya Shkola, Minsk.
4. Morse, Ph.M. and Feshbach, H.: (1953), 'Methods of Theoretical Physics', McGraw-Hill, New York.
5. Sommerfeld, A.: (1950), 'Optik, Vorlesungen über theoretische Physik', Bd. IV, Wiesbaden.

Index

MIX
Papier aus verantwortungsvollen Quellen
Paper from responsible sources
FSC® C105338

If you have any concerns about our products,
you can contact us on
ProductSafety@springernature.com

In case Publisher is established outside the EU,
the EU authorized representative is:
Springer Nature Customer Service Center GmbH
Europaplatz 3, 69115 Heidelberg, Germany

Printed by Libri Plureos GmbH
in Hamburg, Germany